# LABORATORY TECHNIQUES IN SERICULTURE

**R.K, Goel**
Central Silk Board Bangalore

**A.P.H. PUBLISHING CORPORATION**
**4435-36/7, ANSARI ROAD, DARYAGANJ,**
**NEW DELHI 110002**

*Published by*
S.B.Nangia
**A P H Publishing Corporation**
7, Ansari Road, Daryaganj
New Delhi 110002
Ph.: 23274050
E-mail : aphbooks@gmail.com

2026



₹-1500/-

*Printed at*
**Balaji Offset**
Navin Shahdara, Delhi 110032

# PREFACE

Sericulture is a unique combination of art and science for raising of silkworm for production of silk, which is the most precious and glamorous commodity in the world. Basically, silk is a protein being secreted by sericigenous insects belonging to the family Lepidoptera. In fact, all the Lepidopteran caterpillars secrete silk-like filaments but not as strong as one secreted by silkworms. Hence, it remained a challenge to explore secret behind the stoniness and luster of the silk, besides enhancing the production and productivity of silk production through various agronomical, genetical and biotechnological tools.

In last decade, there are several changes in sericulture industry; noteworthy being the teaching and research on various facets of sericulture introduced at graduate and postgraduate levels in colleges and universities. As the sericulture teaching and research was confined to the institutes working under state and central governments, there was a dearth of adequate publications in the form of text books and manuals. Of late, several authors have come forward and published several books on sericulture, which are more or less of theoretical nature or related to practical aspects of silk cultivation and related technologies.

Considering the importance of Biochemistry in the growth of other biological sciences, sericulture is no exception and hence appropriate biochemical methodologies and laboratory techniques will unravel the mysteries behind the process of silk production. There are a number of books available on general and clinical biochemistry as well as laboratory techniques on the study of specialized bio-molecules like proteins, enzymes, nucleic acids etc., their applicability and suitability to sericulture remained a laborious and time-consuming process. Obviously, it was felt necessary to bring out a book, which can provide the desired and relevant methodologies and techniques in a systematic manner and at one place. It is in this context that this book **"Laboratory Techniques in Sericulture"** is presented for education and practical guidance of the students, research workers and scientific staff of Government Institutions, Colleges and Universities engaged in sericulture education and research especially in the field of soil science, plant biochemistry, plant pathology, insect biochemistry and related science subjects.

The entire presentation of the book is based on my practical experience and references collected by me during my research studies. In addition, several related books were referred and the contributions of several scholars were considered while compiling the book.

The book has eight chapters dealing with the techniques involved in soil analysis, soil fertility and application of fertilizers, plant pathology, foliar analysis, histology and histochemistry, biochemical analysis of silkworm biomolecules, ecological studies and biostatistics. Alternative methods have been given for various estimations with a view that based on the lab facilities available, suitable method may be adopted by a researcher. Every effort has been made to bring a understandable body of knowledge before the readers. It is hoped that all the concerned will find this book interesting and useful.

I wish to express my gratitude to all my colleagues and family members who evinced a lot of enthusiasm to see progress in writing the book from time to time. I welcome the constructive suggestions and critical evaluations for further improvement of this book, which will be highly appreciated and thankfully acknowledged.

**R. K. Goel**

# CONTENTS

# 1

# SOIL ANALYSIS

Soil is the uppermost layer of variable depth of the earth's crust consisting of loose material, which is the main support for natural vegetation and other life forms of our planet. Soil is a living and dynamic medium, on and in which almost all life-support processes occur. Strahler and Strahler (1983) defined the soil as ***"a natural surface layer containing living matter and supporting or capable of supporting plants"***. Joffe (1949) stated that ***"the soil is a natural body of mineral and organic constituents differentiated into horizons of variable depth, which differs from the material below in morphology, physical make-up, chemical properties and composition and biological characteristics"***. Soil is the dynamic natural body developed as a result of pedogenic processes during and after weathering of rocks, consisting of minerals and organic constituents, possessing definite chemical, physical, mineralogical and biological properties, having variable depth over the surface of earth and providing a medium for plant growth for land plants (Biswas & Mukherjee, 1992). Thus, *soil is the product of weathering rocks and minerals or debritus at the surface of the earth that has been caused by physical, chemical, mechanical and biological processes or breakdown so that it may support the growth of rooted plants.*

## 1.1 PHYSICAL ANALYSIS OF SOIL

Soil is the storehouse of organic (both living and dead) and inorganic nutrients (minerals) required by the rooted plants, i.e., the main source of plant food. It is generally composed of inorganic matter (45-50% by volume), organic matter (0-5% in the top layer) and pore space or voids (almost 50%), besides microorganisms. The pore space or voids are filled either with gases or water, whose proportions vary with seasons. The inorganic or mineral fraction is insoluble in water and composed of minerals and particles of varying sizes. However, some inorganic compounds, which are soluble in water, are also found in soil. The organic matter consists of decomposed remains of the animals and plants. In addition, the soil contains water and several microorganisms, the quantity of which depends on varying factors.

Physical properties of soil include the texture, structure, density, consistency and colour. **Soil texture** means the size of particles comprising soil while **structure** means the way soil particles are grouped together in the form of peds. **Soil density** is expressed as the weight of cubic centimetre of undisturbed soil or weight of compacted soil. **Soil consistence** is the plasticity or stickiness of soil when wet and depends on the cohesive and adhesive properties of the soil mass. **Soil colour** is also an important distinguishing property.

Soil profile, taken at different depths, thus include the description of soil in respect of the following items.

***Soil Colour:*** *It depends on the content of organic matter in soil. Generally, black-coloured soils have relatively higher contents of organic matter than light-coloured soils. Soil colour usually ranges from very pale brown, through intermediate browns to very dark brown or black as organic matter increases. The red colour of soil is generally attributed to unhydrated iron oxide, the yellow colour to iron oxides, the gray and whitish colour to several substances like quartz, kaolin and other clay minerals, carbonates of lime magnesium, gypsum, various salts and compounds of ferrous iron.*

***Soil Texture:*** *Soil textural class is observed in the field by feeling the soil with fingers. Classes of soil texture are based on different combinations of sand, silt and clay, which can be determined in laboratory by mechanical analysis of soil.*

***Soil Structure:*** *Soil structure plays an important role in soil productivity, soil permeability and root growth. Hence, the description of soil structure should include the shape and arrangement, the size and the distinctness and durability of the visible aggregates or peds.*

***Soil consistence:*** *It is expressed with reference to water content of soil. A soil may be hard when dry, fragile when moist and plastic when wet. Consistence includes stickiness (non-sticky, slightly sticky, sticky, very sticky) and plasticity (non-plastic, slightly plastic, plastic,* *very plastic) when wet. It is characterized when moist by (a) tendency to break into smaller masses rather than into powder, (b) some deformation prior to rupture, (c) absence of brittleness and (d) ability of the material after disturbance to cohere again when pressed together (loose, very friable, friable, firm, very firm, extremely firm). Consistence when dry is characterized by rigidity, brittleness, maximum resistance to pressure more or less tendency to crush to a powder to fragments with rather sharp edges and inability of crushed material to cohere again when pressed together (loose, soft, slightly hard, hard, very hard, extremely hard).*

### 1.1.1 Collection of Soil Samples

Samples are collected from soil survey areas for laboratory measurements for (1) determining fundamental properties of soils in relation to their classification, (2) suggesting their relationship to their environment, (3) checking field measurements and observations of textural class, pH, soluble salts, carbonates, permeability, (4) help in suggesting their response to management practices and (5) help in assessing their physical, chemical and biological properties.

#### *1.1.1.1 Selection of Sampling Site*

The sampling site should be a representative of the soil survey area and many samples should be collected from the sampling site. Wherever conveniently available and unless the samples are specifically taken for a study of cultivated soils, virgin soils are sampled for correlation and analysis. Samples should be taken from a fresh excavation. If the field/sample site is large, then divide the whole area into several sub-areas and collect the soil from each sub-area. Avoid drawing soil samples from spots, which do not represent the field, i.e., bunds and marshy spots. Do not sample a field within three months of application of the lime/gypsum and fertilizers/manures.

#### *1.1.1.2 Preparation of Soil Samples*

A true sample of soil is required to be analysed for study of soils. All the extraneous vegetation is removed from the soil and a representative sample is taken from different areas and depths of the field with the help of a soil auger and heaped together. For this, the soil of different places and depths is spread, mixed thoroughly and divided the soil sample into four quadrates. Soil from two opposite quadrates is removed and other two portions are mixed thoroughly. The process is repeated thrice and the composite soil sample is collected in polythene bag. Normally, the sample should be collected during dry weather conditions. A suitable quantity of soil is further processed for analysis as follows.

a. Spread out the sample on the paper or in shallow trays for at least 24 hours to get air-dry soil.

b. When dry, pulverize the soil with a wooden pestle gently and sieve through a 2.0 mm diameter sieve (about 8 mesh).

c. The soil, which passes through the sieve is called the "fine earth" and the residue on the sieve is taken out; stones are hand-picked, soil particles are carefully broken and again sieved. Care should be taken that no stone is broken. Particles larger than 2.0 mm in diameter are stones and undecayed organic material. The ratio of stones in soil is not uniform and varies from place to place. Accordingly, the soil is classified from its appearance as "stony", "very stony" etc.

d. The process is repeated till all the soil passed through the sieve leaving gravel on the sieve. The sieved soil is well mixed and used for subsequent analysis.

#### *1.1.1.3 Sampling*

An appropriate quantity of soil, which has been subjected for initial analysis as above, is taken as a sample, placed in a container, labelled and stored at a suitable place.

### 1.1.2 Mechanical Analysis of Soil

Soil is a heterogeneous mixture of silicate particles, humus and a variety of insoluble salts and oxides of metals called the solid phase. The silicate particles are of different sizes and irregular shapes. The process of separating these particles into various constituent grades (according to International Classification based on size) viz., sand (fine or coarse), silt, clay etc. and determining their percentage in a given soil sample is called **mechanical analysis of soil.** According to the particle size, the International classification of soils is as follows:

| *Soil Particle* | *Diameter* |
|---|---|
| Gravel | Above 2.0 mm |
| Coarse sand | 2.0 to 0.2 mm |
| Fine sand | 0.2 to 0.02 mm |
| Silt | 0.02 to 0.002 mm |
| Clay | Less than 0.002 mm |

Any fraction of soil, whose particle size is more than 2 mm in diameter is generally called **gravels or stones.** Soil taken for physical and chemical examination is the portion that passes through a 2 mm sieve. The coarse and fine sand together constitute the **coarse fraction of soil,** which is made up of chemically inert silica. It forms a sort of skeleton around the fine fraction of soil, which consists of **silt and clay particles.**

Gravels are separated from the soil by sieving through 2.0 mm sieve. Fine sand is separated by sedimentation after allowing soil suspension to stand for a requisite time while silt and clay are separated by pipetting. The chart for soil classification on the basis of mechanical analysis, as adopted by the Bureau of Soils, is given below.

| *Grading* | *Particle size* | |
|---|---|---|
| | *Modified International Classification* | *Bureau of Soils classification* |
| Gravel | 2 mm | 2 mm |
| Coarse sand | 0.2-2 mm | 0.5-2 mm |
| Medium sand | - | 0.25-0.5 mm |
| Fine sand | 0.02-0.2 mm | 0.05-0.5 mm |
| Silt | 0.002-0.02 mm | 0.005-0.05 mm |
| Clay | 0.002 mm | 0.002 mm |

### *1.1.2.1 Soil Texture*

The soil texture is denoted by the relative proposition of different constituent particles. Normally, soil contains all the particles mentioned under mechanical analysis of soil viz., gravel, coarse sand, fine sand, silt and clay but their quantity varies from soil to soil. These particles are often bound together through the intermediary of water molecules or metal ions or organic matter into aggregates. Depending upon the size and shape of aggregates, soil structure is defined. However, the dominant group of the particles determines the texture of the soil (Fig. 1.1). Based on this and as per the Bureau of Soils, the soil texture may be of the different types as mentioned in Tab. 1.1

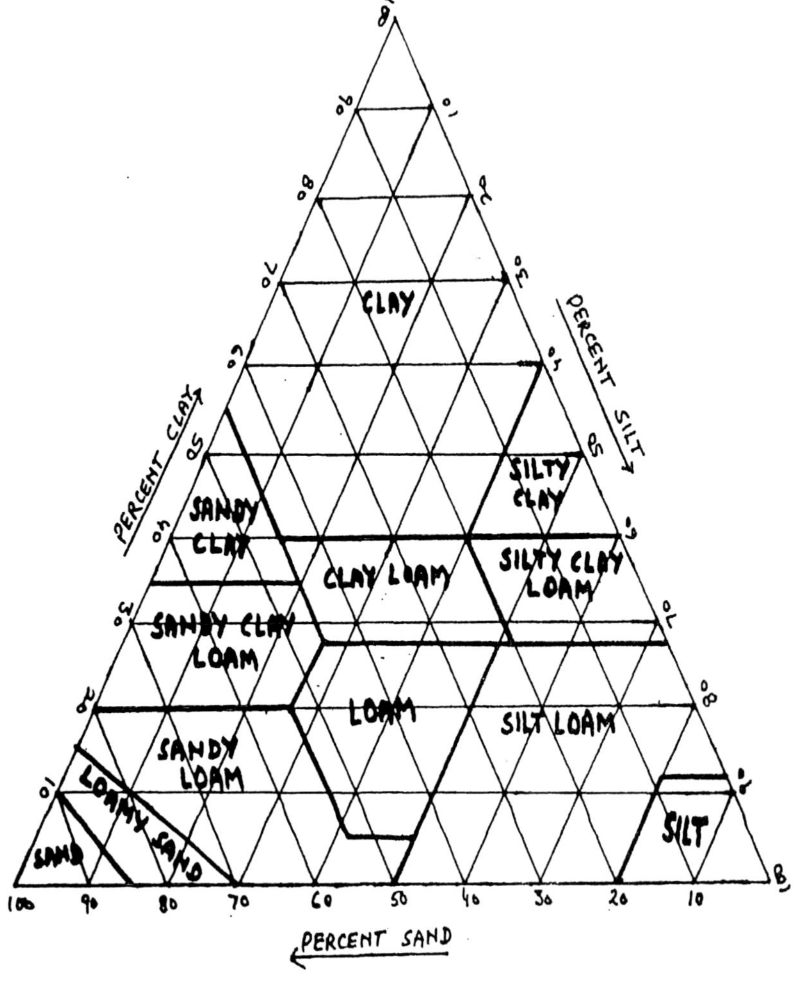


Fig. 1.1 : Triangular Diagram of soil texture

**Table 1.1: Classification of soil texture (as per Bureau of Soils)**

| *Soil Textural Class* | *Conditions* |
|---|---|
| Gravel | Sand and Gravel > 80% Gravel predominates |
| Sandy gravel | Sand > 15% |
| Silty gravel | Silt > 10% |
| Clayey gravel | Clay > 10% |
| Sand | Sand and Gravel > 80% Sand predominates |
| Gravelly sand | Gravel > 15% |
| Silty sand | Silt > 10% |
| Clayed sand | Clay > 10% |
| Loam | Clay > 20%; (Sand & Gravel < 80%. Sand predominates. Sand & Gravel < 70%. Gravel predominates) |
| Gravelly loam | Gravel and sand > 50%. Gravel predominates. |
| Sandy loam | Gravel & sand > 50%. Sand predominates. |
| Gravelly sandy loam | Gravel & sand > 50%. Gravel > 15%. |
| Silty loam | Silt > 50% |
| Clay | Clay > 30% |
| Gravelly clay | Gravel & sand > 50%. Gravel predominates. |
| Sandy clay | Gravel & sand > 50%. Sand predominates. |
| Silty clay | Silt > 50% |
| Loamy clay | Clay < 50%. Sand & gravel < 50%. Silt < 50%. |
| Medium clay | Clay > 50% and < 70% |
| Fat clay | Clay > 70% |

The definition of basic soil textural classes in terms of field experience and feel is given by Shaw (1929), which is as below.

***Sandy soil:*** It is the soil having less than 10 percent silt and clay. Sand is loose and coarse. Hence, the individual sand grains can be seen and felt between thumb and finger easily when rubbed. If a moist sandy soil is squeezed, it forms a cast, which will crumble when touched. A dry sandy soil cannot be squeezed into a cast. This type of soil is available in coastal areas where coconuts grow well.

***Sandy loam:*** The sandy loam is the soil consisting of 10-30 percent silt and clay and more than 50 percent sand and gravel where sand predominates. This soil has much sand and silt, and clay making it somewhat coherent. The individual sand grains can be seen and felt easily. If a dry sandy loam is squeezed, it will form a cast, which crumbles when touched. If a moist sandy loam is squeezed, it forms a cast, which will not break easily.

***Loam:*** It is the soil, which has 30 - 40 percent silt and clay, i.e., this type of soil consists of almost equal quantities of sand, silt and clay in mixture. It is mellow (rich), has somewhat gritty touch yet it is fairly smooth and slightly flexible. If a dry loam is squeezed, it will form a cast, which will bear careful handling.

***Silt loam:*** It is the soil having 40 - 50 percent silt and clay and remaining is fine sand. When dry, the soil appears cloudy (as lumps) but the lumps can easily be broken. When wet, the soil readily runs together and becomes muddy. If dry or moist silt loam is squeezed gently, it forms a cast, which will bear free handling.

***Clay loam:*** A clay loam is a fine-textured soil made up of 50 - 60 percent silt and clay. When a moist clay loam is gently squeezed, it forms a cast, which bears rough handling. If this soil is kneaded in hand, it does not crumble but tends to become a compact mass.

***Clay:*** A clay is the fine textured soil having more than 60 percent clay. When dry, it forms a very hard lump or clod. Slight moisture makes it soft and crumbly. If it is wet, it becomes sticky. When the moist soil is pinched out between thumb and fingers, it will form a long flexible ribbon.

### 1.1.2.2 Principle of Mechanical Analysis

Sieving method has got limited applicability in mechanical analysis of soil. Only gravel/stone content can be determined by the sieving of known quantity of dry soil. However, for fine particles, we have to follow some indirect method of measurement.

Most common method of mechanical analysis of soil is called ***International Pipette method.*** Mechanical analysis of soil consists of two distinct operations.

a. dispersion of soil particles

b. grading of the dispersed particles into different texture units.

***Dispersion*** is effected by suspending a known quantity of soil in water and shaking well in a mechanical shaker. The fractionation of texture units is achieved by using differences in settling velocities. It has been observed that velocity of falling particles in a suspension (i.e. depth of settling) depend upon their size; the other conditions remaining constant. The relationship between particle size and settling velocity is expressed by the Stroke's law, which states -

$$V = \frac{2}{9} \times g \times \frac{(D-d)}{\eta} \times r^2$$

where,

V = Velocity of the falling particles,

r = Radius of the particle,

D = Density of the particle,

d = Density of the liquid,

G = Acceleration due to gravity, and

$\eta$ = Viscosity of the liquid.

Since the density and viscosity of the liquid, acceleration due to gravity and density of the particle are constant, the falling velocity of the particle is directly proportionate to the square of its radius. Larger is the particle, it will settle sooner. In other words, the longest time is taken by clay particles to settle through a certain depth; less time is taken by silt and least by sand particles.

$V = K.r^2$

where K is a constant.

Let us consider a uniform separation of soil particles in water in a cylinder in which particles are settling under action of gravity.

Suppose a single particle with radius r has settled down from the position at A at the surface to that B at a depth of h cm in t seconds, then

$$V = h \div t = Kr^2$$

If h is kept constant at 10 cm, then

$$t = 10 \div K\, r^2 = (1 \div 3470) \times (1 \div 2)$$

the time t required by particles larger than a certain radius r can be calculated.

However, to obtain a uniform suspension of soil mass in water is the most important factor. Sometimes, smaller particles like clay form larger particles of silt or fine sand by sticking together through organic matter or lime and sometimes of iron or silica. In ordinary suspension of soil in water, we get such cemented particles but, to get true suspension, the cemented particles should be broken in original ones. It is achieved by treating soil with hydrogen peroxide and sodium hexametaphosphate solution.

### 1.1.2.3 Procedure for Mechanical Analysis

***Method - I (International Pipette Method):*** 20g of air dry soil (sieved through 2 mm sieve) is taken in a 600 ml beaker and 30 ml of 6% $H_2O_2$ solution (20 volume) is added to it. Reaction starts immediately. The beaker is covered with watch glass and digested on water bath with slow stirring. When the reaction subsides, 20 ml of 6% $H_2O_2$ are added to ensure complete oxidation of organic matter.

The contents of the beaker are cooled and 20 ml of dispersing agent, i.e., sodium hexametaphosphate (51.05 g per litre) and sodium carbonate (13.25 g per litre) solution are added to it. The entire material is collected in a shaking bottle and its volume is made up to 500 ml with distilled water. The bottle is shaken mechanically for three hours. Again, the contents are transferred to a measuring cylinder and the volume is made upto 1000 ml.

The measuring cylinder is shaken well without any whirling motion and kept for sedimentation. At that time, temperature of the solution is noted down and corresponding time for *fine sand sedimentation* is recorded. After the required time, 20 ml of solution is slowly sucked from the cylinder at a depth of 10 cm with the help of ordinary pipette and collected on a weighed nickel crucible ($W_1$g). The solution containing silt and clay is dried on water bath and finally in hot air oven at 105 °C. It is cooled and weighed ($W_2$ g).

The cylinder is again shaken and allowed to settle. The time taken for this case is *silt sedimentation time.* After the lapse of stipulated time, 20 ml of solution now containing only clay is pipetted out from the depth of 10 cm and collected in weighed nickel crucible. It is dried on water bath and finally in hot air oven at 105 °C and weighed ($W_3$ g).

Now the contents in the cylinder are transferred to a 0.2 mm sieve held over the mouth of a 500 ml beaker. The soil material is washed down on the sieve by water till all the particles below the wiremesh of the sieve are passed. The sieve is dried in an oven and the *sand particles* collected and weighed ($W_4$ g).

The solution and other materials collected in beaker are allowed to settle and the supernatant is decanted out. More water is added to that beaker and stirred well. It is again allowed to settle for *sand settling velocity.* The supernatant is decanted. This process is repeated till the water in the beaker becomes transparent. The fine sand is taken in a metal crucible, dried and weighed ($W_5$ g).

***Calculation*** -

Weight of crucible = $W_1$ g

Weight of crucible + silt & clay = $W_2$ g

Weight of crucible + clay = $W_3$ g

Weight of crucible + coarse sand = $W_4$ g

Weight of crucible + fine sand = $W_5$ g

Weight of clay = $(W_3 - W_1)$ g

Weight of silt + clay = $(W_2 - W_1)$ g

Weight of silt = $(W_2 - W_1) - (W_3 - W_1)$ g

Weight of coarse sand = $(W_4 - W_1)$g

Weight of fine sand = $(W_5 - W_1)$g

Percentage of clay = $[(W_3 - W_1) \times 100 \times 1000] \div [20 \times 20]$

Percentage of clay = $\{[(W_2 - W_1) - (W_3 - W_1)] \times 100 \times 1000\} \div [20 \times 20]$

Percentage of coarse sand = $(W_4 - W_1) \times 100 \div 20$

Percentage of fine sand = $(W_5 - W_1) \times 100 \div 20$

## *Method - II*

Place 10 g of air-dried soil in a beaker, add 30 ml of 6% $H_2O_2$ and heat on a water bath for 30-60 minutes. The function of hydrogen peroxide is to oxidize the organic matter. When the frothing ceases, treat with some more $H_2O_2$ and heat again on the top of water bath for 10 minutes. If there is no great frothing at this stage, cool and add conc. HCl drop by drop till the effervescence ceases. The hydrochloric acid removes carbonates and bases combined with clay. Now, add about 150 ml of water and 4 ml conc. HCl or 150-200 ml of 0.2 N HCl. Allow the beaker to stand for about an hour (or leave overnight); and frequently stir the contents. Filter the contents applying suction in a Buchner funnel and wash with hot distilled water till free from acid or chloride. Remove the soil from the Buchner funnel to a 0.2 mm sieve, which should be placed on a large ordinary funnel placed over a large bottle. By jets of water from wash bottle, transfer all the contents passing through the sieve to bottle. To get complete transfer of the soil, use a jet of hot water and finally squeeze the paper between fingers until the liquid runs clear. Note that the volume of liquid in the bottle does not exceed 250-300 ml. The suspension of the soil in the bottle is then treated with 1 or 2 ml of 1 N NaOH, which deflocculates the clay (till the liquid becomes alkaline). The suspension is shaken thoroughly for 24 hours in a mechanical shaker at a speed of 30-40 r.p.m. After it, the volume is made up to 500 ml.

***Determination of coarse sand*** - The residue on the sieve (0.2 mm) is transferred to a weighed crucible, dried at 105 °C in an oven, cooled and weighed. The increase in weight is the weight of coarse sand present in the soil sample.

***Determination of silt and clay*** - The suspension in the bottle consists of fine sand, silt and clay. The bottle is shaken by repeated inversion and then allowed to stand for about 4 minutes and a half (exact time should be taken at a particular temperature from the chart showing sedimentation time). 25 ml of contents are sucked gently with the help of a pipette at a depth of 10 cm and transferred to a weighed crucible. It is dried and weighed. The dry material is a mixture of clay and silt present in 25 ml of suspension ($W_1$ g). The bottle is shaken again and left to stand for about 8 hours (note exact time from the chart at a particular temperature). The 25 ml pipette is lowered to a depth of 10 cm and sucked gently. The contents are transferred to a weighed crucible, dried at 105 °C temperature and weighed. This represents the clay content in 25 ml of the suspension ($W_2$ g). From the difference in weights ($W_1$ - $W_2$), silt content is obtained.

**Table 1.2: Chart showing rate of fall or settling of particles through the depth of 10 cm at various temperatures**

| *Temperature (°C)* | *Silt (0.002 – 0.02 mm)* | *Fine sand (0.02 – 0.2 mm)* |
|---|---|---|
| 17 | 8.41 | 3.02 |
| 18 | 8.18 | 2.94 |
| 19 | 7.98 | 2.87 |
| 20 | 7.77 | 2.80 |
| 21 | 7.57 | 2.73 |
| 22 | 7.38 | 2.66 |
| 23 | 7.21 | 2.59 |
| 24 | 7.03 | 2.53 |
| 25 | 6.86 | 2.47 |
| 26 | 6.71 | 2.41 |
| 27 | 6.54 | 2.36 |
| 28 | 6.40 | 2.31 |
| 29 | 6.25 | 2.26 |
| 30 | 6.12 | 2.21 |
| 31 | 5.98 | 2.16 |
| 32 | 5.86 | 2.11 |
| 33 | 5.74 | 2.06 |
| 34 | 5.62 | 2.02 |
| 35 | 5.50 | 1.98 |

***Determination of fine sand*** - After sampling for clay, most of the suspension is decanted and thrown away. The residue at the bottom is transferred to a beaker.

Water is added in the beaker up to a height of 10 cms. The contents of the beaker are stirred with a rubber tipped glass rode and allowed to stand for 4 minutes and a half. By this time, all the sand particles must have gone below the depth of 10 cm. The suspension is decanted and the residue is again stirred up with water. This process is repeated till the supernatant liquid is clear. The residue is transferred to a weighed crucible, dried and again weighed as fine sand present in 10 g of soil sample.

**Table 1.3: Chart showing time of sedimentation for clay and silt fractions at 10 cm depth**

| *Temperature (°C)* | *Decantation time* | | | |
|---|---|---|---|---|
| | *Clay* | | *Silt* | |
| | *Hrs* | *Min* | *Hrs* | *Min* |
| 8 | 11 | 00 | 6 | 40 |
| 9 | 10 | 40 | 6 | 30 |
| 10 | 10 | 25 | 6 | 20 |
| 11 | 10 | 10 | 6 | 10 |
| 12 | 9 | 50 | 6 | 00 |
| 13 | 9 | 35 | 5 | 50 |
| 14 | 9 | 20 | 5 | 40 |
| 15 | 9 | 05 | 5 | 30 |
| 16 | 8 | 50 | 5 | 20 |
| 17 | 8 | 35 | 5 | 10 |
| 18 | 8 | 25 | 5 | 00 |
| 19 | 8 | 10 | 5 | 00 |
| 20 | 8 | 00 | 4 | 48 |
| 21 | 7 | 50 | 4 | 40 |
| 22 | 7 | 40 | 4 | 30 |
| 23 | 7 | 25 | 4 | 30 |
| 24 | 7 | 15 | 4 | 20 |

| | | | | |
|---|---|---|---|---|
| 25 | 7 | 05 | 4 | 15 |
| 26 | 6 | 55 | 4 | 10 |
| 27 | 6 | 45 | 4 | 05 |
| 28 | 6 | 40 | 4 | 00 |
| 29 | 6 | 30 | 3 | 55 |
| 30 | 6 | 20 | 3 | 50 |
| 31 | 6 | 15 | 3 | 45 |
| 32 | 6 | 05 | 3 | 40 |
| 33 | 5 | 55 | 3 | 35 |

*Calculation*

Coarse sand % = (Weight of coarse sand ÷ weight of soil) × 100

Fine sand % = (Weight of fine sand ÷ weight of soil) × 100

Silt % = (Weight of silt ÷ weight of soil) × (500 ÷ 25) × 100

Clay % = (Weight of clay ÷ weight of soil) × (500 ÷ 25) × 100

### *1.1.2.4 Soil Moisture, Air and Organic Matter*

Besides different sized particles, the soil also contains organic matter, water and air, which play vital role in the soil fertility and productivity.

#### *1.1.2.4.1 Soil Moisture*

In-between soil particles, there are countless interspaces of varying sizes and shapes. The narrow interspaces are called ***capillary interspaces*** while the big ones are the ***non-capillary interspaces***. Their proposition depends upon the soil texture. In black soil (fine textured soil), the capillary interspaces predominate whereas the non-capillary interspaces are found in red soils (coarse textured soil).

The water holding capacity of soil depends upon the capillary interspaces. More the capillary spaces, the more is the water holding capacity of the soil. A soil, which can hold more water is good for plant growth. The soil water has been classified into the followings:

a. Hygroscopic water

b. Capillary water

c. Gravitational water

***Hygroscopic water:*** When a soil is watered to make it just wet, the water is held around the surface of fine particles (silt and clay) by cohesive force as a very thin layer. As this water is held firmly by the soil particles, it is not available to the plants. It is referred as hygroscopic water.

***Capillary water:*** When more water is given to soil, the layer of water around the particles thickens. This is called capillary water or capillary moisture. This water can rise through the capillary interspaces by surface tension and hence is available to plants.

***Gravitational water:*** When still more water is added to soil, the cohesive force weakens and water begins to flow down due to gravitational force till it reaches the water table. This is gravitational water. This water, in many cases, is not available to plants.

## Determination of soil moisture

***Principle:*** Moisture content of soil is defined as the percentage of water retained by a soil when a layer of saturated soil 1 cm deep is centrifuged for 30 minutes at 2440 r.p.m.

Moisture equivalent = (Weight of wet soil after centrifugation ÷ Weight of oven dry soil) × 100

***Method:*** Aluminium centrifuge box is taken and, on its perforated bottom, Whatman No. 1 filter paper is placed. Now, about 30 g of 2 mm sieved air-dried soil is put on filter paper. The box is lightly tapped and kept it on water in a petridish overnight. Water rises from the bottom and saturates the soil. Next day, the centrifuge box is taken out, kept on moist cloth and covered with damp cloth. It is kept like this for 30 minutes to drain out the excess water. The box is then left in centrifuge and centrifuged for 30 minutes at 2440 r.p.m. After that, a known quantity of wet soil is taken in a weighed crucible and dried in hot air oven at 105 °C temperature till the constant weight is obtained (about 16-20 hours). The loss in weight represents the amount of moisture in the soil sample.

### *1.1.2.4.2 Soil Air*

In-between the soil particles, air spaces are found, which constitute the gaseous system of soil. Composition of soil air is almost similar to that of the atmospheric air except that content of $CO_2$ is several times higher. Soil flora and fauna consume oxygen and give out carbon dioxide, which is partially dissolved in soil solution and partly remains in the gaseous phase. Composition of soil air varies with depth.

**Table 1.4: Composition of soil gas in fallow land before and after rain**

| *Depth (cm)* | *Nitrogen (%)* | *Oxygen (%)* | *Carbon di-oxide (%)* | *Argon (%)* |
|---|---|---|---|---|
| Before rain | | | | |
| 0.0-7.6 | 78.45 | 20.44 | 0.12 | 0.988 |
| 7.6-15.2 | 78.44 | 20.30 | 0.25 | 1.007 |
| 15.2-22.8 | 77.08 | 20.29 | 1.65 | 0.976 |
| 22.8-30.5 | 78.21 | 20.56 | 0.29 | 0.937 |
| After rain | | | | |
| 0.0-7.6 | 78.43 | 19.64 | 0.97 | 0.954 |
| 7.6-15.2 | 78.85 | 19.67 | 0.57 | 0.923 |
| 15.2-22.8 | 79.10 | 18.99 | 0.94 | 0.967 |
| 22.8-30.5 | 78.95 | 18.75 | 1.32 | 0.978 |

***Source***: Department of Agriculture in India (Chemical Series), 4:113,1915.

The gaseous exchange between soil and atmosphere is the natural process, by which carbon dioxide in soil is replaced by oxygen or carbon dioxide is washed down beyond root zone by irrigation or rainfall.

### *1.1.2.4.3 Soil Organic Matter*

Organic matter is supplied to the soil in the form of residues of crops that grow on the soil, green manures (residues of green plants and leaves etc.), cattle manure (cow-dung etc.) and compost (mixed manure). Organic matter influences the physical, chemical and biological properties of soil. It can bind sandy soil; thereby increasing its moisture retaining capacity. It improves soil aggregation, which in turn influences infiltration, movement and retention of soil water, soil aeration, soil temperature, soil strength and root penetration. It supplies nutrients to soil and serves as a food for soil microorganisms, which get energy by decomposing the organic matter. It has chelating power for some of the essential trace elements. It has also got substantively the buffering capacity of soil, making it less amenable to pH changes by acids and bases. However, the organic matter content varies from time to time and from place to place in the soil.

During decomposition, the carbohydrates are degraded into carbon dioxide and water; proteins into ammonia and finally into nitrates besides energy. Fats, resins and lignin persist and unite with some proteins and nitrogenous compounds to form soil humus. Thus, *soil humus is a complex organic matter, brown or dark brown in colour formed during microbial decomposition of organic matter in the soil. Its water holding capacity is quite high.*

## Determination of organic matter (Loss of Ignition)

When a soil is ignited, it first loses moisture and then the organic matter contained in it. Loss on ignition is therefore an important index of the organic matter present in the soil.

*Qualitative Estimation* - Take oven-dry soil sample in a beaker and add 100 ml of 6% $H_2O_2$. Keep the beaker on the wire gauge on tripod stand. Warm slightly and notice vigorous frothing. After frothing ceases, add 200 ml of water, shake well and allow the soil particles to settle. Vigorous frothing and pale coloured soil particles are indicative of presence of organic matter in the soil.

*Quantitative Estimation* - The oven-dry soil is transferred to a weighed platinum crucible and ignited at a low red heat for about 5 hours till there is no further change in colour (most soils become black at certain places due to charring of organic matter. When organic matter is completely destroyed, the soil colour is usually uniformly brick red). After ignition, it is cooled in a desiccator and weighed.

If the soil contains appreciable quantities of calcium carbonate, carbon dioxide may be lost during the ignition. To correct this, it is necessary that the ignited soil should be treated with a few drops of ammonium carbonate solution. The moist soil is gently heated to drive off the excess. The dry soil is cooled and weighed.

The loss in weight of the oven dry soil is the loss on ignition of the soil.

### Determination of Loss by Solution

The filtrate after the peroxide-HCl treatment contains in addition to calcium mixed sesquioxides (oxides of aluminium and iron), a small amount of silica, which usually amounts to 2-3% of total soil weight. To determine the dissolved sesquioxides and silica, the filtrate (of mechanical analysis) is brought to the boiling and ammonium chloride and ammonia are added in excess. The precipitate is filtered off and washed with hot water. The precipitate is then dried in steam oven and gently ignited. The weight of the material obtained, expressed as a percentage, is reported as "loss of solution".

The weighed precipitate is then treated with HCl, evaporated to dryness and heated to 150°C for 2 hours. The residue is again moistened with conc. HCl and allowed to stand for 10 minutes. Water is then added to it and silica is filtered off, washed, dried and weighed. The sesquioxides can be obtained by difference.

### *1.1.2.5 Soil Microorganisms*

Soil contains a number of organisms belonging to both animal and plant kingdoms. These are -

#### *1.1.2.5.1 Animal Organisms*

a. Protozoans like rhizopods, flagellates, ciliates etc.

b. Rotifers and nematodes, which feed upon soil protozoans and decay organic matter. Some protozoans are parasitic on plants.

c. Insects, mites, myriapods, burrowing annelids etc., which feed on the microorganisms and decay plant tissue and thus maintain a balance between them.

#### *1.1.2.5.2 Plant Organisms*

a. *Algae* - Green, yellow green and diatoms are present in soil. They form one of the major groups of soil organisms. They add organic matter to soil.

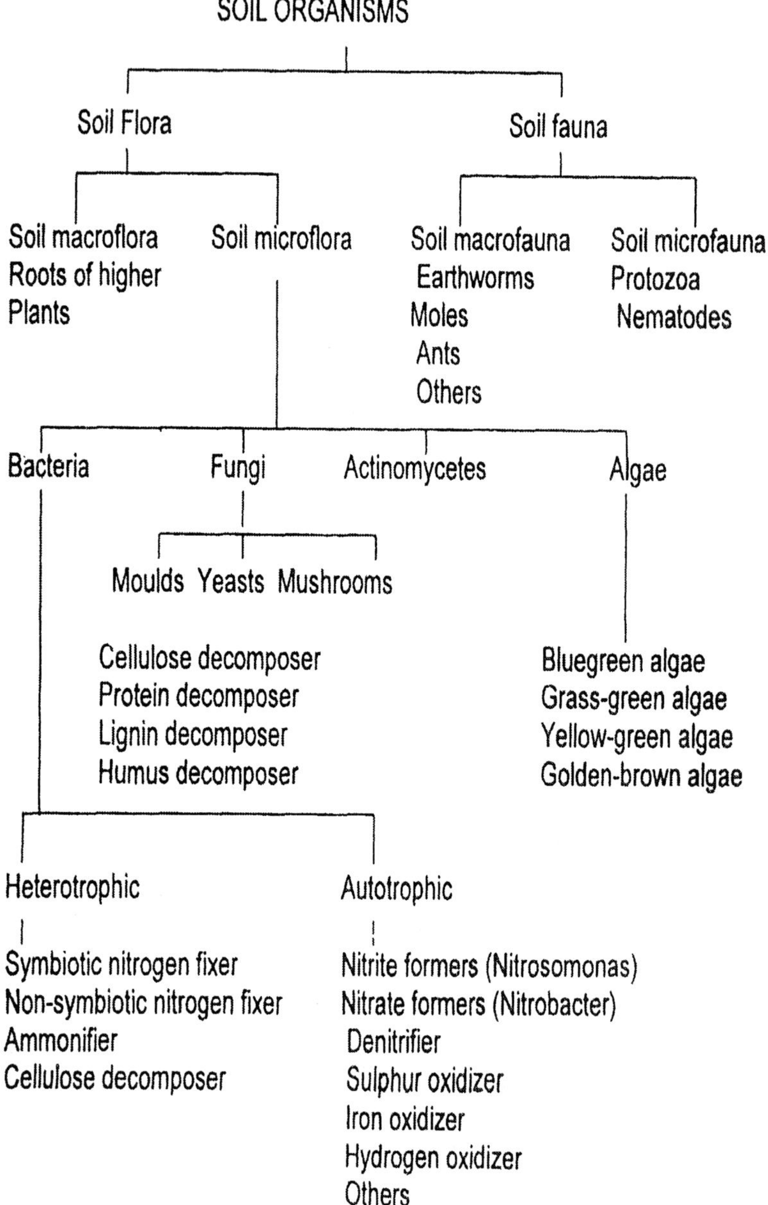


**Fig. 1.2: Classification of Soil Organisms**

b. *Fungi* - It may be saprophytic, symbiotic or parasitic.

*Saprophytic fungi:* Decompose the organic matter of soil and hence are the most useful of all fungi.

*Symbiotic fungi:* Live on the roots of certain plants. Both fungi *and plants gain by this association.*

*Parasitic fungi:* Produce plant diseases e.g. smuts and rusts,

c. *Bacteria* - Of the many bacteria, which live in soil, the important ones are autotrophic and heterotrophic bacteria.

*Autotrophic : bacteria* are the nitrogen fixing bacteria. They fix up nitrogen of the soil and air, which leads to formation of amino acids. When the bacteria and plants die, their proteins are released into the soil.

*Heterotrophic : bacteria* are the bacteria, which decompose the organic matter of the soil to get energy for their activities.

The bacteria are responsible for the chemical processes, which occur in the soil. The beneficial processes are the process of decomposition of organic matter and fixation of atmospheric nitrogen so that they are made available to plants.

## 1.2 SOIL PHYSICS

Soil physics is the interpretation of various soil phenomena in relation to physical parameters. The physical properties of soils are closely related to its chemical and biological properties. As plants require moisture, temperature and air for their best growth and since the soil is the medium for growth of the plants, their influence will be governed by the physical forces related to them. However, some of the physical properties are well defined as single value constants.

### 1.2.1 Single Value Constants

#### *1.2.1.1 Apparent Density*

As the soil is not a continuous mass and consists of particles intervening hollow spaces or pores between them, density of the soil sample will not be actual one but will be higher than the density of the soil mass with pores in-between particles. **Apparent density of**

**soil,** which is indicative of soil condition whether it is loose or compact, is defined as the weight of the soil per unit apparent volume, i.e.,

$$\text{Apparent density} = \frac{\text{Weight of dry soil}}{\text{Apparent volume of that soil}}$$

It varies from 1.1 – 1.6 g $cm^{-3}$ (Mg $m^{-3}$) for fine or compact texture soil to 1.3 – 1.7 g $cm^{-3}$ (Mg $m^{-3}$) for sandy or coarse or loose texture soil. It is also referred as ***bulk density***. High bulk density indicates compactness of soil. Soil organic matter having low density lowers the bulk density of soil.

### *1.2.1.2 Water Holding Capacity (Saturation Capacity)*

**Water holding capacity of the soil is the amount of water required to saturate 100 g of dry soil, i.e.,**

$$\text{Water Holding Capacity} = \frac{\text{Wt. of wet soil} - \text{wt. of dry soil}}{\text{Weight of dry soil}}$$

It is the moisture-saturated stage of soil when all pores and capillaries are filled with water. This is also called as ***saturation*** *capacity of the soil.* If the moisture content is further increased, percolation results, i.e., the excess water begin to flow down under the action of gravity. Normally, it minimizes the influence of gravity. The water holding capacity of soil is of great practical value to agriculturists since it provides a simple means of determining moisture contents required for soils for good plant growth. However, it is unwise to allow a cropped soil to remain at this level (saturation) for a very long period of time because such a saturated condition allows practically no aeration and is harmful to most crop plants. *The maximum water holding capacity of soil can also be determined by the amount of total pore space available in the soil while the available water holding capacity, if the bulk density, moisture content, soil depth etc. are known, by the following formula* -

$$\text{AWHC} = \frac{\text{PVD}}{100}$$

where,

AWHC = Available water holding capacity

P = Present moisture content

V = Bulk density

D = Depth of soil

| *Soil textural class* | *Available Water Holding Capacity* |
|---|---|
| Coarse sand | 0.04 |
| Fine sand | 0.05 |
| Loamy sand | 0.07 |
| Loamy firm | 0.08 |
| Sandy loam | 0.10 |
| Fine sandy loam | 0.13 |
| Sandy clay loam | 0.15 |
| Silt loam | 0.16 |
| Clay loam or clay | 0.17 to 0.18 |

With medium textured soil, good plant growth occurs at moisture content corresponding to 50-70% of total water holding capacity of the soil.

### 1.2.1.3 Particle Density

**Particle density of a soil is the mass per unit volume occupied by the soil particles alone.** Particle density is fixed but bulk density varies with pore volume.

It is determined on the same principle as that of powdered insoluble solid. Small volumetric flasks, say 50 to 100 ml capacity are used in place of a pycometer and water immiscible liquid of known density is used in place of water. Particle density is dependent on the chemical and mineral composition of soil. Because of its variation within narrow limits, the value of 2.65 g $cm^{-3}$ is commonly taken as the value of mineral soils.

### 1.2.1.4 Pore Space

Pore space of a soil mass is the space in-between soil particles occupied by air and water. In dry conditions, the soil is a porous mass. ***Porosity of soil is the volume percentage of the total bulk not occupied by solid particles.*** It is calculated as follows.

Porosity = 100 – (Bulk density ÷ Particle density) × 100

where,

(Bulk density ÷ Particle density) × 100 = solid space %

Porosity generally depends upon the soil texture and amount of different soil fractions viz., sand, silt and clay. A soil with high amount of clay naturally has more porosity than a sandy soil where much a smaller number of particles forms a unit mass or space. These spaces or pores facilitate movement of air and water in the soil. However, it is not always true that a soil with high porosity will allow greater drainage. Normaliy, drainage depends upon the size of the pores. In sandy soil, the pore size is bigger and hence drainage is easier in this soil than in a clay soil.

### *1.2.1.5 Specific Gravity*

**Specific gravity of a soil denotes the average density of minerals particles present in a soil sample.** It is defined as *the weight of soil per unit apparent volume,* i.e.,

$$\text{Specific gravity} = \frac{\text{Wt. Of dry soil in V volume of wet soil}}{\text{V} - \text{Wt. Of water present in V volume of wet soil}}$$

It means that V volume of wet soil contains certain amount of dry soil and water. By eliminating the weight of water from both the sides, we will be able to get exact weight of dry soil and the volume of the soil. In other words, ***specific gravity is the ratio between the weight of a given volume of dry soil particles and the weight of water displaced by these particles.***

Specific gravity of soils generally does not vary much and its average is about 2.65.

### *1.2.1.6 Volume Expansion*

Clay particles, when moist, swell and hydrated clay, when dry, shrink in volume. Volume expansion of soils, when moist, gives a measure of the amount of clay in soil or the mechanical composition of the soil.

### 1.2.2 Determination of Single Value Constants

Air-dry soil of 0.5 mm diameter is taken for this purpose. Whatman No. 1 filter paper is cut and placed on the internal perforated floor of the Keen's box (Keen-Rackzowski). Initial weight of box along with filter paper is recorded ($W_1$ g). The box is then filled with soil by tapping the box briskly and making a convex top of the soil. The convex top of the box is made plain with the help of spatula, wipe off the soil particles from outside and it is weighed again ($W_2$ g). The difference gives the weight of soil taken for analysis. It is now placed in a shallow dish containing distilled water about 1/2 inch deep so that water may rise in the Keen's box through the perforated bottom and moisten the soil to its capacity. It is left overnight in water.

Next day morning, the box is taken out from water and kept on a moist cloth covered with glass jar to avoid evaporation and left for 30 minutes. Meantime, excess water in soil will drain out. The weight of the wet soil and box is taken ($W_3$ g). The portion of soil swollen as a result of water absorption is scraped off to a small weighed watch glass or aluminium box. The Keen's box with remaining soil and the aluminium box/watch glass with excess soil are weighed. Both the boxes are now dried in an hot air oven at 105°C temperature till constant weight is obtained. The volume of the Keen's box is determined by simple formula $\pi r^2 h$.

***Calculation*** -

| | |
|---|---|
| Weight of Keen's box + filter paper | $= W_1$ g |
| Weight of box + filter paper + dry soil | $= W_2$ g |
| Weight of box + wet filter paper + saturated soil | $= W_3$ g |
| Weight of box + wet filter paper + saturated remaining soil | $= W_4$ g |
| Weight of aluminium box/watch glass | $= W_5$ g |
| Weight of aluminium box + excess soil | $= W_6$ g |
| Weight of aluminium box + dry excess soil | $= W_7$ g |
| Weight of box + filter paper + dry remaining soil | $= W_8$ g |

Volume of box (V) $= \pi r^2 h$

Moisture content of filter paper $= x$ g

Apparent density $= (W_2 - W_1) \div V$

Saturation capacity or

Water holding capacity $= [W_3 - (W_2 + x) \times 100] \div (W_2 - W_1)$

Pore space $= \{[(W_4 - W_8) \div x] \div 100\} \times V$

Specific gravity $= (W_8 - W_1) \div [V - (W_4 - W_8)$

Volume Expansion (per 100 ml of soil) $= \{[W_4 - (W_8 + x)]\div V\} \times 100$.

## 1.3 CHEMICAL ANALYSIS OF SOIL

In addition to soil particles, moisture, air and microorganisms, the soil also contains various minerals and organic matter. The percentage of different minerals in a soil is an index of its fertility. **Soil fertility refers to the capacity of soil to supply plant nutrients in available form to the plants.** Determination of soil fertility status helps in getting better plant growth and plantation management. There are two types of minerals present in a soil - macronutrients and micronutrients. However, some of the minerals present in soil are oxygen, carbon, nitrogen, hydrogen, phosphorus, potassium, calcium, magnesium, sulphur, iron, manganese, boron, copper, zinc etc.

### 1.3.1 Determination of Nitrogen

#### *1.3.1.1 Determination of Organic Nitrogen*

##### *1.3.1.1.1 Micro-Kjeldahl Method*

**Theory** : In this method, organic nitrogen of soil is oxidized into ammonia by the action of sulphuric acid. During digestion with sulphuric acid, the water is removed from the organic matter and residual carbon reacts with $H_2SO_4$ to produce sulphur dioxide, which in turn reduces nitrogenous matter. Finally, all the nitrogen is converted into ammonia, which is at once fixed by acid as ammonium sulphate. As the reaction between sulphuric acid and organic matter is slow, the oxidation process is accelerated by addition of catalytic agents like Hg or $CuSO_4$. Normally, copper sulphate is used as catalyst while potassium sulphate raises the boiling point of the acid so as to avoid loss of acid by volatilization.

Since the chemicals may contain nitrogenous impurities, a blank with the same quantities of reagents is carried out. Ammonia content of both the digests is determined by distillation with excess sodium hydroxide and absorption of evolved ammonia in standard $H_2SO_4$. The blank value is usually subtracted from the value of the soil digest.

*Procedure* : 20 g of air-dry soil are taken into a Kjeldahl flask and moistened with 20 ml of water. About 1g copper sulphate and 10g anhydrous sodium or potassium sulphate are added to the soil alongwith 40 ml concentrated, sulphuric acid. After frothing has ceased, the mixture is first heated gently and then strongly till the mass is free from any black particles (about 2-3 hours). After digestion is complete (pale yellow or green colour of the digest), the flask is allowed to cool at room temperature. The contents are diluted with 100 ml of water and then decanted into a 1.0 litre distillation flask. The sandy residue is washed with 50 - 60 ml of water at least 4-5 times and the washings are decanted into the flask after allowing the residue to settle down for few minutes each time. Add a piece of granulated zinc or glass beads to prevent bumping. 150-200 ml of 30% NaOH solution are poured alongwith sides of the flask till the contents are alkaline to phenolphthalein. Do not mix the two solutions until the flask is fitted with delivery tube and condenser, which is connected with a tube dipped in 50 ml of N/10 $H_2SO_4$ solution in a receiving flask containing10-12 drops of methyl red (0.5% in alcohol) since the ammonia gets liberated immediately with the addition of sodium hydroxide solution. With the addition of sodium hydroxide, the contents are heated. The ammonia liberated in the reaction is absorbed in the standard sulphuric acid solution. When about 200 ml of distillate are collected with standard acid (about 20-30 minutes), the distillation is discontinued after testing with red litmus. The tip of the tube dipping into the receiving flask is washed with water and the distillate is now titrated with N/10 NaOH solution till pink colour changes to yellow. The decrease in the strength of the acid gives a measure of the nitrogen content of the soil. Carry out a blank determination in exactly the same manner except using 0.2 g cane sugar in place of the soil so as to correct for any nitrogen present in the reagents.

1 ml of N/10 $H_2SO_4$ = 0.0014 g Nitrogen

*Calculation:* Let $V_1$ ml of N/10 NaOH required for neutralization. Acid used in blank distillation is 10 ml of N/10 $H_2SO_4$ and N/10 NaOH required for neutralization be $V_2$ ml.

Net $N/10 H_2SO_4$ required for soil digest $= (50 - V_1)$ ml

Net $N/10 H_2SO_4$ required for blank digest $= (10 - V_2)$ ml

$\therefore$ Net $N/10\ H_2SO_4$ required by 20 g soil $= (50 - V_1) - (10 - V_2)$ ml

1 ml of $N/10\ H_2SO_4$ $= 0.0014$gN

$\therefore$ $[(50-V_1) - (10-V_2)]$ ml of $N/10\ H_2SO_4$ $= 0.0014 \times [(50-V_1) - (10-V_2)]$

20 g soil contains $0.0014 \times [(50-V_1) - (10-V_2)]$ g N

100 g soil will contain $0.0014 \times [(50 \times V_1) - (10-V_2)] \times 100/20$ g or $\times$ g N

or Nitrogen content of the soil = x %

### *1.3.1.1.2 Winkler's Modification*

*Theory* : In this method, unsaturated solution of boric acid is used instead of absorbing the ammonia in standard $H_2SO_4$ or HCl. Since boric acid is a very weak acid, the absorbed ammonia can be determined by titration with standard HCl with use of suitable indicator (bromophenol blue). This is direct determination requiring only one standard solution. Moreover, a sufficiently large excess of boric acid can be used so as to ensure complete absorption of the ammonia.

*Procedure* : Proceed as before but collect the distillate in 50 ml of 4% boric acid solution. When the distillation is complete, add 3 drops of bromophenol blue indicator solution and titrate the ammonia absorbed with $N/10\ H_2SO_4$.

Nitrogen content of soil = (A-B) $\times$ N $\times$ 0.14 where,

A = Volume of standard acid used in the actual titration

B = Volume of standard acid used in blank titration.

N = Normality of standard solution.

### *1.3.1.2 Determination of Total Nitrogen*

(Modified Kjeldahl method)

*Principle* : In Micro-Kjeldahl method as described above, most of the nitrogen of nitrates is lost during digestion and hence don't represent the whole of the nitrogen of the soil. Accordingly, nitrate nitrogen is fixed by salicylic acid. Sodium thiosulphate is added with sulphuric acid to reduce the nitrates.

*Digestion* -

$$2\,NaNO_3 + H_2SO_4 \longrightarrow Na_2SO_4 + 2\,HNO_3$$

$$\underset{\text{Salicylic acid}}{C_6H_6OHCOOH} + HNO_3 \longrightarrow \underset{\text{Nitrosalicylic acid}}{C_6H_3NO_2(OH)COOH} + H_2O$$

$$Na_2S_2O_3 + H_2SO_4 \longrightarrow H_2S_2O_3 + Na_2SO_4$$

$$H_2S_2O_3 \longrightarrow H_2SO_3 + S$$

$$C_6H_3NO_2(OH)COOH + 3H_2SO_3 + H_2O \longrightarrow \underset{\text{Aminosalicylic acid}}{C_6H_3(NH_2)(OH)COOH} + H_2SO_4$$

$$2\,C_6H_3(NH_2)(OH)COOH + 27\,H_2SO_4 \longrightarrow (NH_4)_2SO_4 + 26\,SO_2 + 14\,Co_2 + 30\,H_2O$$

*Distillation* -

$$(NH_4)_2SO_4 + 2NaOH \longrightarrow Na_2SO_4 + 2\,NH_3 + 2\,H_2O$$

*Absorption* -

$$2\,NH_3 + H_2SO_4 \longrightarrow (NH_4)_2SO_4 \text{ (conical flask)}$$

*Titration* -

$$H_2SO_4 + 2\,NaOH \longrightarrow Na_2SO_4 + H_2O$$

*Procedure* : Weigh 10 g of the soil and place it into Kjeldahl flask. Dissolve 1.0 g salicylic acid in 30 ml of conc. Sulphuric acid and add the mixture to the flask. Shake well until mixed thoroughly, allow to stand at least 30 minutes with frequent shaking and then add 5-7 g sodium thiosulphate. Keep the whole mixture and shake it occasionally for another 30 minutes. Digest the mixture first at low flame and then strongly for 5 minutes. Then add 10 g potassium sulphate and a pinch of copper sulphate and heat more strongly until a clear mixture is obtained. Allow to cool, dilute with water and transfer it to a distillate flask. Add about 120 ml of 30% NaOH solution and connect up to a condenser dipping into conical flask containing N/10 $H_2SO_4$. Suitable indicator is added to the contents of the conical flask. Distil until 150 ml passes over. Back titrate the excess of the acid in the receiver with N/10 NaOH. Then calculate the percentage of nitrogen.

1 ml of N/10 $H_2SO_4$= 0.0014 g Nitrogen

### 1.3.1.3 Determination of Available Nitrogen

*Reagents* : 0.32% $KMnO_4$, 2.5% NaOH, 0.02N $H_2SO_4$, Methyl Red

*Procedure* : Take 20 g soil, 20 ml water, 100 ml $KMnO_4$ and 100 ml of 2.5% NaOH in a distillation flask and fit it up in distillation apparatus. Pipette out 20 ml of 0.02N $H_2SO_4$ in the conical flask and dip the end of delivery tube in it. Distil and collect about 30 ml of the filtrate. Now, add 5 drops of methyl red and titrate with 0.02 N NaOH solution.

*Calculation* :

Weight of soil = 20 g

Volume of 0.02N $H_2SO_4$ taken for absorption = 20 ml

Volume of 0.02N NaOH required for titration = x ml

Volume of 0.02 N acid used for $NH_3$ = (20 – x) ml

Available Nitrogen = (20 – x) × 28 pounds

### 1.3.1.4 Determination of Ammonical Nitrogen

(McLean and Robinson method)

*Principle:* Organic nitrogen in soil is gradually converted into ammonical nitrite and nitrate nitrogen by microbiological agencies. Organic nitrogen is itself of little use for plants since plants cannot take it as such. It is therefore essential to estimate the different forms of mineralized or available nitrogen. The nitrite and nitrate nitrogen hardly comprise more than 1.0% of total nitrogen in ordinary soil.

Although the amount of ammonia in mineral soils is usually very small, but in acid peat soils, pasture soils, heavily dunged soils or in soils after application of ammonical fertilizers, the amount may be much higher. It is retained in the exchangeable form and this method is based on leaching the soil with sodium chloride; thus replacing the ammonium ions by sodium ions.

*Procedure:* 20 g air-dry soil are treated with about 100 ml of 1N NaCl, stirred well and left overnight. Some soils, which are alkaline, give a strongly coloured extract on leaching with sodium chloride because of dispersion of humified organic matter. With these soils, better results are obtained if the leaching is carried out with NaCl solution containing 10 ml of 1N HCl per litre. The mixture is filtered through Whatman Filter paper no. 40 and the residue is leached with excessive quantities of 100 ml of 1N NaCl four to five times till the filtrate measures one litre. 500 ml of the filtrate are transferred to a distillation flask and distilled with 20 ml of 30% NaOH. The distillate is collected in a measured quantity of N/10 sulphuric acid. After the distillation is complete, the acid in the receiving flask is taken out and excess acid is titrated against N/10 NaOH solution with methyl red or bromocresol green indicator. Bromocresol green indicator is better than methyl red since it is least affected by carbon dioxide. The decrease in strength of N/10 acid gives a measure of ammonical nitrogen present in 20 g air-dry soil.

*Calculation:* Let 20 g soil are leached with N NaCl till the leachate is 1.0 litre. 500 ml of the leachate are distilled with excess caustic sauda solution, collecting the distillate in 25 ml of N/10 $H_2SO_4$. If the latter requires 23.2 ml of N/10 NaOH for neutralization, the ammonical nitrogen will be determined as follows.

500 ml of leachate require (25 – 23.2) or 1.8 ml N/10 $H_2SO_4$

1000 ml of leachate will require (1.8 × 1000) ÷ 500 or 3.6 ml of N/10 $H_2SO_4$

i.e. 20 g soil will require 3.6 ml N/10 $H_2SO_4$

100 g soil will require (3.6 × 100) ÷ 20 or 18 ml N/10 $H_2SO_4$

Now,

| | |
|---|---|
| 1 ml of N/10 $H_2SO_4$ | = 0.0014g $NH_4$–N |
| 18 ml of N/10 $H_2SO_4$ | = (0.0014 × 18) g |
| | or 0.0252 g $NH_4$–N |
| i.e. $NH_4$–N content | = 0.0252% |

### *1.3.1.5 Determination of Nitrite Nitrogen*

(Greiss Colorimetric method)

*Principle:* In the process of determination of nitrites, the sulphuric acid is converted by nitrous acid into the corresponding diazo compound and the latter reacts with $\alpha$ - naphthylamine-*p*-azobenzene-

*p*-sulphonic acid, a red azo dye. The intensity of this colour is used to measure the concentration of the nitrite.

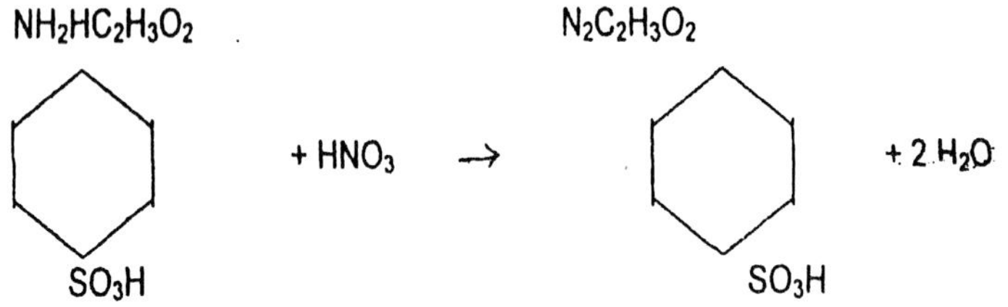

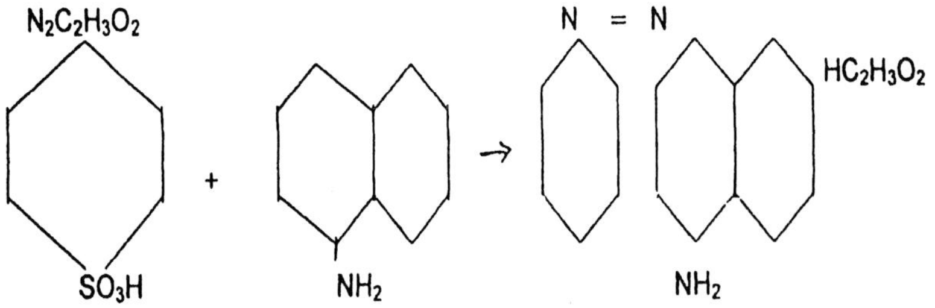

$NH_2HC_2H_3O_2$ (ring) $SO_3H$ + $HNO_3$ → $N_2C_2H_3O_2$ (ring) $SO_3H$ + 2 $H_2O$

$N_2C_2H_3O_2$ (ring) $SO_3H$ + (naphthalene ring) $NH_2$ → N = N (rings) $HC_2H_3O_2$ $NH_2$

Red Azo Dye

***Reagents:***

1. Solution A - 0.5% solution of sulphanilic acid in 30% acetic acid.
2. Solution B - 0.5% solution of α-naphthylamine in 30% acetic acid. Dissolve with the aid of heat, filter and keep in dark-coloured bottle.
3. Solution C - When required, mix equal parts of solution A and solution B.

*Procedure:* 20 g of air-dry soil are treated with about 1g pure calcium sulphate and 100 ml distilled water. Calcium sulphate helps in clearing filtration. The mixture is shaken for about half an hour and filtered through a dry filter paper. To 5 ml of clear filtrate, 1 ml of solution A and 1 ml of solution B are added. The mixture is shaken well and made up to a suitable volume (50 ml or 100 ml) so that no precipitation occurs and reading at the Tintometer should be between 1 and 2. The colour developed is compared with the red slides of a Lovibond Tintometer previously calibrated against colours developed in a similar way in solution of $NaNO_2$ of known strength. In a particular Tintometer,

1°R in 1/2" cell = 0.00324 mg $NO_2$–N

A series of standards is made by introducing 0.2, 0.4, 0.6, 0.8 and 1.0 ml of standard silver nitrite solution (Dissolve 0.22 g silver nitrite in hot distilled water, decompose with slight excess of NaCl, cool and dilute to 1000 ml. Allow the precipitated AgCl to settle, remove 5 ml of clear solution and made up to 1.0 litre. It contains 0.0001 mg of nitrogen per ml.) into Nessler tubes. Each is diluted to 100 ml with distilled water and colour is developed as above.

*Calculation:* Suppose 20 g of soil are treated with 100 ml distilled water and filtered. 5 ml of the filtrate are mixed with the reagents and finally made up to 100 ml. The colour is matched against red slides of a Lovibond Tintometer. Let the reading be 1.2°R in 1/2"cell.

5 ml of filtrate give a reading of 1.2 °R in 1/2" cell

100 ml of filtrate will give a reading of (1.2 × 100) ÷ 5 or 24 °R in 1/2" cell

i.e., filtrate of 20 g soil give a total reading of 24 °R in 1/2" cell

100 g soil will give a total reading of (24 x 100) ÷ 20 or120°R in 1/2"cell

1 °R in 1/2" cell = 0.00324 mg NO2-N

120 °R in 1/2" cell = 0.00324 × 120 mg

= 0.388 mg $NO_2$–N

or 0.000388 g $NO_2$–N

$NO_2$–N content = 0.0004%

### 1.3.1.6 Determination of Nitrate Nitrogen

*(by Colorimetric method)*

In soils and manures, an analysis of nitrate nitrogen helps in determining the availability of nitrogen. The amount is determined by the Colorimetric method.

*Principle:* The method is based on the principle that dried residues of a solution containing nitrates give a yellow colour with phenol-disulphonic acid on making the solution alkaline with ammonia. The yellow colour is developed due to the formation of tri-ammonium salt of nitrophenol disulphonic acid.

$C_6H_5OH + H_2SO_4 \longrightarrow C_6H_3(OH)(SO_3H)_2 + 2H_2O$

$H_2SO_4 + 2\ KNO_3 \longrightarrow 2\ HNO_3 + K_2SO_4$

$C_6H_3(OH)(SO_3H)_2 + HNO_3 \longrightarrow C_6H_2(OH)(SO_3H)_2(NO_2) + 3\ H_2O$

$C_6H_2(OH)(SO_3H)_2(NO_2) + 3\ NH_4OH \longrightarrow C_6H_2(ONH_4)(SO_2ONH_4)_2(NO_2) + 3\ H_2O$ (Triammonium salt of nitrophenol disulphonic acid; a yellow coloured compound)

However, in presence of chloride, the nitrate is lost as follows.

$NaCI + H_2SO_4 \longrightarrow HCl + NaHSO_4$

$NaNO_3 + H_2SO_4 \longrightarrow HNO_3 + NaHSO_4$

$HNO_3 + 3\ HCl \longrightarrow 2\ H_2O + Cl_2 + NOCl$

$HNO_3 + \text{Heat} \longrightarrow 4NO_2 + 2\ H_2O + O_2$

*Reagents :*

1. Phenoldisulphonic acid - Dissolve 25g of pure white phenol in 150 ml of conc. $H_2SO_4$ (AR). Add 75 ml of fuming sulphuric acid and heat to 100°C on water bath for two hours. Cool and preserve the solution in brown stoppered bottle in a dark place.

*Procedure* : Shake 20g air-dried soil with 100 ml of distilled water and stir well for half an hour with addition of 1g calcium sulphate (AR). Filter through a dry filter paper and filtrate is made up to 100 ml. Evaporate 10 ml of filtrate to dryness, cool the residue and treat with 2 ml of phenol-disulphonic acid reagent. Dissolve the residue in a few millilitres of distilled water and transfer to a 100 ml measuring flask. Treat the liquid with excess of $NH_4OH$ (in 2 : 1 ratio). Yellow colour is developed, which indicates the presence of nitrates. It is made up to the mark with distilled water. The reading should preferably be between 1 and 2. The colour is then matched against standard yellow slides of a Loviband Tintometer; calibrated against colours developed with solutions of known quantities of potassium nitrate (0.722 g of oven-dried potassium nitrate are dissolved in water and the solution is diluted to 1.0 litre; 100 ml of this solution is are further diluted to 1.0 litre with water. Each millilitre of the final solution contains 0.01 mg of nitrogen. Or, 0.85 g $NaNO_3$ are dissolved in distilled water. Dilute the solution to 1.0 litre. Take 10 ml of this solution and dilute to 100 ml. Each millilitre contains 0.014 mg $NO_3$-N.).

1° y in 1/2" cell ≡ 0.05 mg $NO_3$–N

### 1.3.2 Determination of Organic Carbon

Organic carbon present in the soil is responsible for many of its physical properties like structure, water holding capacity, micro-biological activities etc. To keep a soil in good agricultural condition, it must be adequately supplied with organic matter.

There are several methods for determination of organic carbon, such as, dry combustion method, Robinson's method, Walkley and Black's method etc. However, rapid titration method of Walkley and Black is more accurate. It is not affected by the presence of calcium carbonate in the soil and is capable of discriminating between elementary carbon and soil organic matter properly.

**Walkley and Black's Method**

*Principle:* Walkley and Black's rapid titration method for determination of organic carbon depends on the oxidation of organic carbon by potassium dichromate.

$$K_2Cr_2O_7 + 4\ H_2SO_4 \longrightarrow K_2SO_4 + Cr_2(SO_4)_3 + 4\ H_2O + 3O'$$

$$\text{Org } 3\ C + 6O' \longrightarrow 3CO$$

i.e. each millilitre of normal $K_2Cr_2O_7$ corresponds to 3 mg of carbon. The excess of dichromate (chromic acid), not reduced by the organic matter is determined by titration with standard $FeSO_4$ solution.

$$2\ FeSO_4 + H_2SO_4 + 3O' \longrightarrow Fe_2(SO_4)_3 + 3H_2O$$

294 g $K_2Cr_2O_7$ which gives 3O' will combine 18 g Carbon

1 g $K_2Cr_2O_7$ will combine (18 ÷ 294) g Carbon

49 g (eq. Wt.) $K_2Cr_2O_7$ will combine [(18 × 49) ÷ 294]g C or 3 g Carbon

or 1 ml of 1N $K_2Cr_2O_7$ = 0.003 g Organic Carbon

But, Walkley found a recovery up to an average of 77%, i.e., 100/77 or a correction factor of 1.3.

Hence, 1 ml of 1 N $K_2Cr_2O_7$ = 0.003 × 1.3 g Organic Carbon

= 0.0039 g Organic carbon

The indicator used for the detection of end point is diphenylamine.

| $C_6H_5NHC_6H$ | → | $(C_6H_5NHC_6H)_2$ | → | $C_6H_5N{:}C_6H_5N{:}C_6H_5$ |
|---|---|---|---|---|
| Diphenylamine | | Diphenylbenzidine (Colourless) | | Diphenylbenzidine (Violet colour) |

A sharp end point (an intense blue-violet colouration) is obtained when the titration is carried out in presence of phosphoric acid or sodium fluoride. It helps in formation of a colourless complex with ferric ions thereby reducing the concentration of the latter and consequently the actual potential of the ferric-ferrous system is kept well below the normal value of 0.76.

*Reagents* :

1. 1.0N $K_2Cr_2O_7$ – Dissolve 49.04 g of AR grade potassium dichromate in water and dilute it to 1 litre.
2. Sulphuric acid - not less than 96%.
3. Phosphoric acid - 85%.
4. 1 N Ferrous ammonium sulphate - Dissolve 292.13 g of ferrous ammonium sulphate in water, add 15 ml of conc. Sulphuric acid and dilute to 1 litre. It should be standardized against 1N $K_2Cr_2O_7$ solution.
5. 0.5% Diphenylamine - Dissolve 0.5 g diphenylamine in a mixture of 100 ml conc. Sulphuric acid (AR grade) and 20 ml of distilled water.

*Procedure* : For determination of organic carbon, the soil should be so grounded as to pass through a 0.5 mm sieve. Take 2 g of sieved soil in a 500 ml conical flask. Add 10 ml of 1N $K_2Cr_2O_7$ solution and 20 ml of cone. Sulphuric acid. Shake for a minute and then leave it for 30 minutes on asbestos sheet. Add about 200 ml distilled water, 10 ml phosphoric acid or 98% hydrogen fluoride or 5 g sodium fluoride and 2-3 drops of 0.5% diphenylamine solution. Shake well and titrate against 1N ferrous ammounium sulphate solution. At the end point, the colour changes abruptly from blue black to green. Then add 0.5 ml of 1 N $K_2Cr_2O_7$ solution to restore an excess of dichromate and complete the titration by adding 1N ferrous ammounium sulphate solution drop by drop until blue colour disappears and changes to green. A blank should also be run side by side.

*Calculation* :

1 ml of 1N $K_2Cr_2O_7$ = 0.003 g Organic Carbon

Organic carbon content = $[(V_1 - V_2) \times 0.003 \times 100] \div W$

where,

$V_1$ – Volume of 1 N $K_2Cr_2O_7$

$V_2$ – Volume of 1 N ferrous ammonium sulphate required for titration.

W – Weight of soil taken.

## 1.3.3 Determination of Calcium Carbonate

### *1.3.3.1 Rapid Titration Method*

*Principle* : When the soil is treated with hydrochloric acid, the calcium carbonate reacts with acid and carbon dioxide gas is liberated.

$$CaCO_3 + 2HCl \longrightarrow CaCl_2 + H_2O + CO_2$$

Therefore, this method involves the neutralization of acid by calcium carbonate and the amount of standard acid neutralized is determined by titration against standard alkali.

$$NaOH + HCl \longrightarrow NaCl + H_2O$$

Reagents:

1. 0.5N HCl – Dilute 40 ml of conc. HCl to 1 litre.
2. 0.1 N NaOH – Dissolve 4 g sodium hydroxide pellets in distilled water and diluted to 100 ml.
3. Phenolphthalein - Dissolve 0.1 g phenolphthalein in 100 ml of 50% alcohol.

*Procedure:* Take 5 g air-dry soil in 250 ml conical flask and add excess of 0.5 N HCl. Boil gently for about 5 minutes and then cool. Add about 50 ml distilled water and 1 ml of phenolphthalein indicator. Titrate it against 0.1N NaOH solution. Near the end point, the soil flocculates rapidly leaving a clear supernatant liquid. At the end point, the liquid is coloured pale pink.

*Calculation:*

$CaCO_3$ content = [($V_1$ – $V_2$) × 0.025 × 100] + W

where,

$V_1$ = Volume of 0.5N HCl required for soil treatment.

$V_2$ = Volume of 0.1 N NaOH required for neutralization of excess acid. W - Weight of soil taken .

### 1.3.3.2 Collin's Calcimeter Method -

*Principle:* This method is based on the same principle as that of Rapid Trtration method except that the carbon dioxide is directly estimated using Calcimeter.

$$CaCO_3 + 2\ HCl \longrightarrow CaCl_2 + H_2O + CO_2$$

*Reagents:* Diluted hydrochloric acid (1:3 ratio)

*Procedure:* 0.2 g calcium carbonate is taken in a 100 ml conical flask and a tube containing 1:3 diluted HCl is placed in it. On shaking, gas is evolved, which is measured on Calcimeter.

Then, 0.2g soil (70 mesh sieved) is taken in a 100 ml conical flask and a tube containing diluted hydrochloric acid is kept in it. The flask is connected with Calcimeter and then shaken. On acid action, carbon dioxide gas is evolved, which is directly measured on Calcimeter.

*Calculation:* Let X g $CaCO_3$ gives $V_1$ ml of carbon dioxide at room temperature. Now, X g of soil sample is taken and by the. action of diluted acid, $V_2$ ml of carbon dioxide is produced.

$V_1$ ml $CO_2$ is produced from X g $CaCO_3$

$V_2$ ml $CO_2$ is produced from [(X ÷ $V_1$) × $V_2$ g $CaCO_3$

X g soil contains [(X × $V_i$) × $V_2$ g $CaCO_3$

100 g soil will contain [(X × $V_1$) × ($V_2$ ÷ X) g $CaCO_3$

or (V2 ÷ VO × 100 g $CaCO_3$

or ($V_2$ × A) g

where, A (100 ÷ $V_2$) is a constant.

## 1.3.4 Determination of Soil Reaction or pH -

According to Bronsted et. al., acids are proton donors while bases are proton acceptors. In other words, acids yield hydrogen ions and alkali the hydroxyl ions. Generally, the hydrogen ioji concentration of most solutions is extremely low. Sorenson (1909) used the term "pH" as a tool to express hydrogen ion concentration. **It is defined as the negative logarithm of the hydrogen ion activity or concentration, i.e.,**

1 ml of 0.5N HCl = 0.025 g $CaCO_3$

$$pH = -\log_{10} [H+]$$

Water is very weakly ionized and at 25°C temperature the concentration of hydrogen ions is only $10^{-7}$ mol/1.

$$H_2O \longleftrightarrow H^+ + OH^-$$

The equilibrium constant for dissociation of water is given by

$$K = \frac{[H^+] [OH^-]}{[H_2O]}$$

or $K = [H^+] [OH^-]$ since $[H_2O]$ is constant.

The ionic product of water at 25°C temperature is $10^{-14}$ and as the concentration of $H^+$ and $OH^-$ ions is equal

$$[H^+] = [OH^-] = 10^{-7}$$

$$pH = -\log_{10} [H^+] = 7$$

pH 7 represents neutrality. Solutions with pH less than 7 are acidic and above 7 are alkaline. Soil pH ranges between 2 and 9 with neutral value at 7.

Soil pH is the most important factor, which determines the chemical nature of soil. Acidic soils are high in exchangeable hydrogen and iron (Fe) and manganese (Mn) are abundant. In alkaline soils, exchangeable bases are quite high and calcium (Ca), molybdenum (Mb), potassium (K), sodium (Na) and magnesium (Mg) are abundant in these soils. Soil pH varies with the soil water ratio and it controls availability of plant nutrients (Fig. 1.3).

The hydrogen ion concentration of soils can be determined either by measuring the potentials developed in a suitable electrical half-cell or by the colour given with suitable indicator solution (Reid & Cummings, 1945).

#### *1.3.4.1 Electrometric Method*

The most convenient and reliable method for measuring pH is the use of a pH meter, which measures e.m.f. of a concentration cell formed from a reference electrode, the test solution and a glass electrode sensitive to hydrogen ions.

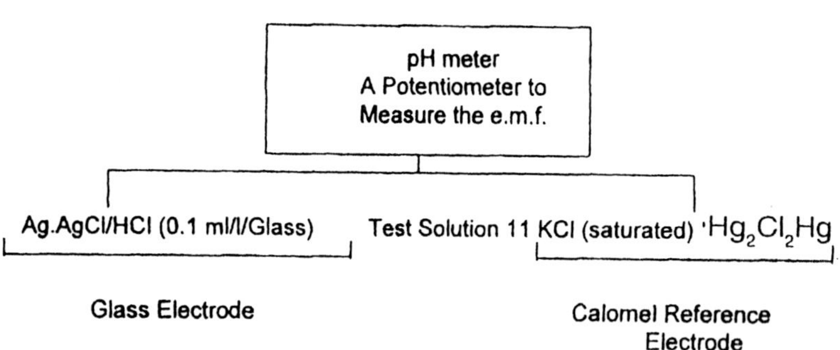


*Reagents:*

1. Standard buffer solution of pH 4.0 (0.05M) - Dissolve 10.207 g recrystallized potassium hydrogen phthalate in water and make up to 1.0 litre.
2. Sorenson's Phosphate buffer solution - Dissolve the appropriate amounts of potassium dihydrogen phosphate and sodium phosphate as indicated below in water and make up to 1.0 litre.

| pH | $KH_2PO_4$ | $Na_2HPO_4 \cdot 2H_2O$ |
|---|---|---|
| 6.24 | 7.262 g | 2.375 g |
| 7.17 | 2.723g | 8.313g |
| 7.73 | 0.908g | 10.688 g |

3. Sodium borate buffer - Dissolve 4.768 g of sodium borate ($Na_2B_4O_7 \cdot 10\ H_2O$) and 3.728 g potassium chloride in water and diluter to 1.0 litre.

*Procedure:* Weigh about 10g of air-dry soil and transfer it to a 125-ml capacity bottle. Add 100 ml of aerated water, fix the stopper and shake it mechanically for about one hour.

Now, remove the glass electrode of Bechman pH meter from the buffer solution and thoroughly rinse the outside with distilled water. Then, dip it in distilled water and flick off any remaining drop. Wipe the upper part of the stem carefully and stir gently. Standarize the pH meter with the buffer solution.

Rinse the stem and ground glass cap of the positive Calomel reference electrode with water and dry lightly with a piece of filter paper. Turn the lower stopcock to the ON position, loosen the ground cap and allow a few drops of KCl solution to escape, in order to renew the liquid junction. Replace the cap and then close the stopcock. Wipe the outside of the cap with a piece of filter paper as before. Shake the bottle with soil sample and bring the wide mouthed bottle containing soil suspension and glass electrode into position. Dip the electrode into the soil suspension and make the pH measurement after 60 seconds.

#### *1.3.4.2 Universal Indicator Method*

*Procedure:* Take about 10 g soil, add pinch of barium sulphate and shake with 50 ml of water for about 30 minutes. If there is not a clear supernatant liquid, transfer the soil suspension to a centrifuge tube and centrifuge it till clear supernatant is obtained. Now, pipette out 10 ml of clear suspension, add 10 drops of Universal indicator and the colour produced is compared with the "colour chart" or in a Loviband Comparator. This gives an approximate pH of soil solution.

### 1.3.5 Water Extract of Soils

When the soil is treated with water, the salts soluble in water come into the solution. Water-soluble salts accumulate in soils under arid and semi-arid conditions. It is necessary to examine salinity of soils for reclamation measures and irrigation purposes. The main constituents of water extract of soils are total soluble salts, carbohydrates, bicarbonates, chlorides, sulphates, nitrates, calcium ($Ca^{++}$), magnesium ($Mg^{++}$) and potassium fig 1.3.

*Procedure:* Take 100 g air-dry soil and 500 ml of distilled water in a flask and shake it for one hour in a shaking machine. Leave it overnight and, next day, filter through a Pasteur Chamberlain pressure filter or Buchner funnel and filtrate is diluted to a desired volume.

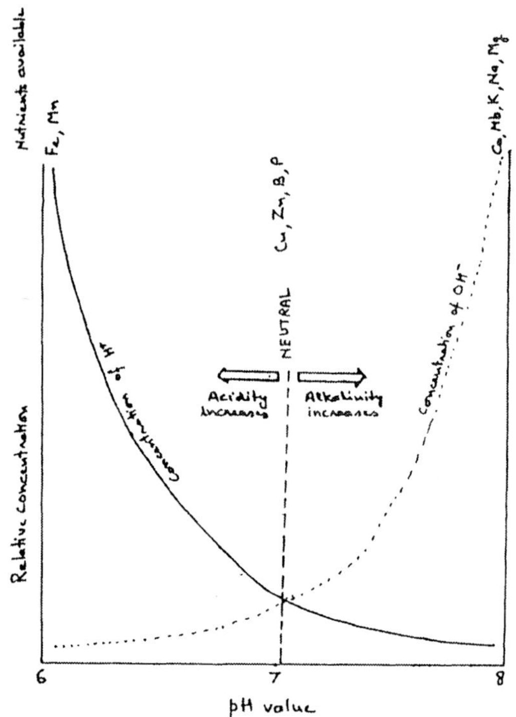


**Fig. 1.3: Soil pH and its relationship to availability of plant nutrients**

### *1.3.5.1 Determination of Total Soluble Salts*

Take 100 ml of water extract in weighed beaker and evaporate to dryness on a water bath. When the volume is reduced to about 5-10 ml, add 2 ml of 20 volume $H_2O_2$ to oxidize organic matter. Evaporate to dryness in an oven at 110°C for 1 hour. Cool the residue in desiccator and weigh it. Calculate the amount of soluble salts as follows.

$$\text{Soluble salt content} = \frac{W_1 \times V}{100 \times W_2} \times 100$$

where,

$W_1$ - weight of residue
$W_2$ - weight of soil
$V$ - volume of water soluble extract taken

### 1.3.5.2 Determination of Calcium Bicarbonate

*Principle* : When a mixture of carbonates and bicarbonates is titrated with standard sulphuric acid, the following reaction takes place.

$$Na_2CO_3 + H_2SO_4 \longrightarrow NaHSO_4 + NaHCO_3$$

$$2\ NaHCO_3 + H_2SO_4 \longrightarrow Na_2SO_4 + 2\ CO_2 + 2\ H_2O$$

Phenolphthalein gives colour so long as $CO_3^{2-}$ is there. In absence of it, the colour is discharged. $HCO_3^-$ can be titrated in presence of methyl orange indicator.

*Procedure* : Take 100 ml of the water extract and titrate it with 0.1N $H_2SO_4$ using phenolphthalein as indicator till the pink colour is just discharged. This end point corresponds to the neutralization of carbonates to the bicarbonates stage.

1 ml of 0.1N $H_2SO_4$ = 0.006 g $CO_3^{2-}$

The liquid from the titration of carbonate is now treated with 1-2 drops of methyl orange indicator and titrated with 0.1N $H_2SO_4$. At the end point, the yellow colour of the solution abruptly changes to orange denoting complete neutralization of the bicarbonates present.

1 ml of 0.1 N $H_2SO_4$ = 0.0061 g $HCO_3^-$

### 1.3.5.3 Determination of Chlorides

*Principle* : In the titration for chlorides, the chlorides are precipitated as AgCl.

$$AgNO_3 + NaCl \longrightarrow AgCl + NaNO_3$$

Potassium dichromate is used as indicator, which reacts with $AgNO_3$ forming a chocolate-coloured $Ag_2Cr_2O_4$

*Procedure* : The aliquot, after neutralization with methyl orange as indicator in the above calcium bicarbonate determination, is treated with 1 ml of 1% potassium chromate solution and titrated with 0.1N $AgNO_3$ solution. At the end point, the colour of the liquid turns to a faint reddish tinge from yellow. The volume of 0.1 N $AgNO_3$ solution required refers to the chlorides present.

1 ml of 0. 1 N $AgNO_3$ = 0.003546 g $Cl^-$

### 1.3.5.4 Determination of Sulphates

*Principle*: Sulphates present in the water extracts are precipitated as barium sulphate by a solution of barium chloride.

$$Na_2SO_4 + BaCl_2 \longrightarrow 2\ NaCl + BaSO_4$$

*Procedure:* Take 100 ml of water extract in a beaker and add few drops of conc. HCl to remove carbonate and bicarbonates. Then, add excess of barium chloride solution and boil till the precipitation is complete. Then, the beaker is kept on a water bath till the precipitate settles down. The liquid is filtered through a 9 cm Whatman No. 44 filter paper. The precipitate is washed repeatedly with hot water till free from chlorides. The precipitate is dried in an oven at 105°C and transferred to a weighed platinum basin and ignited to whiteness. Cool it in a desicator and weigh. The increase in weight of platinum basin gives the weight of barium sulphate.

$$1\ g\ BaSO_4 = 0.412\ g\ SO_4^{2-}$$

If the presence of silica is suspected, purify the partly ignited precipitate by treatment with a little of hydrofluoric acid and a drop of sulphuric acid. Carefully fume off the acids, complete the ignition and weigh.

### 1.3.5.5 Determination of Nitrates

*Reagents:*

1. Phenol Disulphonic acid - Dissolve 25 g of pure white phenol in 150 ml of conc. $H_2SO_4$ (AR grade). Add 75 ml of fuming sulphuric acid and heat to 100°C on water bath for 2 hours. Cool and preserve in brown bottle.

*Procedure:* Evaporate 10 ml of water extract to dryness on a water bath and the residue is treated with 1 ml of phenol-disulphonic acid reagent. Transfer it to 100 ml measuring flask, add excess of $NH_4OH$ and make up to a volume of 100 ml. The yellow colour is matched with standard slides in a Loviband Tintometer.

$$10 \text{ yellow in } \frac{1}{2}\text{" cell} = 0.05 \text{ mg N} = 0.222 \text{ mg } NO_3^-$$

### 1.3.5.6 Determination of Calcium

*Principle:* Calcium is precipitated from the extract by ammonium oxalate as calcium oxalate. Calcium oxalate forms oxalic acid when treated with sulphuric acid and it is oxidized by potassium permagnate, which is the basis of estimation of calcium.

$$CaCl_2 + (NH_4)_2(COO)_2 \longrightarrow Ca(COO)_2 + 2\ NH_4Cl$$

$$Ca(COO)_2 + H_2SO_4 \longrightarrow CaSO_4 + (COOH)_2$$

$$2\ KMnO_4 + 3\ H_2SO_4 \longrightarrow K_2SO_4 + 2\ MnSO_4 + 3\ H_2O + 5\ O'$$

$$(COOH)_2 + O' \longrightarrow 2\ CO_2 + H_2O$$

*Procedure:* Take 100 ml of water extract in a beaker, add about 1 g ammonium chloride, 3-4 drops of methyl red and ammonia till the solution is alkaline, which is indicated by appearance of yellow colour in solution. Then, add few drops of acetic acid till the solution is just acidic (appearance of pink colour). Bring it to boiling now, add 10 ml or more of standard ammonium oxalate solution till the precipitation is complete. Boil for few minutes and then put on a water bath for settling down of precipitate. The solution is now filtered through a 11 cm Whatman No. 44 filter paper. The precipitate is washed several times with hot water till free from chlorides (Filtrate is kept for determination of magnesium).

The precipitate is completely transferred to a beaker by puncturing the filter paper with glass rod and washing the precipitate down by blowing jets of hot water. Wash the filter paper with a few ml of 10% sulphuric acid and then wash again with hot water. If the precipitate of calcium oxalate remains undissolved, add a few more milliliters of sulphuric acid, warm at 60-70°C temperature and titrate it with 0.1N $KMnO_4$ solution till the pink colour is developed. At this stage, the filter paper can be dipped so that unwashed calcium oxalate gets reacted. Finally, complete the titration till a permanent pink colour persists. The volume of 0.1 N $KMnO_4$ solution required corresponds to the amount of calcium present in the water extract.

1 ml of 0.1N $KMnO_4$ = 0.002 g Ca = 0.0028 g CaO

### 1.3.5.7 Determination of Magnesium

*Principle:* In the estimation of magnesium, magnesium is precipitated as magnesium ammonium phosphate by $Na_2HPO_4$ solution and excess of ammonia. Alkalinity of the precipitate is determined by titration with standard acid.

$$MgCl_2 + NH_4OH + Na_2HPO_4 \longrightarrow MgNH_4PO_4 + 2\ NaCl + H_2O$$

$$2\ MgNH_4PO_4 + H_2SO_4 \longrightarrow (NH_4)_2SO_4 + 2\ MgHPO_4$$

*Procedure:* Filtrate of calcium estimation is concentrated and evaporated to dryness. The residue is ignited till the ammonium salts are decomposed. Now, the residue is treated with a few drops of HCl and water and filtered through a 9 cm Whatman No. 44 filter paper. To the filtrate, add 10 ml of 10% $Na_2HPO_4$ solution and then excess of ammonia till the distinct smell of ammonia is found. Stir well and keep overnight. In the meantime, magnesium is precipitated as $MgNH_4PO_4$ The solution is now filtered and the precipitate is first washed with dilute ammonium hydroxide (1:37) till free from chlorides and then with alcohol till free from ammonium. The precipitate along with filter paper is transferred to a beaker and added distilled water and a few drops of methyl red. Titrate with 0.1N NaOH solution. The net amount of alkali required to neutralize the $MgNH_4PO_4$ corresponds to the amount of magnesium present.

1 ml of 0.1 N NaOH = 0.0012 g Mg = 0.002 g MgO

### 1.3.5.8 Determination of Potassium

*Principle:* In the cobaltinitrite method, the potassium is precipitated as $K_2NaCo(NO_2)_6 \cdot H_2O$ when potassium salts are treated with $NaNO_3$ and $Co(NO_3)_2$

$$3\ NaNO_3 + Co(NO_3)_2 \longrightarrow Na_3Co(NO_2)_6$$

$$Na_3Co(NO_2)_6 + 2\ KCl \longrightarrow K_2NaCo(NO_2)_6 + 2\ NaCl$$

*Reagents:*

1. Sodium Nitrite Solution (35%) - Dissolve 35 g of sodium nitrite in water and dilute to 100 ml.
2. Cobalt Nitrate solution (20%) - Dissolve 20 g cobalt nitrate in water and dilute it to 100 ml.
3. Saturated solution of sodium chloride
4. Sodium sulphate solution (2.5%)
5. Glacial acetic acid

*Procedure:* The residue obtained in the estimation of total soluble salts is treated with 1.5 ml of glacial acetic acid and 10 ml of saturated sodium chloride solution. Stir well and, after about 5 minutes, add 5 ml each of sodium nitrite and cobalt nitrate solutions stirring all the time. Stir for about 1 minute and allow the solution to stand overnight in a cool place. The mixture is then filtered, preferably in a Gooch crucible, and beaker and the precipitate is washed repeatedly with 2.5% solution of sodium sulphate. When the filtrate is no longer yellow coloured, transfer the contents to the beaker, add excess of 0.1 N $KMnO_4$ solution with about 20 ml of 10% sulphuric acid and heat nearly to boiling. Allow the reaction to proceed for 20 minutes and then decolourize $KMnO_4$ by measured quantity of excess of 0.1 N oxalic acid solution. Titrate the excess of oxalic acid with 0.1N $KMnO_4$ solution. The difference between the net volume of permagnate required and oxalic acid solution refers to the amount of permagnate reduced by the precipitate $K_2NaCo(NO_2)_6$ from which the amount of K can be calculated indirectly.

1 ml of 0.1 N $KMnO_4$ = 0.00083 g $K_2O$ = 0.0007 g K

### *1.3.5.9 Determination of Sodium*

*Principle:* When a solution of uranyl magnesium acetate in acetic acid is added to water extract, the precipitate of triple salt, sodium uranyl magnesium acetate $[Na(UO_2)_3Mg(CH_3COO)_9\ H_2O]$ is formed.

*Reagent:*

1. Uranyl magnesium acetate solution - Dissolve 3.2 g sodium free crystalline uranyl acetate and 10 g magnesium acetate in water by warming, add 2 ml of glacial acetic acid and 50 ml of 90% alcohol and dilute to 100 ml. Leave in a cool dark place for 1-2 days and filter off any precipitate of sodium uranyl magnesium acetate and store in Pyrex bottle.
2. Alcohol saturated with triple salt.

*Procedure:* Take 6 ml of water extract, add 15 ml of uranyl magnesium acetate and stir well. Cover and keep it for about 1/2 hour but not more than 2 hours otherwise a small amount of potassium salt will be precipitated. Filter through a small Gooch crucible charged with asbestos. Wash the precipitate of triple salt 3-4 times with alcohol saturated with triple salt. Dry the precipitate in an oven at 105°C temperature for one hour, cool and weigh as triple salt. The Gooch crucible is now washed with hot water several times when the triple salt goes into solution. The Gooch crucible is then dried in oven and weighed again. The difference in two weights gives the weight of triple salt formed from sodium in 6 ml of water extract.

1 g of triple salt = 0.015 g sodium

## 1.3.6 HCl Extract of Soils

Generally, digestion of soils with hydrochloric acid dissolves almost whole quantities of nutrients, which are not readily available to plants. If the soil is fused with sodium carbonate, even more constituents come out in solution. Analysis of HCl extract is quite helpful in determining comparative quantities of lime, phosphate and potash in different soils.

Preparation of HCl extract –

*Reagents:*

1. Constant Boiling HCl (Sp gr. 1.125). Dilute 675 ml of conc. HCl to 1.0 litre with water.

*Procedure:* 10 g air-dry soil (2 mm sieved) are taken in a conical flask and treated with 25-50 ml of constant boiling HCl. Put the flask over a sand bath and a funnel on to the mouth of the flask to prevent evaporation of acid. The digestion of soil is continued for 1 hour and when the digestion is complete, add 50 ml of water to the flask and filter the contents. Wash the residue on the filter paper with hot water several times till free from acid. Make up the volume of filtrate to 250 ml with water.

### *1.3.6.1 Determination of Insoluble*

*Principle:* The residue of acid digestion is known as acid insoluble, which mainly consists of silica and resistant rock minerals.

*Procedure:* The residue of acid digestion is dried in an oven and transferred to a weighed platinum crucible. Ignite it, cool and weigh again. The final weight of residue represents the amount of insoluble; calculated as follows.

$$\text{Insoluble content} = \frac{W_1 - W_2}{W} \times 100$$

where,

$W_1$ = Weight of crucible with residue,
$W_2$ = Weight of crucible, and
$W$ = Weight of soil sample taken.

### *1.3.6.2 Determination of Phosphorus - (Volumetric method)*

*Principle:* To check loss of phosphorus through volatilization, magnesium nitrate is added, which destroys organic matter.

$$MgNO_3 + heat \longrightarrow MgO + NO_2$$

$$2\ NO_2 + C \longrightarrow N_2O + CO_2 + O'$$

Simultaneously, the following reactions also take place.

$$2\ P_2 + 5\ O_2 \longrightarrow 2\ P_2O_5$$

or

$$P_2O_3 + O_2 \longrightarrow P_2O_5$$

$$P_2O_5 + 3\ H_2O \longrightarrow 2\ H_3PO_4$$

$$2\ MgO + 2\ H_3PO_4 \longrightarrow Mg_2P_2O_7 + 3\ H_2O$$

Magnesium pyrophosphate is non-volatile and insoluble in water.

*Reagents:*

1. Conc. Nitric acid
2. Dilute nitric acid (1:3)
3. Concentrated hydrochloric acid
4. Magnesium Nitrate solution - Dissolve 346 g magnesium nitrate in 1385 ml of distilled water. Each millilitre contains 0.2 g magnesium nitrate.
5. Ammonium hydroxide
6. Ammonium nitrate solution - Dissolve 400 g of ammonium nitrate in distilled water and dilute to 1.0 litre. 5 ml of this solution contains 2.0 g of the salt.
7. Ammonium molybdate solution - 20 g salt are shaken with 80 ml water till nearly dissolved. Add 20 ml ammonium hydroxide, shake well and filter. The reagent is used in $HNO_3$ medium. For each estimation, take 10 ml of conc. $HNO_3$ in a beaker and add 10 ml of molybdate solution shaking the beaker all the time. A clear warm solution is obtained.

*Procedure :* Take 50 ml of HCl extract in an evaporating dish, add the equivalent of 1.5 - 2.0 g magnesium nitrate (7.5-10 ml of magnesium nitrate solution), evaporate to dryness on a steam bath and ignite to a dull cherry red until the organic matter is destroyed.

15 ml of conc. $HNO_3$ and 5 ml of conc. HCl are added to the ignited sample, covered with a watch glass and digested on the steam bath or over a low flame for 30 minutes. If the flame is used, it should be so adjusted that the solution will be as close as possible to the boiling point without actually boiling. This converts all the phosphorus compounds to orthophosphoric acid, which is soluble in water. Add 20-30 ml of distilled water and allow to stand for sometimes so that all the dehydrated silica may be settled down. Filter and wash with hot water using such an amount that the filtrate is 50-70 ml.

50 ml of this filtrate are taken in a beaker and boiled with 1-2 drops of conc. $HNO_3$. Now, add excess of ammonium hydroxide till the liquid is distinctly alkaline. Filter and wash the residue with hot water till free from chlorides. Transfer the precipitate to the beaker. Dissolve it in dilute nitric acid and filter if the liquid is cloudy. The solution should be colourless at this stage. It is collected in a clean beaker and added to it 5 g of ammonium nitrate. It increases the concentration of nitrate ions, which helps in precipitation of $P_2O_5$ as ammonium phosphomolybdate in granular form and keep molybdic acid in solution. The solution is then warmed and treated with 20 ml of ammonium molybdate solution in $HNO_3$. The amount of molybdate solution required will vary according to the phosphate content of the sample to be analyzed. Avoid too great excess as it merely wastes the reagent. (The precipitating power of 10 ml of 20% ammonium molybdate is 0.04 g of $P_2O_5$). A yellow precipitate of ammonium phosphomolybdate is formed. The liquid is stirred well and kept at a warm place (40-50°C) overnight. A complete precipitation will give a colourless liquid at the top with the yellow precipitate at the bottom at this stage. This is filtered in a Gooch crucible and washed thrice with about 1 ml of diluted $HNO_3$ (to dissolve out any molybdic acid formed during precipitation) and then with 3% $KNO_3$ solution (cold) till free from acid. $H_2O$ is not used as it induces deflocculation of the precipitate and the later may then pass through the filter paper.

The precipitate is now transferred to the beaker and dissolved in excess of 0.1N sodium hydroxide (pink colour with phenolphthalein indicator). Titrate back with 0.1N $HNO_3$. The net volume of standard alkali required corresponds to the amount of $P_2O_5$ present in 50 ml of the extract taken for estimation of $P_2O_5$.

1 ml of 0.1N NaOH = 0.0003088 g $P_2O_5$

*Calculation:* Let 10 g soil be digested with HCl and filtered. The volume of the filtrate is made up to 250 ml. 50 ml of the filtrate are treated with excess ammonium hydroxide and filtered. The washed precipitate is dissolved in dil $HNO_3$ and filtered. The filtrate when warm is treated with ammonium molybdate solution. The washed phospho-molybdate precipitate is dissolved in 20 ml of 0.1N NaOH solution. The alkaline solution required $V_1$ ml of 0.1 N $HNO_3$ for neutralization.

50 ml of the extract have $P_2O_5$, which requires $(20–V_1)$ ml NaOH

250 ml extract will have $P_2O_5$ $[(20\text{-}V_1) \times 250] \div 50$ NaOH

10 g soil have $P_2O_5$, which requires $[(20–V_1) \times 250] \div 50$ ml 0.1 N NaOH

100g soil have $P_2O_5$, which requires $[(20–V_1) \times 250 \times 100] \div 50$ ml 0.1N NaOH

1 ml of 0.1 N NaOH = 0.0003088 g $P_2O_5$

$[(20–V_1) \times 250 \times 100] \div$ 50ml C.1N NaOH = 0.0003088 $\{(20\text{-}V_1) \times 250 \times 100] \div 50$ g $P_2O_5$

### 1.3.6.3 Determination of $K_2O$ and $Na_2O$

*Principle :* HCl extract contains mostly chlorides of iron (Fe), aluminium (Al), calcium (Ca), magnesium (Mg), sodium (Na) and potassium (K). If the solution is made alkaline, iron and aluminium are precipitated and a further treatment with ammonium carbonate $[(NH_4)_2CO_3]$ precipitates calcium and magnesium. The filtarte now contains sodium and potassium, which can be removed by ignition with an excess of ammonium chloride.

*Procedure :* 50 ml of HCl extract are taken in a conical flask and carefully neutralized with ammonium hydroxide after treating with a few drops of conc. $HNO_3$ and solid ammonium chloride. At this stage, the extract will appear just cloudy. Now, treat it with 10 ml of a saturated solution of ammonium carbonate (free from K salt), boil for 2-3 minutes and while boiling add a little solid ammonium oxalate. It facilitates removal of excess barium, calcium and magnesium by precipitation while ammonium, sodium and potassium will remain in the solution. Filter off the precipitate and wash well with hot distilled water. Evaporate the filtrate to dryness and ignite in a platinum crucible to volatilize ammonium salts. The residue is re-dissolved in 10 ml of HCl, evaporated to dryness and extracted with water. The filtrate is taken in a weighed platinum crucible, evaporated and dried in an oven and weighed. The weight gives the quantity of potassium and sodium salts.

The residue of Na & K salts is transferred, by solution, to a beaker and then re-evaporated to semi-dryness. The dry residue is treated with 1.5% acetic acid, 10 ml of a saturated solution of sodium chloride, 5 ml of 35% sodium nitrite and 5 ml of 20% cobalt nitrate stirring all the time. Stir for about 1 minute and allow to stand overnight in a cool place. The mixture is then filtered in a Gooch crucible. The precipitate is repeatedly washed with 2.5% sodium sulphate solution. When the filtrate is no longer yellow-coloured, transfer the precipitate or residue in Gooch crucible to the beaker, add excess of 0.1N $KMnO_4$ solution alongwith about 20 ml of 10% sulphuric acid solution. Heat nearly boiling and allow the reaction to go for 20 minutes. Warm if necessary until all oxides of manganese have been dissolved. Decolourize the $KMnO_4$ by measured quantity of excess of 0.1 N oxalic acid solution. Titrate the excess of oxalic acid with 0.1N $KMnO_4$ solution. The difference between net volume of permagnate required and the oxalic acid refers to the amount of permagnate reduced by the precipitate $K_2NaCO(KO_2)_6$ from which the amount of K can be calculated indirectly.

1 ml of 0.1 N $KMnO_4$ = 0.00083 g $K_2O$

or = 0.0007 g K

The figure for $K_2O$ can be converted to KCl by multiplying the latter with a factor 1.585. This can be subtracted from the combined weight of chlorides to get the weight of NaCl. From NaCl, the percentage of $Na_2O$ can be calculated.

1 g NaCl = 0.53g$Na_2O$

*Calculation :*

$Na_2O$ content $= (W_1 \times 0.53 \times V_1 \times 100) \div (V_2 \times W_3)$

where,

$W_1$ = Weight of NaCl

$V_1$ = Total volume of HCl extract

$V_2$ = Volume of extract taken for estimation

$W_3$ = Weight of soil, and

$W$ = $W_1 + W_2$

where,

$W_1$ = Weight of NaCl + KCl

$W_2$ = Weight of KCl

While,

$$W_2 = V \times 0.00083 \times 1.585$$

where, V is the volume of 0.1N $KMnO_4$ required.

### *1.3.6.4 Determination of Ferric Oxide and Alumina*

*Procedure:* Take 50 ml of HCl extract in a beaker, add about 4-5 g solid ammonium chloride and 1-2 drops of conc. $HNO_3$. Boil and then cool for 2-3 minutes. Add excess of ammonium hydroxide and boil again. Allow the precipitate to settle down. Filter and wash it with hot water till free from chlorides. Preserve the filtrate for estimation of calcium and magnesium.

The precipitate is dried and ignited in a weighed platinum crucible. Cool in a desiccator and weigh. It represents the amount of ferric oxide, alumina and $P_2O_5$ in 50 ml of HCl extract. Transfer the ignited material to a conical flask by jets of water. Add 10 ml of conc. $H_2SO_4$ and boil gently till all the precipitate is dissolved. Cool, dilute and add 5-10 g metallic zinc. Leave the flask till the nascent hydrogen produced reduces all the iron to the ferrous state. Reduction should continue till a drop of the supernatant liquid gives no red colour with $NH_4CNS$ solution. Filter off rapidly the excess zinc in a Buchner funnel, wash well with cold water and titrate the filtrate immediately with 0.1N $KMnO_4$ solution till a permanent pink colour remains.

1 ml of 0.1N $KMnO_4$ = 0.08 g $Fe_2O_3$

Once the $Fe_2O_3$ and $P_2O_5$ are determined, their sum can be subtracted from the weight of $Fe_2O_3 + Al_2O_3 + P_2O_5$ to get the amount of $Al_2O_3$.

### 1.3.6.5 Determination of CaO and MgO

*Procedure :* The filtrate after precipitation of $Fe_2O_3$ + $Al_2O_3$ + $P_2O_5$ is concentrated to about 100-150 ml. Add 2 drops of methyl red and then acetic acid drop by drop till the colour of the solution changes from yellow to pink. The acidification of solution helps in checking the precipitation of magnesium along with calcium. The liquid is now boiled with 20 ml of a saturated solution of ammonium oxalate. Boil it again and put it on a water bath till the precipitate of calcium oxalate assumes a granular form at the bottom. It is then filtered and the precipitate is washed with hot water till free from chlorides. The precipitate is now transferred with hot water to a beaker, dissolved in dilute $H_2SO_4$, warmed to about 70°C and titrated with 0.1N $KMnO_4$ till the colour of the solution remains pink. At the end point, permanent pink colour is developed.

1 ml of 0.1 N $KMnO_4$ = 0.0028 g CaO

The filtrate after the precipitation of calcium is evaporated to dryness and ignited to remove the ammonium salts. The residue is dissolved in a few drops of HCl and 10 ml of water and filtered. The filtrate is treated with 10 ml of 10% sodium phosphate solution ($Na_2HPO_4$) and then excess of ammonia. The liquid is stirred well and left overnight for completion of precipitation of $MgNH_4PO_4$. The solution is filtered and the precipitate washed with dilute $NH_4OH$ (1:19) and later with alcohol. The precipitate is transferred to the beaker, treated with excess 0.1N $H_2SO_4$ and the excess acid is titrated back with 0.1N NaOH with methyl red indicator.

1 ml of 0.1 N $H_2SO_4$ = 0.002 g MgO

*Calculation :* Let 10 g soil be digested with HCl and filtered. The filtrate is made up to 250 ml. 50 ml of the filtrate are treated with excess $NH_4OH$ and filtered. The filtrate is made acidic with acetic acid and boiled with ammonium oxalate. The washed precipitate is dissolved in sulphuric acid, warmed and titrated with 0.1N $KMnO_4$ solution. Let 0.1N $KMnO_4$ solution required be $V_1$ ml. Filtrate from the above is evaporated to dryness and ignited. The residue is dissolved in HCl and filtered. The filtrate is treated with $Na_2HPO_4$ and excess of ammonium hydroxide. The washed precipitate of $MgNH_4PO_4$ is dissolved in 15 ml 0.1N $H_2SO_4$. The volume of 0.1N NaOH required to neutralize the acid be $V_2$ ml.

50 ml of extract requires $V_1$ ml 0.1 N $KMnO_4$

250 ml of extract will require ($V_1 \times 250$) ÷ 50 ml of 0.1 N $KMnO_4$

i.e. 10 g soil require ($V_1 \times 250$) ÷ 50 ml of 0.1 N $KMnO_4$

100 g soil will require ($V_1 \times 5 \times 100$) ml. of 0.1 N $KMnO_4$

1 ml of 0.1 N $KMnO_4$ = 0.0028 g CaO

($V_1 \times 5 \times 100$) ml 0.1 N $KMnO_4$ = ($V_1 \times 5 \times 100$) × 0.0028 g CaO

CaO content = (($V_1 \times 5 \times 100$) × 0.0028 %

50 ml extract require $V_2$ ml of 0.1 N $H_2SO_4$

250 ml extract will require ($V_2 \times 250$) ÷ 50 ml of 0.1N $H_2SO_4$

i.e. 10 g soil requires ($V_2 \times 250$) ÷ 50 ml of 0.1 N $H_2SO_4$

100g soil will require ($V_2 \times 5 \times 100$) ml of 0.1 N $H_2SO_4$

1 ml of 0.1N $H_2SO_4$ = 0.002 g MgO

($V_2 \times 5 \times 100$) ml of 0.1N $H_2SO_4$ = ($V_2 \times 5 \times 100$) × 0.002 g MgO

MgO content = ($V_2 \times 5 \times 100$) × 0.002 %

### 1.3.7 Citric Acid Extract of Soils

HCl extract of soils do not give the amount of $K_2O$ and $P_2O_5$ immediately available to the plants. Extraction of sap from plant roots indicated that their acidity might be equal to that of a 1% solution of citric acid. It has therefore been considered customary to extract a soil in the 1:10 ratio with 1% citric acid solution to estimate available $K_2O$ and phosphorus.

Preparation of Citric Acid Extract (Dyer's method) -

Weigh 100 g of air-dry soil and transfer it to a Winchester quart bottle, add 1.0 litre of water and 10 g citric acid. Allow the solution to remain in contact with the soil for 7 days, shaking a number of times each day. If a shaker is available, 24 hours continuous shaking is sufficient. Filter through a Buchner funnel. Evaporate 500 ml of the extract to dryness and ignite gently until all the organic matter is destroyed. The residue is cooled, treated with strong HCl, evaporated to dryness and heated for some hours at about 120°C to render silica insoluble. Extract the residue with hot water to dissolve potassium chloride. From the residue, available $P_2O_5$ is determined and the filtrate is used for determination of available $K_2O$.

### 1.3.7.1 Determination of Available Potassium ($K_2O$)

*Procedure:* The filtrate is evaporated to dryness and the residue is treated with 1.5 ml of glacial acetic acid, 10 ml of saturated solution of NaCl, 5 ml of 35% $NaNO_3$ and 5 ml of 20% Cobalt nitrate. Keep the solution overnight and then filter through a Gooch crucible charged with asbestos pulp. The precipitate is washed with 2.5% $Na_2SO_4$ solution till the filtrate is no longer yellow coloured. Transfer the precipitate to the beaker and dissolve it in excess of 0.1N $KMnO_4$ and $H_2SO_4$, then bring to boiling and allow the reaction to go for 20 minutes. Decolourize the $KMnO_4$ with known quantity of 0.1N oxalic acid and titrate again with 0.1N $KMnO_4$ till the permanent pink colour is developed.

1 ml of 0. 1 N $KMnO_4$ = 0.00093 g $K_2O$

### 1.3.7.1 Determination of Available Phosphorus

*Procedure:* The residue of the above is transferred to a silica basin and ignited. The residue is treated with 1 ml HCl and 10 ml $HNO_3$. The liquid is evaporated to dryness and treated with 10 ml of dilute $HNO_3$. This process is repeated twice. The final residue is treated with 50 ml of dilute $HNO_3$ and digested for 2-3 minutes till clear solution is obtained. The filtrate is warmed and treated with 20 ml of ammonium molybdate solution. The solution is left overnight and filtered. The precipitate is washed with 3% $KNO_3$ solution till free from acid. It is now transferred to a beaker and dissolved in excess of 0.1 N NaOH with phenolphthalein as indicator.

1 ml of 0.1N NaOH = 0.0003088 g $P_2O_5$

*Calculation:* Let 100 g soil be shaken with 1 litre of 1% citric acid solution for 24 hours and filtered. 500 ml extract is evaporated to dryness and ignited. The residue is digested with water and filtered. The filtrate is evaporated to dryness and $K_2O$ estimated in the residue. Let net volume of 0.1 N $KMnO_4$ solution required be $V_1$ ml.

The residue is dissolved in HCl and $HNO_3$. The liquid is evaporated to dryness to remove silica. The residue is finally dissolved in dilute $HNO_3$. Available phosphorus is estimated in the filtrate. Let net volume of 0.1N NaOH required be $V_2$.

1 ml of 0.1 N $KMnO_4$ = 0.00093 g $K_2O$

$V_1$ ml of 0.1N $KMnO_4$ = $V_1 \times 0.00093$ g $K_2O$

500 ml of the extract contains $V_1 \times 0.00093$ g $K_2O$

1000 ml of the extract will contain $(V_1 \times 0.00093 \times 1000) \div 500$ g $K_2O$

or $(V_1 \times 0.00093 \times 2)$ g $K_2O$

Available $K_2O$ content of soil = $(V_1$ $0.00093 \times 2)\%$

1 ml of 0.1 N NaOH = 0.00031 g $P_2O_5$

$V_2$ ml of 0.1 N NaOH = $V_2 \times 0.00031$ g $P_2O_5$

500 ml of the extract contains $V_2 \times 0.00031$ g $P_2O_5$

1000 ml of the extract will contain $(V_2 \times 0.00031 \times 1000) \div 500$ g $P_2O_5$

or $(V_2 \times 0.00031 \times 2)$ g $P_2O_5$

Available $P_2O_5$ content of soil = $(V_2 \times 0.00031 \times 2)$ %

### 1.3.8 Base Exchange Capacity and Exchangeable Bases of Soil

When a solution of salt like potassium chloride is leached through a soil, it is observed that a certain amount of potassium has been removed from the solution and an equivalent quantity of other bases such as calcium, magnesium or sodium has come into solution. **The capacity of a soil to exchange bases from solution and to retain them against loss by drainage is known as the Base Exchange capacity. It is a property possessed by the finest constituents in the soil, i.e., the clay and organic colloidal matter. This phenomenon is referred as ionic exchange.**

It is not restricted to cations but both cation and anion exchanges are found in the soil. Cation exchange has been commonly known as **Base Exchange.** Since base replacement is a reversible reaction, complete replacement can only be effected if the replaced ions are removed from the sphere of action. Suppose a solution of ammonium acetate (leaching solution) is leached through a soil, the $NH_4^+$ ions replace the other cations viz., Ca, Mg, K, Na etc and after several washings the soil is free from all the cations except $NH_4^+$. The excess of $NH_4^+$ ions is replaced by alcohol (wash the soil with 60% ethyl alcohol). From the leachable solution, exchangeable $Na^+$ and $K^+$ can be easily estimated by flame photometer. The $NH_4^+$ present in the form of ammonium salt can be estimated after evolution of ammonia gas on base reaction and followed by acid-base titration.

Suppose the photometer is adjusted for

40 ppm solution = 100, i.e. 20 ppm = 50

Amount of soil taken = 10 g

Total volume of ammonium acetate leachate = 250 ml

Let the reading of unknown solution be R'. Then,

$$\text{Ions in g percent} = R^2 \times \frac{40}{100} \times \frac{1}{10^6} \times \frac{250}{1} \times \frac{100}{10}$$

$$= R \div 1000$$

$$K^+ \text{ ions} = (R \div 1000) \times (1000 \div 39.1) = 0.026\ R \text{ m.e. } \%$$

$$\text{m.e. \% of C.E.C.} = \frac{\text{Volume of acid used} \times L \times N \times 100}{\text{weight of soil}}$$

If the acid used is of 0.1 N normality and soil taken is 10g, then

$$\text{m.e. \%} = L\ X \times \frac{1}{10} \times \frac{100}{10} = X \times 1$$

i.e., the titration reading directly gives m.e. % (milligram equivalent percent).

An acidic soil has lesser amounts of exchangeable bases like Ca, Mg, K & Na and more of $H^+$, which alter the reaction of the soil. A soil, which has got high amount of exchangeable $Ca^+$, flocculates quickly, have good tilth and permits good drainage. An alkali soil has high amount of exchangeable $Na^+$ and hence it is deflocculated possessing high pH and drainage is difficult.

### *1.3.8.1 Estimation of Base Exchange Capacity and Total Exchangeable loss/bases*

*Reagents :*

1. 1.0 N ammonium acetate solution (pH 7.0): Prepare 1.0 N acetic acid by dissolving 115 ml of glacial acetic acid (Sp. Gr. 1.052) to 1.0 litre and 2.0 N ammonia by diluting 108 ml of conc. $NH_4OH$ (Sp. Gr. 0.88) or 150 ml of $NH_4OH$ (Sp. Gr. 0.91) to 1.0 litre. Mix both the solutions and allow to stand

for some time. Adjust the pH of the solution to 7.0 by $NH_4OH$ if pH is below 7.0 or acetic acid if the pH is above it.

2. 60% alcohol: Add 500 ml of water in 1.0 litre of neutral alcohol.

*Procedure:*

*Cation Exchange capacity* - 10g of air-dry soil is taken in a 500 ml beaker and treated with 100 ml of 1N ammonium acetate solution. Stir the solution for an hour and keep overnight. Next day, the solution is filtered through Whatman filter paper. The residue is washed repeatedly with ammonium acetate solution till the leachate is about 1.0 litre. The leachate is preserved for estimation of total exchangeable loss/bases.

The residue is washed thoroughly with 60% alcohol till the filtrate gives negative to $NH_4^+$ ions. The residue is then transferred to a distillation flask and added 20 ml of 30% NaOH solution, few drops of phenolphthalein and a glass bead. The flask is heated and the distillate is collected in a receiving flask containing 25 ml of 10% boric acid and a few drops of mixed indicator (if pH 4.5, 0.5% Bromocresol green and 0.1% methyl red dissolved in 100 ml of 95% ethanol and pH is adjusted to 4.5 with dilute HCl or NaOH). When the distillate is about 200 ml, it is disconnected and titrated against 0.1N $H_2SO_4$ solution. The total amount of 0.1 N $H_2SO_4$ required to neutralize ammonia evolved for 100 g soil is the total base/cation exchange capacity of soil.

1 ml of 0.1N $H_2SO_4$ = 0.1 m. e. of any base

*Total Exchangeable Loss/Bases* - The leachate of the above is evaporated to dryness on a steam bath and ignited carefully, first over a small flame and then strongly. After cooling, the residue is treated with excess of 0.1N HCl (excess denoted by red colour with methyl red indicator) till a clean pink colour results. It is now titrated back with 0.1N NaOH till the colour changes from pink to yellow.

1 ml of 0.1 N HCl = 0.1 m.e. of any base

### *1.3.8.2 Determination of Exchangeable Calcium and Magnesium (EDTA method)*

*Reagents :*

1. $NH_4Cl$-$NH_4OH$ buffer pH 10.0: Dissolve 67.5 g of $NH_4Cl$ and 570 ml of conc. $NH_4OH$ in 1 liter of water.

2. 10% KOH
3. 0.01N calcium chloride solution
4. 0.01N EDTA
5. EBT & Calcone indicator: 0.4% Erichrome Black T in methanol containing 4 g hydroxylamine hydrochloride.
6. 1N NaCl: Dissolve 58.5 g NaCl in 1 litre water and adjust pH to 7.0 by HCl or NaOH.

*Procedure :* 10 g of air-dry soil (10 mesh) is taken in a 500 ml beaker and added to it 100 ml of 1N NaCl with stirring for 5 minutes. It is left overnight. Next day, the solution is filtered through a Whatman No. 42 filter paper. The filtrate is collected in a 250 ml measuring cylinder. The residue on filter paper is leached repeatedly with 1N NaCl and the filtrate is made up to 250 ml.

*Calcium* - Pipette out 5 ml of aliquot in 50 ml porcelain dish and dilute to a volume of 25 ml with water. Add few drops of 10% KOH and two drops of Calcone indicator. Then titrate it against EDTA using 10 ml micropipette. At the end point, colour changes from red lavender to purple.

$$\text{Ca m.e \%} = \frac{X \times 250 \times 100}{5 \times 10} \times \text{Normality of EDTA}$$

$$\text{or} = X \times A$$

where,

X – Volume of EDTA used in titration

A – 500 × Normality of EDTA

*Calcium & Magnesium* - Pipette out 5 ml of aliquot into a 100 ml conical flask and dilute to 25 ml. Five drops of $NH_4Cl$-$NH_4OH$ buffer and 1-2 drops of EBT indicator are added to it. Titrate it against EDTA solution until colour changes from red to blue or green.

$$\text{Ca \& Mg m.e. \%} = \frac{Y \times 250 \times 100}{5 \times 10} \times \text{Normality of EDTA}$$

or = Y × A

where,

Y - Volume of EDTA used in titration

A - 500 × Normality of EDTA

Mg m.e. % = Y.A-X. A

### 1.3.8.3 Determination of Exchangeable Sodium and Potassium

*Procedure:* The leachate of method for estimation of cation exchange capacity is evaporated to dryness and ignited. The residue is moistened with a few drops of saturated ammonium carbonate solution. It is digested with warm water and filtered in a weighed platinum crucible. The filtrate is evaporated to dryness, ignited, cooled and weighed. Increase in weight of crucible represents the weight of sodium and potassium carbonates.

Transfer the dry residue of the above in a beaker with a few ml of dil. HCl and evaporated to dryness again. The residue is treated with glacial acetic acid (1.5 ml), 10 ml of saturated NaCl solution, 5 ml of 35% $NaNO_2$ and 5 ml of 20% cobalt nitrate solution. The liquid is stirred well and left overnight. Next day, it is filtered through Gooch crucible charged with asbestos. The precipitate is washed with 2.5% $Na_2SO_4$ solution till the filtrate is no longer yellow. The precipitate along with crucible is now treated with dil. $H_2SO_4$ and excess of 0.1 N $KMnO_4$ and warmed gently. Add a further amount of standard permagnate solution if there is colour discharge. Heat upto nearly boiling. After a few minutes, add a sufficient measured quantity of 0.1 N oxalic acid to decolourize the solution. Warm for few minutes and then titrate the excess of oxalic acid with 0.1 N $KMnO_4$.

1 ml of 0.1N $KMnO_4$ = 0.0007 g K or 0.018 m.e. K

The amount of K can be converted into $K_2CO_3$ by multiplying the K value by1.77. The amount of $K_2CO_3$ can be substrated from the total weight of $Na_2CO_3$ and $K_2CO_3$ and thus the amount of $Na_2CO_3$ is obtained. From $Na_2CO_3$, the amount of Na is obtained by multiplying with 0.44.

*Calculation:* Suppose total weight of $Na_2CO_3$ and $K_2CO_3$ is W g and volume of 0.1N $KMnO_4$ required is V ml. Then,

$$\text{Weight of K} = \frac{W \times 0.0007 \times 100}{10}$$

$$= 0.007\ W$$

or weight of $K_2CO_3$ $= 0.007\ W \times 1.77$ g

or say, $W_1$ g

Therefore,

Weight of $Na_2CO_3$ = $(W - W_1)$ g

$$\text{Exchangeable Na} = \frac{(W - W_1) \times 0.44 \times 1000}{23} \text{ m. e. \%}$$

$$\text{Exchangeable K} = \frac{0.007W \times 1000}{39} \text{ m. e. \%}$$

### 1.3.9 Determination of Available Phosphorus

#### 1.3.9.1 Bray's Method for Acidic Soils

*Reagents:*

1. ***Bray's Extract No. 1 :*** 12.2 g $NH_4F$ is dissolved in 200 ml distilled water and filtered. 18 litres of water containing 40 ml conc. HCl are added to the filtrate and the volume is made up to 20 litres with distilled water.
2. ***Molybdate Reagent :*** Exactly 15 g AR grade ammonium molybdate is dissolved in about 300 ml of distilled water, warmed to 45°C and then filtered to remove any sediment. The molybdate solution is cooled and 350 ml of 10N HCl are added slowly with rapid stirring. When the solution has cooled down to room temperature, it is diluted with distilled water to exactly 1000 ml. It is mixed thoroughly and stored in an amber-coloured glass stoppered bottle.
3. ***Stannous chloride :*** 10 g $SnCl_2 . 2H_2O$ is dissolved in 25 ml of conc. HCl. A piece of pure metallic tin is added to the solution and kept in an amber-coloured bottle. 0.5 ml of this stock solution is diluted to 66 ml by distilled water just before use.

*Preparation of Standard Curve* : 0.1916 g of pure dry potassium dihydrogen phosphate are dissolved in 1 litre distilled water. This solution contains 0.1 mg $P_2O_5$ per ml. 2 ml of this stock solution are diluted to 100 ml with distilled water. It will contain 2 μg $P_2O_5$ per ml. 0.5, 1.0, 2.0, 4.0, 6.0 and 8.0 ml of this solution are separately taken in 25 ml volumetric flasks. 5 ml of extract solution are added to each flask followed by 5 ml of molybdate solution. The solution is diluted to 20 ml with distilled water. 1 ml stannous chloride solution is added, shaken well and diluted to 25 ml mark by distilled water. After 10 minutes and before 20 minutes, the blue colour is developed, which is read on photoelectric colorimeter using 660 mμ red filter (Klett No. 66). Graph is plotted with meter reading against μg $P_2O_5$.

*Procedure* : 50 ml of the Bray's extract no. 1 are mixed with 5 g soil sample. Shake for 5 minutes and filtered with Whatman filter paper. 2.5 ml of aliquot are taken into 25 ml volumetric flask and the process is repeated as in case of standard curve.
Let there be A μg $P_2O_5$ as per standard curve. Then,

$$\frac{A}{1000} \times \frac{50}{5} \times \frac{100000}{5} \times \frac{1000}{1000} = 2\,A \text{ (kg } P_2O_5 \text{ per acre)}$$

### 1.3.9.2 Olsen's Method for Neutral or Alkaline Soils

*Reagents:*

1. Bicarbonate Extractant: Dissolve 8.4 g sodium bicarbonate in 20 litres of distilled water and adjust the pH to 8.5 by NaOH or HCl.
2. Active Carbon (Darco 60)
3. Molybadte solution: same as Bray's Method except that instead of 350 ml of 10N HCl, 400 ml of 10N HCl are added.
4. Stannous chloride: same as Bray's Method

*Procedure* : Use multiple dispenser to add 50 ml of the bicarbonate extractant to 2.5 g measured soil sample. Add 1 g measured quantity of decolouring carbon, shake for 30 minutes on mechanical shaker and filter. Develop colour and read on colorimeter as described in

Bray's Method. Calculate kg $P_2O_5$ per acre by multiplying concentration in µg with 4.

## 1.4 SOIL MANAGEMENT

Productivity of some soils is lowered by unfavourable soil conditions viz., salt content, soil reaction etc. These soils are referred as **problem soils**, which require remedial measures and management practices for satisfactory crop production. There are two major categories of problem soils - salt affected soils (saline soils and sodic soils) and acidic soils.

### 1.4.1 Salt-affected Soils

#### *1.4.1.1 Saline soils*

These soils contain excess of neutral salts dominated by chlorides and sulphates and adversely affect plant growth. The main soluble salts that accumulate in these soils are of sodium, calcium and magnesium as cations and chlorides and sulphate as anions. The concentration of potassium carbonates, bicarbonates and nitrates is comparatively low. Total concentration of salts is estimated in the soil solution either from weight after evaporation or from electrical conductivity.

Crop plants differ in their tolerance to salinity. Removal of excess salt to a desired level in the root zone is the basic principle of reclamation of saline soil, which is done either by leaching with good quality water or adequate drainage. Leaching is a process of transportation of soluble salts by downward movement of water in soil by application of water. Drainage is the removal of excess water from soil. It may be surface drainage or subsurface drainage. However, besides water management, cropping system is very important for crop production in saline soils. For optimum crop production in saline areas, selecting crop and crop varieties, adoption of best suited cultural and fertilizer practices are essential. Cultivation of salt tolerant crops is the best management practice.

#### *1.4.1.2 Sodic soils*

*Sodic soils are characterized by high exchangeable sodium percentage, sufficient to interfere with plant growth, high sodium absorption ratio (SAR) of saturation extract, presence of a large concentration of sodium carbonate type salts, and low permeability. Such soils are also referred as alkali soils.* High pH (more than 8.5) and high sodicity (ESP more than 15) of sodic soil are the main causes of unproductiveness. Reclamation of these soils is done by the ameliorative measures, which replace the exchangeable sodium by calcium and the exchangeable sodium released as sodium salt is leached out of root zone. Gypsum is widely used for reclamation of sodic soils, the requirement of which is determined by Schoonover's method. In this method, a weighed quantity of sodic soil (5 g) is shaken with 100 ml of a saturated gypsum solution of known calcium concentration and filtrate is analyzed for $Ca^{++}$ plus $Mg^{++}$. If $C_1$ is the concentration (m.e. / l) of $Ca^{++}$ in gypsum solution and $C_2$ is the concentration of $Ca^{++}$ and $Mg^{++}$ in filtrate, then the gypsum requirement is given by

$$\text{Gypsum requirement} = \frac{(C_1 - C_2) \times 100 \times 100}{1000 \times 5}$$

$$= (C_1 - C_2) \times 2 \text{ m. e. of } Ca^{++} \text{ per 100 g soil}$$

### 1.4.2 Acidic soils

Soil pH is the measure of soil acidity. Soils with low pH contain relatively high amounts of exchangeable $H^+$ and $Al^{++}$. **Soils, which have pH less than 5.5 in 1:1 soil-water suspension, are known as acidic soils.** Such soils are also characterized by high exchangeable hydrogen and almunium concentration and managed in two ways - by growing crops suitable for a particular soil pH or by ameliorating the soils through application of amendments like lime, which raises soil pH.

The lime requirement of soil is determined by the Shoemaker's buffer method. In this method, 5 g soil is treated with 5 ml of distilled water and 10 ml of the extractant buffer (1.8 g nitrophenol, 2.5 ml triethanolamine, 3 g potassium chromate, 2 g calcium acetate, and 53.2 g calcium chloride dehydrate are dissolved in 1 litre of water and the pH is adjusted to 7.5 with dilute NaOH solution), stirred continuously for 10 minutes and the pH of the suspension is determined. Lime requirement is read from the table given below.

**Tab. 1.5: Determination of Lime requirement of soils**

| *pH of soil-buffer suspension* | *Lime required to bring the soil to indicated pH (in tons/acre of pure calcium carbonate)* | | |
|---|---|---|---|
| | pH 6.0 | pH 6.4 | pH 6.8 |
| 6.7 | 1.0 | 1.2 | 1.4 |
| 6.6 | 1.4 | 1.7 | 1.9 |
| 6.5 | 1.8 | 2.2 | 2.5 |
| 6.4 | 2.3 | 2.7 | 3.1 |
| 6.3 | 2.7 | 3.2 | 3.7 |
| 6.2 | 3.1 | 3.7 | 4.2 |
| 6.1 | 3.5 | 4.2 | 4.8 |
| 6.0 | 3.9 | 4.7 | 5.4 |
| 5.9 | 4.4 | 5.2 | 6.0 |
| 5.8 | 4.8 | 5.7 | 6.5 |
| 5.7 | 5.2 | 6.2 | 7.1 |
| 5.6 | 5.6 | 6.7 | 7.7 |
| 5.5 | 6.0 | 7.2 | 8.3 |
| 5.4 | 6.5 | 7.7 | 8.9 |
| 5.3 | 6.9 | 8.2 | 9.4 |
| 5.2 | 7.4 | 8.6 | 10.0 |
| 5.1 | 7.8 | 9.1 | 10.6 |
| 5.0 | 8.2 | 9.6 | 11.2 |
| 4.9 | 8.6 | 10.1 | 11.8 |
| 4.8 | 9.1 | 10.6 | 12.4 |

# 2

# SOIL FERTILITY AND APPLICATION OF FERTILIZERS

Soil fertility status and application of fertilizers are closely related aspects for better farm management. **Soil fertility refers to the capacity of the soil to supply plant nutrients in available form to the plants.** Although a fertile soil may or may not be productive depending upon several factors, but every productive soil must be fertile. Therefore, it is very essential for a farmer to know the soil fertility status so that he may manipulate it with the addition of fertilizers so as to get maximum production. Otherwise, the imbalanced fertilizer application may produce the following ill effects

* Excess of elements may be toxic to the protoplasm of the plant,
* Excess of an element may cause deficiency of another e.g.
  - Excess of nitrogen causes potash starvation in some crop plants,
  - Excess of potash causes deficiency of manganese in tomato,
  - Excess of phosphorus causes deficiency of zinc in most of the crops.

Apart from these, imbalanced fertilization increases cost of production because excess of any nutrient does not help in higher production at all.

## 2.1 SOIL FERTILITY

The soil fertility status is evaluated by several methods as discussed below:

### 2.1.1 Chemical Analysis of Soil

A composite soil sample collected randomly is subjected to chemical analysis for a quantitative determination of available nitrogen, phosphorus, potash and other plant nutrients. It will give the amount of nutrients present in soil, i.e., soil analysis reveals its fertility. The data is useful for recommending the type and amount of fertilizers and other amendments for increased and profitable crop production. The detailed methods are discussed in Chapter 1.

***Interpretation of Soil Test Values*** : Different soils with different soil test values differ in their capacity to supply nutrients to crops. The crops vary in their nutrient requirement for a certain level of yield. The lower the soil test value for a particular nutrient, the higher is the response to the fertilizer nutrient. Assessment of nutrient requirement for various crops or crop sequences for different soils is essential for rational use of fertilizers. It is done by soil test calibration method, which has got two approaches.

***Soil analysis-correlation approach*** - In this approach, a group of soils ranging in fertility from high to low in respect of a particular nutrient are selected and a crop is grown on these soils with varying doses of that nutrient over a basal dose of other nutrients in adequate amount. The soil test values are plotted against the percentage yield and correlation coefficient between soil test values and percent yield response is calculated.

Percent yield =

$$100 \times \frac{\text{(Crop yield with adequate nutrients)} - \text{(yield of control)}}{\text{Crop yield with adequate nutrients}}$$

Control has all the nutrients except the one under study. Based on the contents of available nutrients and soil test values (NPK), the soils are grouped into classes such as low, medium and high.

*Nutrients*

| *Nutrient* | *Low* | *Medium* | *High* |
|---|---|---|---|
| Organic carbon (%) for nitrogen | Below 0.5 | 0.5-1.0 | Above 1.0 |
| Available Nitrogen (kg/ha) | Below 250 | 250 - 500 | Above 500 |
| Available phosphorus (kg/ha) | Below 25 | 25-50 | Above 50 |
| Available potash (kg/ha) | Below 125 | 125-250 | Above 250 |

*pH/soil reaction*

| *pH* | *Reaction* |
|---|---|
| <7.0 | Acidic |
| 7.0 | Neutral |
| >7.0 | Alkaline |

*Electrical conductivity*

| *EC (dS/m)* | *Extent of suitability* |
|---|---|
| <1.0 | Suitable for mulberry growth |
| 1.0-2.0 | Critical for rooting of mulberry |
| 2.0-3.0 | Critical for growth of mulberry |
| >3 | Injurious to mulberry |

***Critical soil test level approach*** - As per this approach, the critical soil test level is the level of nutrient below which the application of that nutrient gives positive response and above which the response is low. In this approach, the soils collected from each field are analysed, field experiments are conducted with the application of graded doses of fertilizers and the response curves are fitted. A scattered diagram of percentage yield (y-axis) and soil test value (x-axis) is plotted. It is divided in four quadrants as shown positioning the lines in such a way that the number of points in the upper left and lower right quadrants is minimum. The point where the vertical line parallel to the y-axis crosses the x-axis is defined as the 'critical soil test value'. The critical levels vary according to the nature of soil and crop.

***Fertilizer recommendation*** - Fertilizer recommendation for efficient and economic crop production is based on soil testing and the economic return-profitability of fertilizer application. Outputs for different doses of fertilizers are obtained from the response curve. On the basis of costs of inputs and outputs, the required information may be derived on maximum yield, maximum profit per hectare, maximum rate of return per rupee of investment and minimum profitable application. A farmer may make his choice accordingly.

***Soil Test Report, Soil Test Summaries, Nutrient Index and Soil Fertility Map*** - Soil test information is compiled soil-wise or area-wise in the form of soil test summaries. The soil test summary of an area indicates the number of samples falling under the categories of Low, Medium and High NPK content. It is useful for ascertaining the logistics of fertilizer distribution and consumption.

Average soil fertility is expressed by soil's nutrient index (NI) calculated as per the following formula.

$$\text{Nutrient Index} = \frac{N_l + 2N_m + 3N_h}{N_l + N_m + N_h}$$

where, $N_l$, $N_m$ and $N_h$ are the number of soil samples falling in the category of low, medium and high nutrient status and are given weightages of 1, 2 and 3 respectively.

A soil fertility map is prepared by plotting the value of the NI on an outline map of any area.

### 2.1.2 Field Trials for Fertility Evaluation

Field trials are conducted just to determine the response pattern of the plants to the nutrients applied for ascertaining the fertilizer scheduling. According to the nature of experiments, the trials are of two types.

*Simple Trial* - This type of trials are conducted to find out as to what element is the limiting factor in foliage production. Different combinations of N, P and K are tried and compared with control.

*Complex Field Trials* - In this type of experiment, more than one factor is always studied and the results explain about the whole of the fertilizer scheduling. The most suited design are split plot design or complex randomized block design.

## 2.2 IMPORTANCE OF PRIMARY ELEMENTS OF FERTILIZERS

In general, nitrogen, phosphorus and potassium are the most essential elements for growth of crop plants and their rational proposition in fertilizer application ensures luxurious growth. So far as sericulture is concerned, muga and oak tasar silkworms are reared on the existing natural flora of their food plants while mulberry, tropical tasar and eri silkworms are reared on the food plants, which are propagated. So, application of these three elements is a pre-requisite for having good foliage harvest. The quality and quantity of foliage in mulberry, tropical tasar and eri food plants depend upon the quantity of these elements in the fertilizers applied. If the quantity of nitrogen fertilizers applied to mulberry plants is large, the water and nitrogenous matter increase while the content of carbohydrates, phosphorus, potassium and lime decreases in the leaves. On the other hand, large amount of phosphorus fertilizers leads to low nitrogenous matter content and in case the amount of potassium fertilizers and lime is increased, the same phenomenon is observed. However, protein content in leaves increases with the increase in water content of soils. In short, application of fertilizers will be governed by the nutrient requirements of the silkworm larvae.

## 2.3 FERTILIZER APPLICATION

The importance of fertilizer application for both increased productivity and improved quality of leaves has been well recognized. Soil fertility status has to be enriched with fertilizers and manures. Of the three major elements, nitrogen is the most important and vital for increased production. However, balanced use of NPK fertilizers gives best results as compared to individual fertilizer application. Compound fertilizers are applied as basal doses and the straight fertilizers containing nitrogen are applied in split dose as top dressing. In India, for irrigated mulberry crops, ratio of NPK is 2.5:1:1 and for rain-fed 2:1:1.

### 2.3.1 Dosages of Fertilizers and manures

Quantity of fertilizers and manures depend on several factors like soil type, removal of plantation, rainfall distribution pattern, spacing, pruning, harvesting method and climate etc. In addition, it also depends on the availability of a particular nutrient in the particular fertilizer or manure.

In India, dosages of fertilizers vary from region to region and with intensity of plant cultivation. Under rainfed conditions, organic manure is applied at the rate of 5-10 MT per hectare of mulberry plantation as basal dose at the time of annual pruning and fertilizers at the rate of 100 kg N, 50 kg $P_2O_5$ and 50 kg $K_2O$ per hectare in two split doses. Under irrigated conditions, 10-20 MT of organic manure per hectare is applied as a basal dose and 250 kg N, 100 kg $P_2O_5$ and 100 kg $K_2O$ in 4-5 split doses in a year per hectare. In temperate climate, fertilizers are applied during the growing season while in tropical climate fertilizers are applied during monsoon season under rainfed conditions.

### 2.3.2 Methods of Application of Fertilizers and manures

Effective utilization of fertilizers and manures depends upon several factors, like, soil factors (fertility, soil reaction, physical condition of soils), climatic factors (temperature, rainfall distribution and day length), crop factors (nature of crop, variety of crop) and agronomic factors (fertilizer responsiveness of crops and their varieties, timely sowing, proper spacing, proper dose, time and method of application). The method of application may be in solid form or in liquid form and also may be broadcasting, placement and foliar application depending upon the overall situation.

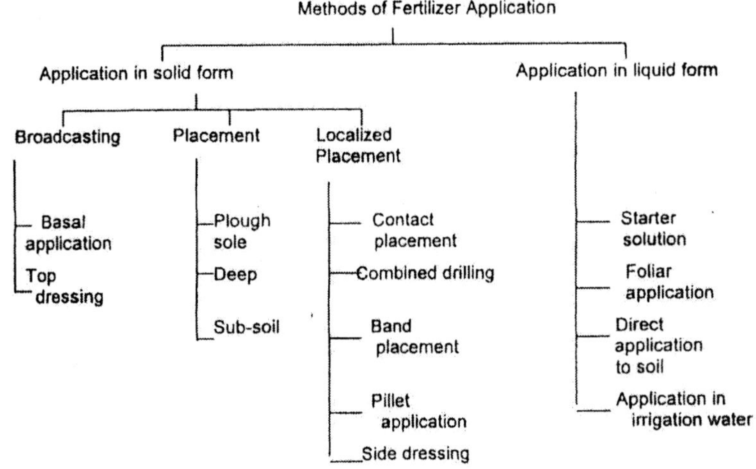


## 2.4 ANALYSIS OF MANURES AND FERTILIZERS

Manures and fertilizers are applied to soils to augment the supplies of nutrients by foliage crops to grow as these are regularly taken by the plants from the soils. A manure or fertilizer is generally tested for the following constituents.

a. Organic matter

b. Nitrogen

c. $P_2O_5$

d. $K_2O$

e. CaO

### 2.4.1 Analysis of Nitrogenous Fertilizers

*Analysis of Nitrogen in Ammonium Sulphate* - Weigh accurately 1 g of ammonium sulphate and transfer it to a 250 ml flask. Make up the volume with water. Shake well and pipette 25 ml of the solution in a distillation flask. Distil the contents of flask with 10 ml of 30% NaOH solution; keeping 25 ml of $H_2SO_4$ with a drop of methyl red indicator in a receiving flask. Ammonia liberated during distillation process will be absorbed by the acid in the receiving flask. When about 200 ml of the distillate are collected, disconnect the flask after washing the tip of condenser and titrate the acid with 0.1N NaOH till the pink colour just changes to yellow. The difference in the volume of acid taken and the volume of alkali required to neutralize the excess acid after distillation gives the amount of acid neutralized by ammonia.

1 ml of 0.1N $H_2SO_4$ = 0.0014g N
= 0.0017g $NH_3$
= 0.0066 g $(NH_4)_2SO_4$

Suppose,

0.1N $H_2SO_4$ taken = $V_1$ ml

0.1N NaOH required to neutralize the excess acid = $V_2$ ml

Volume of 0.1N $H_2SO_4$ required for 25 ml of ammonium sulphate solution = $(V_1-V_2)$ml

Therefore,

1 g of $(NH_4)_2SO_4$ contains = $(V_1 - V_2) \times 0.0014$ g N
= $0.0014\ (V_1 - V_2)$g N

Nitrogen content of sample = $0.0014\ (V_1 - V_2) \times 100$

Now,

$(NH_4)_2SO_4 = 2\ NH_3$ = 2 N

Therefore,

132g of $(NH_4)_2SO_4$ = 28 g of N

| | |
|---|---|
| 100g of $(NH_4)_2SO_4$ | = (28 × 100) ÷ 132 |
| or | = 21.21% N |
| Purity of sample | = [0.0014 ($V_1$ − $V_2$) × 100] × 100 ÷ 21.21 % |

*Analysis of Nitrogen in Sodium Nitrate* - Weigh 5 g fertilizer and transfer it to a beaker. Dissolve in 50 ml water and filter in 250 ml flask. Wash the residue 5-6 times and make up the volume of filtrate and washings to 250 ml. Transfer 10 ml of the solution to a distillation flask and dilute to 600 ml with water. Add 5 g of Devarda's alloy and 10 ml of 30% NaOH solution. Distil and collect the distillate in 50 ml of 0.1N $H_2SO_4$ having a few drops of methyl red as indicator. Discontinue distillation when about 200 ml of the distillate is collected. Titrate back the excess acid by 0.1 N NaOH, the end point becoming yellow at this stage. The net amount of 0.1 N acid required refers to the amount of nitrogen present in the material taken.

| | |
|---|---|
| 1 ml of 0.1 N $H_2SO_4$ | = 0.0014g N |
| Suppose, | |
| 0.1N $H_2SO_4$ taken | = $V_1$ ml |
| 0.1N NaOH required to neutralize the excess acid | = $V_2$ ml |
| Volume of 0.1N $H_2SO_4$ required | = ($V_1$ − $V_2$) ml |
| 10 ml of the solution contains | = ($V_1$ − $V_2$) × 0.0014 g N |
| 250 ml of the solution contains | = ($V_1$ − $V_2$) × 0.0014 × 25 g N |
| or | = 0.035 ($V_1$ − $V_2$) g N |
| 4.75 g $NaNO_3$ contains | = 0.035 ($V_1$ − $V_2$) g N |
| Therefore, | |
| Nitrogen Content | = 0.035 ($V_1$ − $V_2$) × 100 % |

### 2.4.2 Analysis of Phosphatic Fertilizers

*Analysis of $P_2O_5$ in Superphosphate -*

*Water soluble* $P_2O_5$: 2 g powdered superphosphate fertilizer are taken in a beaker. Add 25 ml of distilled water, stir well and filter. Wash the residue 4-5 times with hot water. Collect the filtrate and washings in 250 ml flask and make up the volume. 50 ml of the extract are taken in a basin. Add 10 g solid $NH_4NO_3$ and 20 ml conc. $HNO_3$. The solution is then warmed to 80°C and 40 ml of 20% ammonium molybdate solution are added till the precipitation is complete. Keep the mixture in a warm place overnight. Filter it through Whatman No. 42 Filter Paper and wash with cold 3% $KNO_3$ solution till free from acid. The precipitate is now transferred to the beaker, dissolved in excess of 0.1 N NaOH and the excess alkali determined by titration with 0.1N $HNO_3$ using phenolphthalein as indicator.

1 ml of 0.1 N NaOH = 0.00031 g $P_2O_5$

Suppose,

0.1N NaOH taken = $V_1$ ml

0.1N $HNO_3$ required to neutralize the excess alkali = $V_2$ ml

Volume of 0.1 N NaOH required for 50 ml of extract = $(V_1 - V_2)$ml

Volume of 0.1N NaOH required for 250 ml of extract = $(V_1 - V_2) \times (250 \div 50)$ ml

$P_2O_5$ present in 2 g = $(V_1 - V_2) \times 50 \times 0.00031$ g

Therefore,

$P_2O_5$ Content = $0.00155\ (V_1 - V_2) \times 100$ %

*Acid soluble* $P_2O_5$: 50 ml of the extract are evaporated to dryness on water bath and treated with 10 ml of dilute $HNO_3$ and re-evaporated. The process is repeated twice and then the residue is treated with 50 ml of 10% $HNO_3$, shaken with a glass rod and warmed for a minute and then filtered. The filtrate is warmed and then treated with 20 ml of 20% freshly prepared ammonium molybdate solution till the precipitation is complete. The mixture is shaken well and left at a warm place overnight. It is filtered through Whatman No. 42 Filter paper and washed with dilute $HNO_3$ twice followed by washing with cold 3% $KNO_3$ solution till free from acid. The precipitate is now transferred to a beaker, dissolved in standard 0.5N NaOH and the excess of the alkali titrated against 0.1N $HNO_3$ using phenolphthalein as indicator.

1 ml of 0.1 N NaOH = 0.00031 g $P_2O_5$

Suppose,

0.5 N NaOH taken = $V_1$ ml

0.1N $HNO_3$ required to = $V_2$ ml

neutralize the excess alkali

Then, $V_1$ ml of 0.5N NaOH = 5 $V_1$ ml of 0.1 N NaOH

Volume of 0.1N NaOH required = (5 $V_1$ - $V_2$) ml

HCl soluble $P_2O_5$ content =

$$\frac{(5V_1 - V_2) \times 250 \times 0.00031 \times 100}{50 \times 2}$$

or = $(5V_1 - V_2) \times 250 \times 0.00031$ %

### 2.4.3 Analysis of Potash Fertilizers

Weigh 1 g fertilizer and transfer it to a beaker. Dissolve the substance in about 50 ml of water and filter into 250 ml flask. Wash the residue thrice and make up the volume. Pipette 10 ml of the filtrate into a beaker and evaporate to dryness. Add 1.5 ml glacial acetic acid, 10 ml of saturated solution of NaCl, 5 ml of 35% $NaNO_3$ solution and 5 ml of 20% cobalt nitrate solution. The mixture is shaken well and left overnight in a cool place. It is then filtered; the precipitate and beaker washed repeatedly with 2.5% $Na_2SO_4$ solution till the filtrate is no longer coloured. The precipitate is transferred to the beaker, excess 0.1N $KMnO_4$ and 25 ml of dilute 10% $H_2SO_4$ are added, heated and allowed to stand for 20 minutes. Then, the permanganate is decolourized with excess 0.1N oxalic acid. The excess oxalic acid is warmed at 60-70°C and titrated with 0.1N $KMnO_4$ till faint permanent pink colour remains. The net amount of 0.1N $KMnO_4$ required refers to the amount of $K_2O$ present.

1 ml of 0.1 N $KMnO_4$ = 0.00083 g $K_2O$

Suppose,

0.1 N $KMnO_4$ taken = $V_1$ ml

0.1N oxalic acid used = $V_2$ ml

0.1 N $KMnO_4$ used to complete titration = $V_3$ ml

Then,

$V_1$ ml of 0.1N $KMnO_4$ used = $(V_1 + V_3) - V_2$ ml

$K_2O$ in 100 g fertilizer =

$$\frac{V_1 + (V_3 - V_2) \times 250 \times 0.00083 \times 100}{10 \times 1}$$

or = $(V_1 + V_3) - V_2) \times 25 \times 0.00083$ %

### 2.4.4 Analysis of Organic Manure

*Determination of Moisture* - Weigh 2-5 g of the powdered air dry material in a weighed platinum basin and keep it in an oven at 105°C temperature for 4 hours. Cool in a desicator and weigh. Loss in weight represents moisture present in sample.

*Determination of Organic Matter* - The platinum basin containing the sample is heated first gently and thereafter strongly till a white ash is obtained. The loss in weight represents the organic matter present in the manure.

*Determination of Sand* - Transfer the ash to a beaker, warm the basin with few ml of dil HCl and transfer the acid to the beaker. Add another 25 ml of dil. HCl and digest on a sand bath for 30 minutes. Filter and wash the residue several times with hot water till free from acid. Dry the filter paper in an oven, ignite on a weighed basin, cool and weigh. Increase in the weight of basin refers to the sand insoluble matter in the sample.

*Determination of $K_2O$* - The filtrate and washings of the above are collected in a 250 ml flask and made up to 250 ml with water. 50 ml of the solution are evaporated to dryness and $K_2O$ content is estimated as detailed in 2.4.3.

*Determination of $P_2O_5$* - Evaporate 50 ml of the extract to dryness. Treat the residue with 10 ml of 10% $HNO_3$ and evaporate to dryness again. Then, the dry residue is warmed with 10g solid $NH_4NO_3$ and treated with 20 ml of 20% ammonium molybdate solution till the precipitation is complete. The mixture is shaken well and left overnight in a warm place. It is then filtered through It is filtered through Whatman No. 42 Filter paper and washed with dilute $HNO_3$ twice followed by washing with cold 3% $KNO_3$ solution till free from acid. The precipitate is now transferred to a beaker, dissolved in standard 0.1N NaOH and the excess of the alkali titrated against 0.1N $HNO_3$ using phenolphthalein as indicator. From the net amount of 0.1N NaOH required, the amount of $P_2O_5$ present in the sample is calculated.

1 ml of 0.1 N NaOH = 0.00031 g $P_2O_5$

Suppose,

0.5 N NaOH taken = $V_1$ ml

0.1N $HNO_3$ required to neutralize the excess alkali = $V_2$ ml

Volume of 0.1 N NaOH required = $(V_1 - V_2)$ ml

$(V_1 - V_2) \times 250 \times 0.00031 \times 100$

$$P_2O_5 \text{ content} = \frac{(V_1 - V_2) \times 250 \times 0.00031 \times 100}{50 \times 4.5675}$$

*Determination of Nitrogen* - Take 1 g powdered air-dry manure in a Kjeldahl flask, add about 20 ml water, 25 ml conc. $H_2SO_4$, 1 g $CuSO_4$ and 10 g $K_2SO_4$. Digest the mixture till it is clear. Cool and dilute to about 500 ml. Transfer the contents of flask to one litre distillation flask and treat with 150 ml of 30% NaOH and boil. Collect the distillate in a conical flask containing 10 ml of $H_2SO_4$ with a few drops of methyl red as indicator. When about 200 ml of the distillate are collected, discontinue distillation and the excess of acid in the flask is titrated with 0.1N NaOH solution. Run a blank along side.

1 ml of 0.1 N $NaOH_4$ = 0.0014 g N

Suppose,

0.1N $H_2SO_4$ used = $V_1$ ml

0.1N NaOH required to neutralize the excess acid = $V_2$ ml

Blank for $H_2SO_4$ used = $V_3$ ml

Net volume of 0.1N $H_2SO_4$ required = $(V_1 - V_2)$ ml

Therefore,

Nitrogen Content = $0.0014 (V_1 - V_2) \times 100$ % on air dry basis
or = $[0.0014 (V_1 - V_2) \times 100] \div 91.01$ % on oven dry basis

**3**

# TECHNIQUES IN PLANT PATHOLOGY

The word 'Pathology' means the study of sufferings and main objectives of plant pathology are to study etiology, pathogenesis, epidemiology and control of plant diseases. The pathogens responsible for causing diseases of plant belong to viruses and viroids, bacteria, fungi, algae, nematodes and insects. On the basis of the extent to which diseases are associated with plants, diseases are classified as ***localized,*** limited to a particular part of a plant, or ***systematic,*** spread throughout the entire plant. Diseases may be grouped as ***soil-borne, air-borne*** **and** ***seed-borne*** on the basis of their mode of perpetuation and their mode of primary infection. Diseases are also classified on the extent of occurrence and geographical distribution as ***endemic*** when a disease is more or less constantly present from year to year in a moderate to severe form in a particular part of the earth or country; ***epidemic*** **or** ***epiphytotic,*** when a disease occurs relatively for a short period of time but in a severe state involving major areas of the crop; and ***sporadic,*** when a disease occurs occasionally. **Disease signs** may be noted by the presence of distinct characteristic features of the pathogen on the host tissues while the **disease symptoms** are the evidences of disease characterizing the type or reaction or disease syndrome. These symptoms may of the following types.

a. Spots, lesions, shredding, vein clearing or drying of leaves or exudation on leaves.
b. Spots, lesions, cankers, galls, gummation breaking, discoloration and rotting of stems.
c. Spots, malformation, rotting and dropping of fruits,
d. Blackening, rotting, swelling and drying of roots.

## 3.1 COLLECTION AND CULTIVATION OF MICROORGANISMS

### 3.1.1 Collection of Samples

Type of sample to be collected depends upon the part of the plant infected. Leaf or leaves samples are collected in case of leaf spots, rust, smut, powdery or downy mildew diseases; leaf drying, curling, twisting, bending, thickening or shortening; mosaic, vein clearing, streaks, yellowing, malformation; resetting, burning, chlorosis etc.; gall formation etc. Stem samples are collected in case of stem spots, pustules, rotting, necrosis; rust or smut like symptoms; discoloration of stems, tumor or gall on stems; browning or drying of stem; cankering, scabbing, bark shredding, rotting or gummosis etc. Root samples are collected in case of root rotting, blackening, drying etc.; bulging of roots, gall formation, flattening, clubbing or thickening etc. Seed and fruit samples are collected in case of spots, malformation, rotting and dropping of fruits etc. Samples are collected in large number, dipping the cut end in water or soil as case may be.

### 3.1.2 Cultivation Techniques of Microorganisms

Microorganisms are ubiquitous in our environment. They occur in air, soil, water, food, sewage, decomposing matter, living plants and on body surfaces. Microbes include viruses, bacteria, mycoplasmas, algae, fungi, protozoa and worms. They may be saprobic and/or parasitic, capable of living within or outside the cells of higher forms of life. They are isolated from mixed population to individual species for study. **A culture containing a single unadulterated species of cells is called a pure culture** and there are several isolation techniques to get a pure culture.

#### *Serial Dilution-Agar Plating Method*

*Principle:* The plate count technique is one of the most commonly used procedures to culture microorganisms. This method is based on the principle that when material containing bacteria are cultured, every viable bacterium develops into a visible colony on a nutrient agar medium. The number of colonies, therefore, is same as the number of microorganisms in the sample. In this method, a small measured volume/amount (10 ml or 10g) of the material is mixed with a large volume of sterile water or saline called the **dilution blank** or **diluents** to make microbial suspension. Dilutions are usually made in multiple of ten. A single dilution is calculated as follows.

$$\text{Dilution} = \frac{\text{Volume of the sample}}{\text{Total volume of sample and diluent}}$$

Serial dilutions are later prepared by transferring a known volume (10 ml or 1 ml) of the dilution to second dilution blank and so on. Once diluted, the specified volume of the dilution sample from various dilutions is added to sterile Petri plates (in triplicate for each dilution) to which molten and cooled (45-50°C) suitable agar medium is added. The colonies are counted on a Quebec colony counter. The number of organisms developed on the plates after an incubation period of 24-48 hours per ml is obtained by multiplying the number of colonies obtained by plate by dilution factor, which is the reciprocal of the dilution.

$$\text{Number of cells per ml or g} = \frac{\text{Number of colonies (average)}}{\text{Amount plated} \times \text{dilution}}$$

*Procedure:* Collect the sample at random and mix them thoroughly. Label 90 ml sterile water blanks as 1, 2, 3, 4, 5, 6 and 7 and sterile Petri dishes as $10^{-2}$ (3 plates), $10^{-3}$ (6 plates), $10^{-4}$ (9 plates), $10^{-5}$ (9 plates), $10^{-6}$ (6 plates) and $10^{-7}$ (3 plates) with a wax pencil. Place 10g or 10 ml sample into numbered 1 water blank and make a microbial suspension of $10^{-1}$ dilution. Vigorously shake the dilution on a magnetic stirrer for 20-30 minutes to obtain uniform suspension of microorganisms. Transfer 10 ml of suspension from water blank 1 into water blank number 2 with a sterile pipette to make a 1:100 ($10^{-2}$) dilution and shake well for about 5 minutes. Prepare another dilution 1:1000 ($10^{-3}$) by pipetting 10 ml of the suspension into water blank number 3, using a fresh sterile pipette and shake it. Make further dilutions $10^{-4}$ to $10^{-7}$ by pipetting 10 ml of suspension into water blank number 4 to 7 as prepared above. Transfer 1 ml of aliquots each from $10^{-2}$ dilution blank into 3 sterile Petri dishes, from $10^{-3}$ dilution blank to 6 petri dishes, from $10^{-4}$ to 9 petri dishes, from $10^{-6}$ to 6 petri dishes and from $10^{-7}$ to 3 petri dishes. Add approximately 15 ml of the cooled medium (45°C) to each Petri dish and mix the inoculum by gentle rotation of the Petri dish. The three media are to be added to various dilutions as follows.

a. For bacteria - nutrient agar medium to 12 plates with $10^{-4}$ to $10^{-7}$ dilutions,

b. For actinomycetes - glycerol yeast agar medium supplemented with aureomycin to plates with $10^{-3}$ to $10^{-6}$ dilution,

c. For fungi - sabouraud agar medium supplemented with streptopenicillin to plates with $10^{-2}$ to $10^{-5}$ dilutions.

Upon solidification of the media, incubate all the plates in an inverted position at 25°C for 2-7 days. Observe the plates for number and distribution of colonies of bacteria, fungi and actinomycetes from each dilution (Fig. 3.1). Select the plates, which contain colonies in range of 30-300 and make plate counts using a colony counter. Determine the average microbial count and record the results in chart (Tab. 3.1).

**Table 3.1: Number of colonies of bacteria, actinomycetes and fungi in each dilution plate**

| Organism | Dilution | Dilution factor | Number of colonies per plate | | | Average number of colonies per dilution | Organisms per g of soil |
|---|---|---|---|---|---|---|---|
| | | | I | II | III | | |
| Bacteria | $10^{-4}$<br>$10^{-5}$<br>$10^{-6}$<br>$10^{-7}$ | $10^{4}$<br>$10^{5}$<br>$10^{6}$<br>$10^{7}$ | | | | | |
| Actinomycetes | $10^{-3}$<br>$10^{-4}$<br>$10^{-5}$<br>$10^{-6}$ | $10^{3}$<br>$10^{4}$<br>$10^{5}$<br>$10^{6}$ | | | | | |
| Fungi | $10^{-2}$<br>$10^{-3}$<br>$10^{-4}$<br>$10^{-5}$ | $10^{2}$<br>$10^{3}$<br>$10^{4}$<br>$10^{5}$ | | | | | |

***Calculation*** : Calculate the number of organisms (bacteria, actinomycetes and fungi) per g of sample by applying the following formula.

$$\text{Viable cells/g sample} = \frac{\text{Mean plate count} \times \text{Dilution factor}}{\text{Amount / weight of sample plated}}$$

**Streak-plate Method**

*Principle:* This streak-plate method is the most practical method of obtaining discrete colonies and pure cultures. In this method, a sterile loop or transfer needle is dipped into a suitable diluted suspension of microorganisms, which is then streaked on the surface of an already solidified agar plate to make a series of parallel, non-overlapping streaks, with the aim to obtain colonies of microorganisms that are pure i.e. growth derived from a single cell/spore.

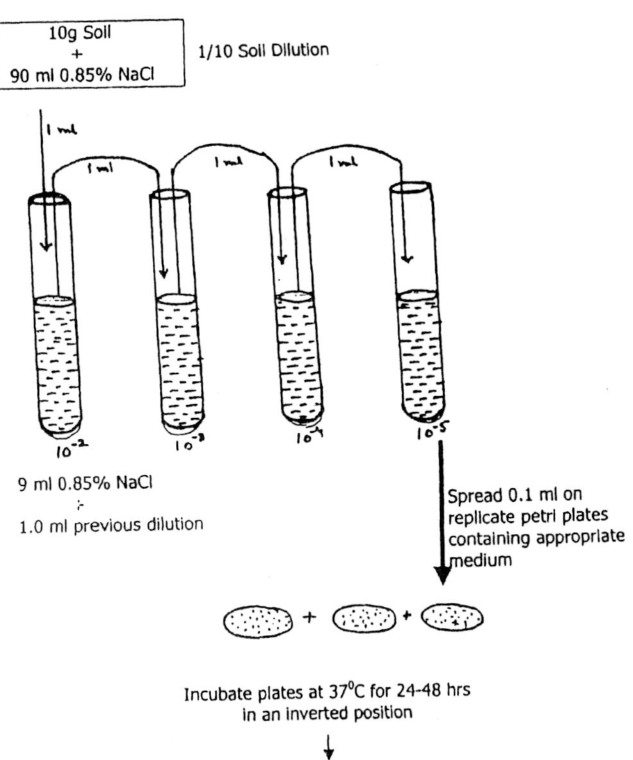


**Fig. 3.1: Procedure for estimation of number of micro-organisms by Serial Dilution Agar Plating Method**

*Procedure* : Label all the plates, on the bottom, with the name of the organism(s) to be inoculated. Hold the tube containing broth in the left hand and sterilize the loop holding in right hand. Remove cotton wool plug and immediately flame the mouth of the tube. Insert the loop into the broth and withdraw one loopful of culture. Flame the mouth of the tube, replace the cotton wool plug and place the tube in the test tube rack. Lift the Petri dish/plate cover and hold it at an angle of 60°. Place the inoculum on the agar surface at the farthest edge and streak the inoculum from side to side in parallel lines across the surface of the area. Reflame and cool the loop and turn the Petri plate to 90°. Touch the loop to a corner of the culture in area 1 and streak the inoculum across the agar in area 2 as above. The rest of the agar surface is now used to complete the streaking. Replace the cover of the plate and incubate at 25°C in an inverted position for 48-72 hours. A confluent growth will be seen where the initial streak was made, the growth is less dense away from the streak and discrete colonies farthest away from the streak. Select an isolated colony from the plate and record their features.

**Pour-plate Method**

*Principle:* In this technique, successive dilutions of the inoculum (serially diluting the original sample) are added into sterile Petri plates to which is poured melted and cooled (42-45°C) agar medium and thoroughly mixed by rotating the plates, which is then allowed to solidify. After incubation, the plates are examined for the presence of individual colonies growing throughout the medium. The pure colonies, which are of different size, shape and colour are isolated and transferred to test tube culture media for making pure cultures.

*Procedure:* Place the seven agar deep tubes into boiling water bath for melting. Allow these to cool to 45°C. Label the empty sterile culture tube as number 1 and six water blanks as number 2 to 7. Also label Petri plates 1 to 7. Place the labelled tubes in a test tube rack. 1 ml of the bacterial suspension is transferred aseptically from tube number 1 to water blank number 2 and returns the pipette to tube number 1. Shake the tube number 2 and, with a sterile pipette, transfer 1 ml to tube number 3. Return the pipette to tube number 2. Make the serial dilutions till the six water blanks (number 2 to 7) are used by repeating the earlier steps. Transfer 1 ml of the bacterial suspension each from tube numbers 1 to 7 to Petri plate number 1 to 7 using respective pipettes. Remove the nutrient agar tube from water bath and pour the medium into plate 1 and rotate the plate gently to ensure uniform distribution of the cells in the medium. Repeat this step for addition of medium to plates 2,3,4,5,6 and 7. Allow the medium to solidify and incubate at 37°C in an inverted position for 24-48 hours. Observe the plates for appearance of individual colonies growing throughout the agar medium. The progressively poured plates will have lesser and lesser number of colonies, which will be distributed more or less sparsely in the plates (Fig. 3.2). These colonies may be transferred to other media or agar slants for further study.

- Transfer with 1 ml pipette, using a fresh sterile pipette for each dilution

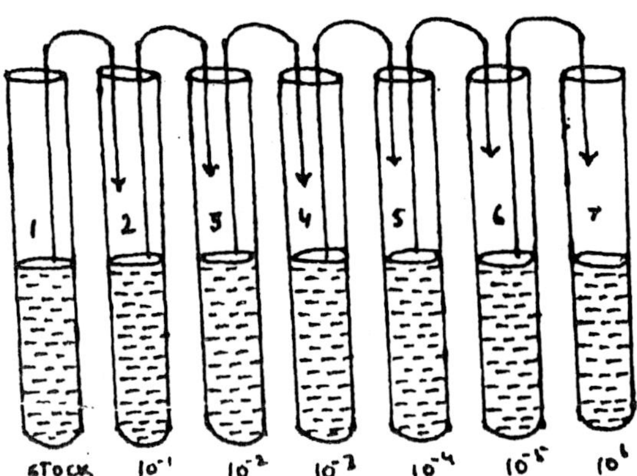


- Add 1 ml of suspension to respective labeled plates using respective pipettes
- Pour molten, cooled nutrient agar medium from the agar deep tubes and rotate the plates gently to ensure uniform distribution of cells/spores

Allow medium to solidify

↓

Incubate at 37$^0$C for 24-48 hours in an inverted position

↓

Observe appearance of individual isolated colonies

**Fig. 3.2: Purification of a culture by pour-plate method**

**Broth Cultures**

*Principle* : Broth is a liquid formulation used for growing microorganisms. The media termed broths are made by dissolving various solutes in distilled water and later sterilizing it. Appropriate physical (like temperature) and chemical (like oxygen) environments are required for growing microorganisms in a broth. For temperature, incubators are used while oxygen is provided to broth cultures through a process called aeration with the help of roller drum apparatus or fermentors. Growth of the microorganisms occurs throughout the container and may present a dispersed cloudy or particulate appearance.

*Procedure* : Label each nutrient broth tube with the name of organism. Thoroughly mix the broth culture by thumping the bottom of the tube. Hold both the culture and broth tube in the left hand and sterilize the loop, remove the cotton plugs from these tubes and alternately flame the mouths of these tubes. Transfer a loopful of culture into the sterile broth with the sterile loop. Flame the mouth of the tubes and replace the plugs. Flame sterilize the inoculating loop immediately after transfer. Inoculate each organism into a separate tube of sterile nutrient broth following earlier steps. Incubate all the tubes, including control at 30°C in an incubator for 24 hours. Observe the broth cultures for growth of microorganisms without disturbing them. Record any change in appearance (growth) 24 hours after incubation.

## 3.2 DIRECT OBSERVATION

The collected samples are divided in several parts for the following observations.

### 3.2.1 Slide Mounting

*Reagent* :

*Lactophenol solution* - Warm phenol crystals (20 g) with distilled water till entire phenol dissolves. Add lactic acid (20 ml), glycerol (40 ml) and 0.05 g cotton blue powder. Mix well.

*Crystal Violet Solution* : Dissolve 2 g crystal violet and 1 g ammonium oxalate in 100 ml distilled water.

*Burke's Iodine* - Dissolve 2.5 g safranin in 10 ml 95% ethanol and dilute with 100 ml distilled water.

*Method :* Take a small piece of diseased portion and put it in a drop of lactophenol solution on the slide, leave it for 5 minutes and then tease the piece with the help of two sterilized dissecting needles. After thorough teasing, put the cover slip on the material. Pass the slide over the flame 2-3 times for few seconds. Observe under a light microscope and study the presence of fungal bodies. If fungal mycelia are not found, then put the cut end of the sample into 1 ml of 1% sucrose or 1% peptone solution for one hour. Take a drop from the solution on the microscope slide and prepare a smear with the help of another slide. Warm over the flame till the smear dries and stain with crystal violet solution for 30 seconds. Rinse the stain, cover with Burke's iodine and leave for 30 seconds. Rinse with 80% ethanol till the colour ceases to come out. Wash the slide with water. Counterstain with Safranin for 1-2 minutes. Wash with water, gently dry and observe under microscope.

### 3.2.2 Moist Chambering

Sterilize the Petri dishes, blotting paper of its size and water by appropriate technique (in hot air oven at 180°C for 4-5 hours or by UV radiation). Select the specimen with some healthy portion and sterilize it surface by dipping in 0.1% $HgCl_2$ for 1 minute. Take all the materials in transfer chamber, put 3-4 pieces of diseased specimen on the microscope slide placed in petridish. Pour sterilized distilled water in such a way that the bottom filter paper remains moist. Cover the top lid, keep the petridish in B.O.D. incubator at 25 ± 2°C for 48-72 hours, scrap the diseased tissue with a needle and observe under microscope.

Prepare the desired agar media and take only 5 ml of the medium into the sterilized culture tubes, which are then plugged with sterilized non-absorbent cotton tightly. Autoclave in moist heat at 15-20 lbs pressure (140-160°C) for 20 minutes. Remove the tubes and keep them in a standing position and medium is allowed to solidify. The slant should be prepared in such a way that maximum portion is covered with agar medium but it must not touch the cotton plug. The slants are kept in erect position. The media plates are prepared by pouring the agar medium in the sterilized Petri dishes and allowing it to solidify. Now, take some diseased specimen with the help of inoculating needle and put it on the slant or Petri-plate and culture the micro-organism in moist chamber for 4-5 days.

### 3.2.3 Detection of Plant Pathogen

Boil the specimen for 1-2 minutes in ethanol-water (1:2; v/v) and store overnight at room temperature. Wash the specimen with 50% ethanol for 15 minutes twice and again with 0.05 M NaOH for 15 minutes. Wash four times with distilled water. Put it in 0.1 M Tris-HCl buffer (pH 8.5) for 30 minutes. Stain for 5 minutes with 0.1% solution of Calcofluor white M2R New. Wash four times with water for 10 minutes each and once with 25% aqueous glycerol. Mount on glycerol and observe under microscope.

### 3.2.4 Techniques to Handle Powdery Mildew Disease

*A Whole Mount Technique* - Cut the infected leaves into portions of about 5 mm square and drop them into a fixative containing dioxin and propionic acid (95:5; v/v). Keep the flask in the oven at 80°C temperature for 24 hours. Remove the specimen from fixative and clear with distilled water for at least 3 hours. Place the specimen on slides, a few drops of Haemgum staining mount are added, covered with cover slips and observed under microscope.

*A Rapid Staining Method* - Keep 1-2 cm infected leaf fragments at 70°C temperature for 10-60 minutes or overnight at room temperature in ethanol-chloroform (75:25; v/v) mixture containing 0.15% TCA. Stain the cleared leaves in Coomassie Brilliant Blue R-250 solution (15% TCA in water and 0.6% Coomassie Blue in 60% methanol) for 15-30 minutes. Preserve the stained segments in glacial acetic acid-glycerol-water (5:20:75; v/v). This method may be used for other fungi on different hosts.

### 3.2.5 Techniques to Handle Rust Disease

Collect rust infected leaves, wash with running tap water and blot dry with absorbent tissue paper. Place the leaves either in a sealed plastic bag or in petridish containing filter paper saturated with distilled water and incubate for 2 days at 24°C temperature. Collect fresh uredospores with brush and transfer them to a sterilized beaker. Prepare uredospore suspension from the freshly grown ones with distilled water and Tween-80.

Place the uredospores on the appropriate medium and incubate in the dark in a water saturated atmosphere at 37°C. Germination of uredospores start on 2nd day and branching of mycelium on the 4th day and after 3-4 weeks, full growth takes place.

## 3.3 TECHNIQUES FOR PLANT PATHOGENIC BACTERIA

### 3.3.1 Isolation of Organism in Nutrient Agar or Peptone Beef Extract Agar Medium

*From soil* - To 5g air-dried and sieved soil, add 500 ml of sterilized distilled water and shake in a mechanical shaker for 30 minutes. Pipette out 10 ml of soil water suspension and add 90 ml of distilled water. Shake thoroughly. Take 20 ml of such diluted soil suspension and dilute with equal volume of sterilized water. From each of the above dilutions, spread 1 ml of clear suspensions in the Petri dishes. Add 20 ml of melted but cool Potato dextrose agar medium. Rotate the Petri dishes so that the samples get dispersed in the medium uniformly. Incubate the Petri dishes at 37°C ± 2°C temperature for 5-6 days. Observe these plates regularly.

*From Leaf Surface* - Prepare Petri dishes with GYCA medium (5g glucose, 5g yeast extract, 4g calcium carbonate, 15 g agar and 1 litre distilled water). Collect bacterial disease infected leaves with old brown lesions and tiny black spots. Surface sterilize these leaves by dipping in 0.1% mercuric chloride solution and wash thoroughly 3-4 times with sterilized distilled water. Cut the leaves into small pieces and place the pieces on the GYCA plates. Incubate the plates at 37°C for 4-5 days. After the incubation period, a watery pale yellow growth develops from the cut ends of the leaf pieces. Transfer parts of this growth on fresh plates and incubate. Pure culture of bacteria is obtained in this way.

### 3.3.2 Measurement of Bacterial Growth by Turbidity Measurements (Spectrophotometer Method)

*Principle:* Bacterial growth generally referred to reproduction-increase in population size - rather than to enlargement of mycelial growth. Reproduction in bacteria occurs mainly through binary or transverse fission that results in the formation of two daughter cells. This is repeated at intervals by each new daughter celi in turn or with each successive round of division the population increases. The time required for a complete fission cycle i.e. parent cell to two daughter cells is called the generation time. In bacteria, it is a doubling process in which each new fission cycle or generation increases the population by a factor of 2. The length of generation time is a measure of the growth rate of an organism. The average generation time in bacteria is 30-60 minutes under optimum conditions. Determination of bacterial growth requires inoculation of a sterile broth medium and incubation of the culture under optimum temperature and gaseous conditions. The bacteria will reproduce rapidly under these conditions and the microbial growth may be measured by a variety of techniques viz., dry weight determination, determination of total nitrogen by Kjeldahl method, direct counting of cells under a light microscope or indirectly by viable count (plate count), which detects living organisms by their ability to form colonies on agar surfaces, and turbidity measurements that relate cell number to the turbidity of a broth culture. Optical density (OD) is directly proportional to cell concentration. With the increase of bacterial numbers, the broth becomes more turbid causing the light to scatter and allowing less light to reach the photoelectric cells. The change in direction of light is indicated in the Spectrophotometer as percentage of transmission or %T. Absorbance is a logarithmic value and used to plot bacterial growth curve, which shows four distinct phases of growth–the lag phase, logarithmic phase, stationary phase and death phase.

*Procedure:* Aseptically transfer 5 ml of the broth culture to 100 ml brain-heat infusion broth flask and determine initial OD at 600 nm wavelength. Place the inoculated culture flask in the shaker set at 120 rpm at 37°C for 6 hours. Aseptically transfer, after 30 minutes inoculation, 5 ml of the culture to a cuvette and determine OD of the sample at 600 nm. Repeat these steps at each 30 minutes interval for a period of 6 hours and record the optical density.

**Table 3.2: Determination of generation time by turbidity measurement**

| *Culture (minutes of incubation)* | *Optical Density at 600 nm* |
|---|---|
| 0 | |
| 30 | |
| 60 | |
| 90 | |
| 120 | |
| 150 | |
| 180 | |
| 210 | |
| 360 | |

Plot a graph between optical density and incubation time to prepare the growth curve and calculate the generation time.

### 3.3.3 Staining of a Bacterial Smear

***Simple Staining of Bacteria*** - Prepare bacterial smears on a marked glass slide with the help of flamed loop and fix it by heating over a spirit lamp. Keep the slide on the staining tray and apply about 5 drops of a stain (Crystal violet, Carbol fuschin or Methylene blue) for the specified period. Pour off the stain and wash the smear gently with slowly running tap water. Blot dry the slide using blotting paper. Examine the slide under oil-immersion objectives and write the description of the bacteria on the basis of microscopic examination. The bacteria stain a deep blue. The *Bacillus* is rod shaped with clear areas (i.e. endospores) within them; *S. aureus* cells are spherical, occurring singly, in pairs and irregular clusters; the small rod of *E. coli* frequently gives the appearance of being coccus.

***Gram Staining method*** - Prepare a bacterial smear on glass slide and fix it by heating over a spirit lamp. Stain in crystal violet solution (2g Crystal violet, 1g ammonium oxalate and 100 ml distilled water) for 1 minute. Wash the Crystal violet under running water for 5-10 seconds. Rinse the slide with Burke's iodine solution (1g iodine, 2g potassium iodide and 100 ml distilled water) and allow to stand for 1 minute. Wash with water for few seconds and then add 95% ethanol drop-wise across the smear for 20 seconds. Wash and counterstain with aqueous Safranin (2.5% Safranin solution in 95% alcohol). Wash with water, dry and observe under microscope. Gram positive bacteria takes blue or deep violet colour and Gram negative ones red colour.

***Acid-fast staining of bacteria*** - Prepare a bacterial smear, air dry and heat fix the smear. Flood the smear with Carbol fuchsin, heat the slide for 3-5 minutes and add more stain during heating to prevent the smear becoming dry. Cool the slide and wash with distilled water. Decolourize the smear with acid-alcohol for 10-30 seconds or until the smear becomes faint pink in colour. Wash the slide with distilled water and counterstain the smear with methylene blue for 1-2 minutes. Wash with distilled water, blot dry and observe under microscope. Acid-fast bacteria appear red and non-acid fast bacteria blue coloured.

***Bacteria Cell Wall Staining*** - Prepare a bacterial smear and fix by heat. Apply 3 drops of Cetyl pyridinium chloride (0.34%) and one drop of aqueous congo red solution. Mix with a glass rod for 5 minutes. Flood the smear with congo red. Wash the slide with distilled water and air dry the smear. Counterstain with methylene blue, wash with distilled water, blot dry and observe under microscope. The cell wall stains red and cytoplasm stains blue.

***Bacterial Spore Staining*** - Make smear of a bacterial culture. Air dry and heat fix the smear. Flood the smear with Malachite green (5% aqueous). Heat the slide to steaming and steam for 5 minutes adding more stain to smear from time to time. Wash the slide slowly under running tap water and counterstain with Safranin (0.5% aqueous) for 30 seconds. Wash the smear with distilled water, blot dry and observe under microscope. The endospores of *Bacillus* stain green and the vegetative cells, which are rod shaped, each containing an elliptical, centrally located spore, stain red. *Staphylococcus* cells staining red are spherical, occurring singly, in pairs and irregular clusters.

***Capsule Staining*** - Place several loops full of culture onto a clean slide. Prepare a thick smear. Apply Congo red stain. Allow smear to air dry. Fix the smear with acid-alcohol for 15 seconds. Wash the smear with distilled water and flood it with Acid Fuchsin for 60 seconds. Wash with distilled water, blot dry and observe under microscope. The capsules appear colourless surrounding red cells against a dark blue background.

***Flaqella Staining Method*** - Prepare a bacterial suspension on Casamino agar plate. Take a few drops of culture at one corner of the slide and smear the turbid bacterial liquid with another sterilized slide. Dry the slide by slightly heating on spirit lamp flame and stain the bacteria with Liefson's flagella stain (12 ml of water saturated solution of ammonium aluminum sulphate [$NH_4Al(SO)_4$. 12 $H_2O$], 3 ml of Basic fuchsin [saturated in 95% ethanol], 10 ml distilled water and 15 ml of 95% ethanol) using sufficient amount. Keep for 10 minutes till a distinct iridescent film is formed over the surface. Wash the excess stain with water, blot dry and examine under fluorescent microscope.

### 3.3.4 Chemical Tests for Bacteria

*Whether aerobic or anaerobic* - Dissolve nicotinic acid (0.01g) in 1 litre distilled water and add 1g $NH_4Cl$, 4g $KH_2PO_4$, 4g $K_2HPO_4$,

0.2g $MgSO_4$. $7H_2O$, 0.028g $FeSO_4$, 0.003g $ZnSO_4$ and 0.018g phenol red. Mix well and autoclave 25 ml of the solution for 5 minutes in conical flask (50 ml capacity). Add one drop of bacterium on the surface, shake well and incubate at 30°C for 48 hours. Strict aerobic bacteria will grow on the surface (on the upper layer only), microaerophiles grow just a few minutes below the broth and the strict anaerobic bacteria grow only in the bottom payer or may not grow at all.

*Nitrate Test* - Grow the bacteria in Nitrate broth (containing 10g peptone and 1g nitrite free $KNO_3$ dissolved in 1 litre water) in conical flask attached with a bent glass outlet at the bottom. Collect the bacterial culture along with broth starting from 48 hours after incubation until 5 days. Test for nitrate by draining out 5 ml broth in a test tube and adding a few drops of Sulfanilic (Dissolve 8 g sulfanilic acid in 1 litre of 5N acetic acid) and Naphthyl-amine (Dissolve 5g naphthylamine in 1 litre of 5N acetic acid). Pink or red colour in the tube indicates the presence of nitrite.

*Indole Production* - Multiply the bacteria in Tryptophan broth (10g tryptophan in 1 litre water) and insert a flap of Whatman No. 1 filter paper soaked in saturated oxalic acid solution through the side of the cotton plug. The filter paper should not touch the broth surface. If indole is produced, then the filter paper turns pink.

*$H_2S$ Production* - Grow the bacteria peptone water tubes and like indole production test, insert filter paper strip into the tube pre-soaked in saturated lead acetate solution. In case of $H_2S$ production, the filter paper turns black.

*Casein Hydrolysis* - Prepare milk agar plates (2.5% water agar with 1.25% skimmed milk) and streak the bacterium once across. Look for casein clearing zones at different growth intervals.

*Gelatin Hydrolysis* - Add 0.4% gelatin with nutrient agar, streak across with the test bacterium and incubate at 28°C. After 3-5 days, flood the plate with mercuric chloride and hydrochloric acid solution (15 g $MgCl_2$, 20 ml conc. HCl and 100 ml distilled water). Gelatin hydrolysis is indicated by a clear zone underneath around the growth in contrast to the white opaque gelatin.

## 3.4 TECHNIQUES FOR PLANT PATHOGENIC FUNGI

### 3.4.1 Isolation of Fungal Pathogens from Leaves

*Principle:* On leaves, lesions (spots), blights and mildews are the common symptoms produced by fungi. When the cells are killed in a limited area of the leaf, it results in a lesion. When a lesion becomes necrotic and dead part falls away, it results in a shot-hole effect. When the leaf damage is sudden and serious and dead part turns brown, the blight results. The mildew disease is caused when the pathogen produces a visible growth over the host surface. Most fungal pathogens sporulate on the host and the spores may be collected directly over a Petri-dish containing culture medium. The colonies produced after incubation can be sub-cultured on separate plates.

*Requirements:*

Infected young leaves / aerial part of a plant

Potato dextrose agar plates supplemented with streptomycin sulphate or streptopencillin (5 No.)

Moisat chambers

Disinfectant (10% chlorox, 1% sodium hypochlorite or 0.1% $HgCl_2$)

Forceps and razor blade

Sterile Petri plates (6 No.)

Sterile water

Spirit lamp

*Procedure*:

- Cut across lesions of 5 to 10 mm square, containing both the diseased and healthy looking tissue.
- Surface sterilize the cut portions, by dipping in a surface sterilant solution for different times, varying from 15 to 120 seconds.
- Wash the treated pieces in three changes of sterile water and blot dry on clean sterile paper towels to remove the sterilant.
- Aseptically transfer the pieces on a nutrient medium, usually 3-5 pieces per plate.

- Incubate the inoculated plates in an inverted position at 25°C for 3-5 days.
- In some cases, the infected leaves or portions of leaf are taken in moist chambers and incubated at 25°C for 2 days. The growth of fungus is directly transferred to a nutrient medium.
- If fruiting bodies are present on the leaf, these are picked up, dipped in a surface sterilant for a few seconds and then placed on a nutrient medium.

*Observations and Results*

Observe the incubated plates from the second day onwards for the growth of fungus. Select the plates showing growth of pathogen arising from the infected portions only. Aseptically transfer bits of mycelia from the margin of the colonies on fresh nutrient medium plates for further study. Mycelial growth on the nutrient medium from the infected tissues indicates that the disease may be due to a fungus.

### 3.4.2 Isolation of Fungal Pathogens from Stems, Fruits and Other Aerial Plant Parts

*Principle:* Almost same method used for isolating fungal pathogen from leaves can also be employed to isolate these pathogens from superficial infections of the stems, fruits and other aerial plant parts. The deep seated infections from these tissues are generally isolated by cutting pieces of internal tissue, or by cutting from freshly exposed area of the advancing margin of the infection with a flamed scalpel directly on a nutrient agar medium or on water agar.

*Requirements:*

Infected woody tissue (fruits, stems etc.)

Nutrient agar medium plates (acidified PDA, PDA, malt agar, corn meal)

Sterile Petri plates

Sterilant (70% alcohol)

Scalpel

Sterile water

Cotton

Spirit lamp

*Procedure*

- Wipe the infected woody tissue with a cotton swab dipped in 70% alcohol (ethanol) followed by lightly flaming the tissue.
- Using a flamed scalpel, peel off outside layer of the tissue, exposing tissues not previously exposed to contaminants,
- Dig out small pieces of tissue (1/8 to 1/2" size) with flamed scalpel or forceps from the freshly exposed area of the advancing margin of infection,
- Place 3 to 4 pieces, well separated, per nutrient agar plate.
- Incubate the plates at 25°C, in an inverted position, for 5-7 days.
- If infected material is a root, wash it thoroughly with running tap water and shake dry. Follow step 1 to 5 above.

*Observations and Results*

Observe the incubated plates from the second day onwards for 1-2 weeks for the mycelial growth and spores of fungus. Transfer bits of agar containing mycelium from the edge of the developing colonies to slants or nutrient medium plates for identification and future use. Mycelial growth from the infected tissue on a nutrient agar medium indicates that the disease may be due to a fungal pathogen that can grow on a nutrient medium.

### 3.4.3 Isolation of Fungal Pathogens from Soil

*Principle:* There are several fungal pathogens, which are soil-borne. These survive in soil or on infested plant debris either as resting spores or as mycelial strands or rhizomorphs and cause primary infection under favourable conditions causing diseases like pre-emergence killing and damping off of seedlings, seedling blight, root-rots, vascular wilts etc. Isolation of fungal pathogens from soil is done by serial dilution agar-plating technique.

*Requirements*

Soil sample

PCNB agar, plate

Sterile 90-ml water blanks (4)

Sterile 10-ml pipetes (5)

Sterile 1 ml pipette

Bunsen burner/spirit lamp

500 ml Erlenmeyer flask

Wax marking pencil

*Procedure*

- Take 1g soil sample into 200 ml sterile water taken in a 500 ml Erlenmeyer flask.
- Agitate the soil thoroughly for 1-2 minutes.
- Allow it to settle.
- Decant the suspension.
- Repeat the washing procedure (step 2 to 4 above) for 16 times.
- Make a suspension of the soil for a few seconds in a warring blender.
- Prepare soil dilution following the serial dilution technique (label 90-ml sterile water blanks as 1, 2, 3 and 4 and sterile Petri dishes as $10^{-4}$; transfer 10 ml of suspension into water blanks and shake well for 5 minutes).
- Take out 1 ml aliquots into sterile Petri dishes; add 15 ml of the cooled PCNB medium to each Petri dish and mix the inoculum by gentle rotation of the Petri dish.
- Upon solidification, incubate all the plates in an inverted position at 25°C for 5-7 days.

*Observations and Results :*

Observe the incubated plate for the isolated fungal colonies. Whitish, pinkish or yellowish aerial colonies of fusaria will appear after 5-7 days of incubation.

### 3.4.4 Measurement of Fungal Growth by Mycelial Weight Determination

*Principle :* The direct measurement of an increase in fungal weight can be determined by inoculating culture flasks with a known volume of broth followed by filtering off mycelium, drying at 105°C for 48 hours to a constant weight and weighing.

*Procedure :* Inoculate a Czapek-Dox agar plate with the fungi for preparation of a seed plate. Incubate at 25°C for 96 hours in an inverted position. Pour 20 ml Czapek liquid medium in each flask. Plug the flasks with cotton plugs and autoclave at 15 lb/in$^2$ (121°C) for 15 minutes. Inoculate each flask with 8 mm diameter disc, cut from the periphery of the actively growing colony. Incubate the inoculated flasks at 25°C for 12 days. Prepare three replicates for each observation. After 2 days of incubation, remove three inoculated flasks. Filter the contents through a pre-weighed Whatman filter paper. Dry the filter papers in an oven at 105°C for 48 hours. Re-weigh the dried filter papers along with the mycelium. Repeat these steps at each one-day interval for a period of 12 days. Measure the growth increments at intervals of 2, 3, 4, 5, 6, 7, 8, 9, 10 and 12 days. Calculate the weight of mycelium and prepare a graph illustrating the growth increments in dry mycelial weight of the fungi.

### 3.4.5 Staining of a Fungal Growth

*Lactophenol cotton blue mounting of fungi* - Place a drop of lactophenol cotton blue on a clean slide. Transfer a small tuft of fungus, usually with spores and spore bearing structures, into the drop using a flamed cooled needle. Gently tease the material using the two mounted needles and mix the stain with the mould structures. Place a cover slip over the preparation and examine under low-power and high-power objectives. Describe the type of hyphae, conidiophore, conidiogenous cells, conidia and their arrangement on the conidiophores/conidiogenous cells. The fungal cytoplasm is seen as a lightly stained blue region forming a layer inside the unstained cell wall of the hyphae, conidiophores, phialides and conidia that are surrounded by a light blue background on the slide.

*Nuclear staining in filamentous fungi* - Grow the fungus on thin agar film on new glass slides in a Petri dish lined with moist filter paper. Fix spores and mycelia for 10-12 minutes in a 3:1 mixture of ethyl alcohol and acetic acid or in a 3:1:1 mixture of ethyl alcohol, acetic acid and lactic acid. Rinse the preparation in 95% ethyl alcohol. Transfer the preparation in 70% ethyl alcohol and then rinse in distilled water. Submerge the preparation for 5 minutes in 1N HCl at room temperature followed by its transfer to 1N HCl at 60°C for 7 minutes. Rinse first in water and then in phosphate buffer (pH 6.9), 5 changes in each. Transfer the preparation in dilute Giesma stain (1 ml stain in 15 ml phosphate buffer at 6.9 pH) and stain for 2 hours. Rinse in phosphate buffer followed by water. Allow the preparation to air dry at room temperature and then mount in euparal. Cover the preparation with a cover slip and examine microscopically under high power and oil-immersion objectives. Make observations for color and number of nuclei in spores and hyphal cells. The nuclei are seen as rose red to pink surrounded by colorless or light blue zone of cytoplasm.

*Staining of vesicular arbuscular mycorrhizae (VAM)* - Wash VAM infected roots thoroughly and fix overnight or store in FAA (Formalin-Acetic acid-Alcohol). Remove FAA from the roots by washing with several changes of tap water. Transfer the samples into KOH solution taken in autoclave resistant jars. Clear the samples by autoclaving them at 15 psi for 15 minutes. Rinse with several changes of tap water. Wash with deionized water. Stain roots in a staining solution (made up of equal volumes of 80% lactic acid, glycerine and distilled water with 0.1% w/v chlorazol black E) for 1 hour or longer at 90°C. Keep stained roots into glycerine overnight for de-staining or for storage. Mount roots on clean glass slides using mounting fluid (chloral hydrate 20g, gum Arabic 20g, glycerine 20g, glucose syrup 3 ml, basic fuschin 10 drops, and 35 ml water) and observe slides under bright-field microscope for VAM structures.

*Staining of actinomycetes* - Grow culture of *Streptomyces albus* on the surfaces of sterile cellophane placed on solidified nutrient agar medium. Incubate culture at 30°C till it sporulates. Remove the cellophane bearing growth from the agar surface. Stain for 2 minutes in the mixture of stains [2 parts Bismark brown (0.1% w/v), 2 parts of Toluidine blue (0.1% w/v) and 1 part of saturated ammonium sulphate solution]. Rinse in distilled water and allow it to dry. Mount in Canada balsam and observe microscopically under high-power and oil-immersion objectives. Record the color of aerial sporulating structures and the mycelium. Vegetative mycelium stains light yellow and the spores blue with red granules in the aerial hyphae. The colony of S. *albus* appears granular, powdery, velvety or floccose. The hyphae are thin, slender and coenocytic; the sporulating aerial mycelium forms chains of 3 to many spores.

## 3.5 IDENTIFICATION OF AN UNKOWN MICROORGANISM

**The science dealing with identifying, naming and classifying microorganisms is called Taxonomy.** Classification deals with the separation of living organisms up to groups while identification determines the correct place of an organism in a previously established plan of classification. Once an unknown microbial culture is isolated from a sample, such as, food, water, soil, infected tissue of animal or plant etc., it has to be identified.

### 3.5.1 Identification of an Unknown Bacterium

When an unknown bacterium is isolated in the pure form, it is usually identified by a combination of information viz., microscopic observations i.e. morphology and arrangement of cells, cultural (growth) characteristics on agar and in broth, the gram stain and other staining reactions, absence or presence of motility and biochemical tests, which are detailed below.

*Morphological Observations*

| | |
|---|---|
| Shape: | Spheres, short rods, long rods, filaments, commas, spirals |
| Arrangement: | Single, pair, chains, clusters |
| Capsules: | Present or absent |
| Gram stain: | Positive or negative |
| Spore stain: | Nonspore former, spore former (central, sub-terminal, terminal) |
| Buds or sheaths: | Positive, negative |
| Motility: | Motile, non-motile |

*Cultural Characteristics on Agar plate*

| | |
|---|---|
| Nutrition: | Autotroph, heterotroph |
| Colonies: | Golden, yellow, white, glistering |
| Temperature: | Minimum, Maximum, Optimum |
| Growth: | (agar medium) absent (0), slight (+), moderate (++), abundant (+++) |
| Form: | Circular, irregular, rhizoid |
| Margins: | Entire, lobate, undulate, serrate, filamentous |
| Elevation: | Flat, raised, convex, unibonate |
| Density: | Opaque, transparent |

**Tab. 3.3: Key to Identification of some selected heterotrophic bacteria**

- Cells spherical
  - Gram-positive
    - Catalase produced
      - Cells arranged in tetrads or glucose not fermented
        - Red pigment → *Micrococcus roseus*
        - Yellow pigment → *Micrococcus luteus*
      - Cells not in tetrads or glucose fermented
        - Acid from mannitol → *Staphylococcus aureus*
        - No acid from mannitol
          - Acid from fructose → *Staphylococcus epidermidis*
          - No acid fructose → *Staphylococcus saprophyticus*
    - Catalase not produced
      - Growth in 6.5% NaCl or at pH 9.6 → *Streptococcus faecalis*
      - No growth in 6.5% NaCl or at pH 9.6
        - Acid from glycerol → *Streptococcus equisimilis*
        - No acid from glycerol
          - No acid from inulin → *Streptococcus mitis*
          - Acid from inulin
            - Acid from raffinose → *Streptococcus saliverius*
            - No acid from raffinose → *Streptococcus sanguis*
  - Gram-negative
    - Glucose not fermented → *Neisseria Flavescens*
    - Glucose fermented
      - Nitrate reduced → *N. mucosa*
      - Nitrate not reduced → *N. sicca*

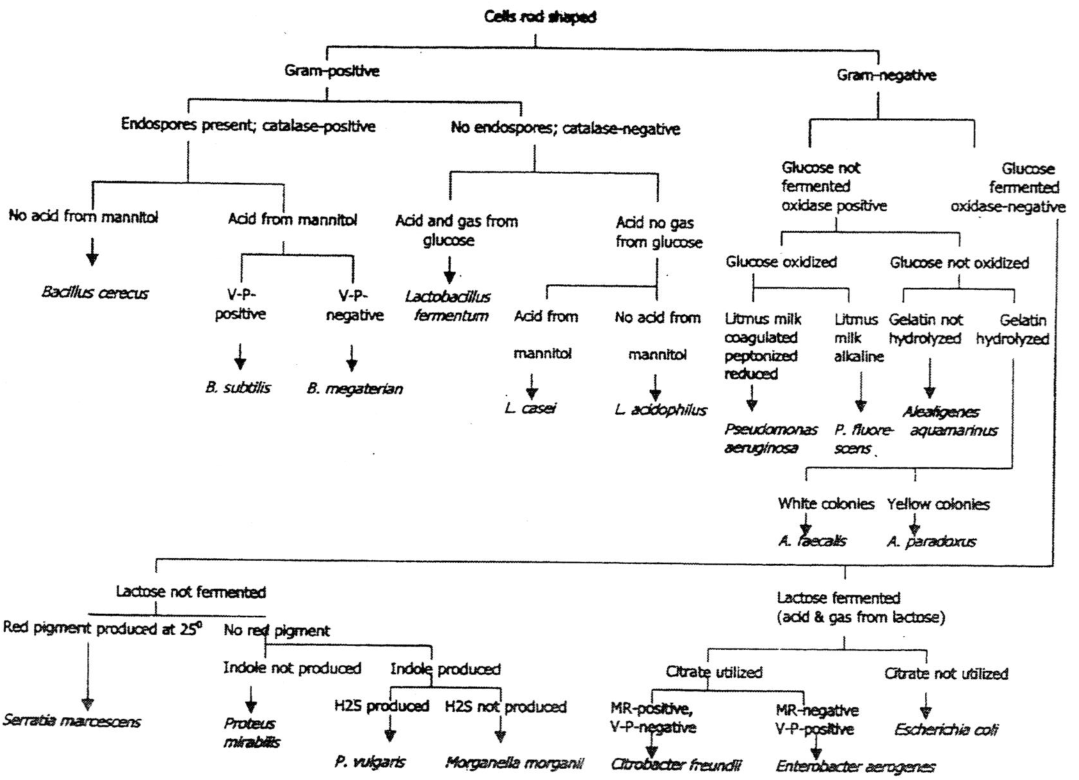
Cells rod shaped
Gram-positive
Gram-negative
Endospores present; catalase-positive
No endospores; catalase-negative
Glucose not fermented oxidase positive
Glucose fermented oxidase-negative
No acid from mannitol
Acid from mannitol
Acid and gas from glucose
Acid no gas from glucose
Glucose oxidized
Glucose not oxidized
Bacillus cerecus
V-P-positive
V-P-negative
Lactobacillus fermentum
Acid from mannitol
No acid from mannitol
Litmus milk coagulated peptonized reduced
Litmus milk alkaline
Gelatin not hydrolyzed
Gelatin hydrolyzed
B. subtilis
B. megaterian
L. casei
L. acidophilus
Pseudomonas aeruginosa
P. fluorescens
Alcaligenes aquamarinus
White colonies
Yellow colonies
A. faecalis
A. paradoxus
Lactose not fermented
Lactose fermented (acid & gas from lactose)
Red pigment produced at 25°
No red pigment
Indole not produced
Indole produced
Citrate utilized
Citrate not utilized
Serratia marcescens
Proteus mirabilis
H2S produced
H2S not produced
MR-positive, V-P-negative
MR-negative V-P-positive
Escherichia coli
P. vulgaris
Morganella morganii
Citrobacter freundii
Enterobacter aerogenes

*Growth on Broth -*

| | |
|---|---|
| Surface growth: | Ring, pellicle, none |
| Clouding: | Slight, heavy, none |
| Sediment: | Abundant, scanty, none, granular, flaky orflocculent |

*Biochemical Test*

Carbohydrate fermentation

| | |
|---|---|
| Glucose: | Red to yellow |
| Mannitol: | Red to yellow |
| Lactose: | Red to yellow |
| Sucrose: | Red to yellow |

| | |
|---|---|
| Lactose fermentation: | purple to pink |
| Gas formation: | separation of curd or development or fissures within curd |

*Litmus Milk Reaction*

| | |
|---|---|
| Litmus reduction: | purple to white or milk coloured |
| Proteolysis: | Deep purple band on the upper position and Whey-like brownish translucent medium |
| Alkaline reaction: | Deep blue |
| Starch hydrolysis: | A clear zone surrounding bacterial growth |
| Gelatin hydrolysis: | Liquidified cultures following incubation |
| Casein hydrolysis: | Clear area surrounding the bacterial growth |
| Lipid hydrolysis: | Clear area surrounding the bacterial growth |
| Hydrogen sulphide production: | Black precipitate |
| Indole production: | Formation of red layer |
| Methyl-red test: | Presence of red colour |

| | |
|---|---|
| Voges-Praskauer (VP) test: | Deep rose (pink) colour |
| Citrate utilization test: | Green to blue |
| Urease test: | Red to deep pink |
| Nitrate reduction test: | Red colour |
| Catalase activity: | Bubbles of free oxygen gas |
| Oxidase test: | Development of pink, the maroon and finally black colour on the bacterial colonies |

*Procedure:* Streak the unknown bacterial culture into agar plates for isolation. Incubate one inoculated plate at 37°C and other plate at room temperature for 24-48 hours in an inverted position. Inoculate two trypticase soy agar slants by means of a streak inoculation and incubate these slants at 37°C for 24 hours for the growth of the bacterium. Keep one culture in the refrigerator as stock and the other culture is used as working culture. Perform a gram staining reaction on the 24-hours old culture. After observing its staining and morphological characteristics, conduct the following biochemical tests with bacterial culture by inoculating it into different media, slants and broths as given below.

| | |
|---|---|
| Carbohydrate fermentation - | phenol red glucose broth, phenol red lactose broth, Phenol red sucrose broth |
| Litmus Milk Reaction - | Trypticase soy agar slant |
| Starch hydrolysis - | Starch agar plate |
| Gelatin hydrolysis - | Nutrient gelatin deep tube |
| Lipid hydrolysis - | Tributyrin agar plate |
| Hydrogen sulphide production- | SIM medium |
| Indole production - | SIM medium |
| Methyl-red test - | MR-VP broth |
| Voges-Praskauer (VP) test - | MR-VP broth |
| Citrate utilization test - | Simmon's citrate agar slant |
| Urease test - | Urea broth |
| Nitrate reduction test - | Trypticase nitrate broth |
| Catalase activity - | Trypticase soy agar slant |
| Oxidase test - | Trypticase soy agar plate |

Incubate all the inoculated deep tubes, slants and broths at 37°C for 24-72 hours and examine them for the cultural characteristics. Record observations and results for various biochemical tests.

*A. Morphological Characteristics*

1. Cell shape
2. Cell arrangements
3. Cell size
4. Endospores
5. Motility

*B. Staining Reaction*

1. Gram stain
2. Other stain reaction

*C. Cultural Characteristics*

(a) On trypticase soy agar plate

1. Diameter
2. Growth
3. Colour
4. Form
5. Elevation
6. Margin

(b) On agar slant

1. Growth
2. Form
3. Colour
4. Consistency

(c) In broth

1. Growth
2. Pellicle
3. Sediment
4. Flocculent

**Table 3.4 Key to indentification of some selected heterotrophic bacteria**

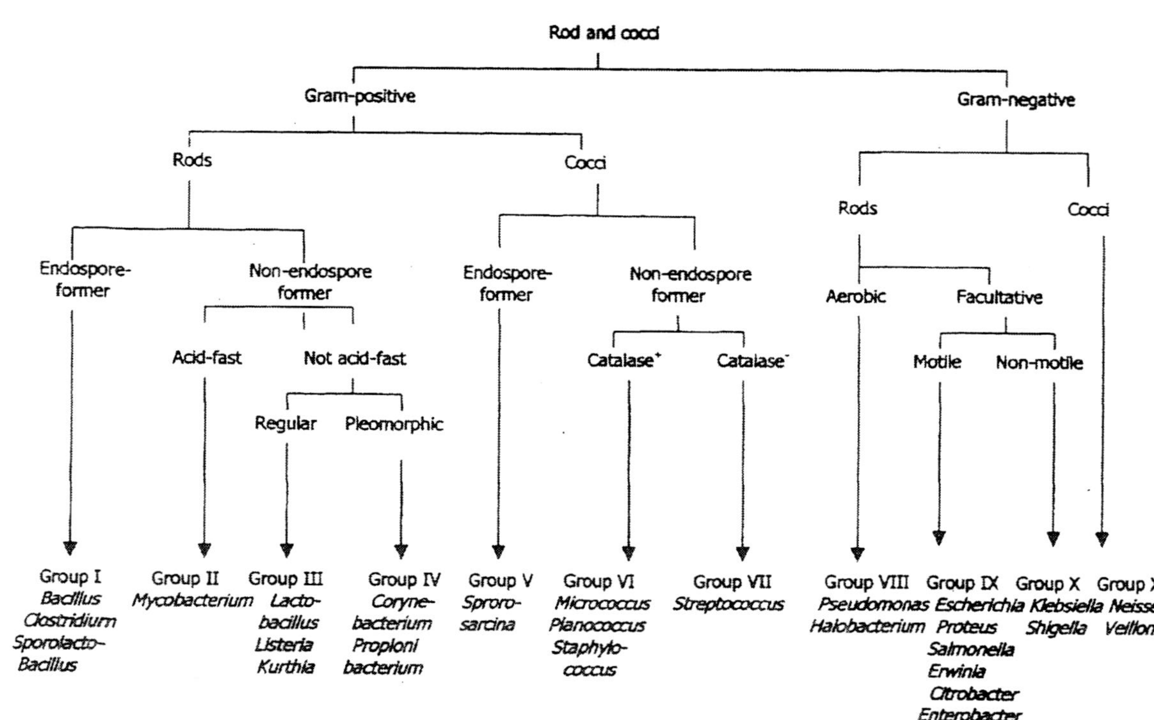

*D. Biochemical Tests*

Carbohydrate fermentation

Litmus Milk Reaction

Starch hydrolysis

Gelatin hydrolysis

Lipid hydrolysis

Hydrogen sulphide production

Indole production

Methyl-red test

Voges-Praskauer (VP) test

Citrate utilization test

Urease test

Nitrate reduction test

Catalase activity

Oxidase test

On the basis of the above results, identify the unknown bacterium as to its genus and species with the help of the key given in Table 3.3.

### 3.5.2 Identification of an Unknown Fungus

The fungi can be differentiated from bacteria or algae on the basis of the facts that fungi have true nuclei, which contain several chromosomes confined in a nuclear membrane. Identification of a fungus is often made based on the following characteristic structures–

| | |
|---|---|
| Colonial morphology - | Waxy, leathery, velvety, fluty |
| Hyphae - | Coenocytic or septate |
| Asexual spores - | Zoospores, sporangiospores, conidia or blastospores |
| Sexual spores - | Ascospores, basidiospores, oospores or zygospores |
| Reproductive bodies- | Ascocarps, basidiocarps, simple conidiospores, synnemata, sporodochia, pycinidia or acervuli |
| Arrangement of conidia - | Solitary, masses, chains (acropetal or basipetal) |

*Procedure:* Inoculate the Sabouraud agar plate with the fungal culture. Incubate the plate at room temperature (25°C) in an inverted position for 5-7 days. Prepare a slide culture following slide culture technique and incubate at room temperature for 2-5 days. Make a lactophenol cotton blue preparation for microscopic examination (Gently heat the cover slip culture in an open flame to fix the fungus to the glass and then put a drop of lactophenol cotton blue on the glass slide. Gently lower the cover slip culture into the drop. Place a second drop of lactophenol cotton blue on the cover slip culture and cover it with an 18 mm square cover slip. Put 4-5 drops of histological stain fluid on the mount and place a second cover slip - 22 mm square -over the culture.). Observe the fungal growth on slide culture, plate culture as well as stained preparation for structure of hyphae and details of sporating structures microscopically. Draw the details and record the results.

*Colony appearance*

Morphology

Hyphae colour

Spore colour

Underside

*Characteristics*

Hyphae

Type of spores

Colour of spores

Arrangement of spores

Type of fruiting bodies

Dimensions of various structures

# 4

# FOLIAR ANALYSIS

Insects feed throughout their life cycle for having adequate supply of essential elements viz., carbon, nitrogen, phosphorus, sulphur, inorganic salts etc. for body maintenance, growth and reproduction. The ultimate energy sources in insect metabolism are carbohydrates, proteins and fats. Besides these, some minerals, water, vitamins and steroids also play significant role in insect nutrition. The substances - organic or inorganic -, which satisfy these needs of phytophagous insects, vary to some extent from plant to plant. Hence, nutritional evaluation of a feed is of vital importance in the metabolism of a phytophagous insect. Foliar analysis or chemical analysis of feeding stuff constitutes quantitative determination of various chemical constituents present in a feed stuff.

## 4.1 PREPARATION OF SAMPLES

Sampling and collection of plants depend on the kind of plants to be sampled, determinations to be made and overall objectives of analysis. The preparation of sample is as important as the analytic procedure and hence it should be done with proper care and due precautions. It is desirable to collect enough individual plant or plant parts.

- Separate the plant material by using sharp knife or dissecting scissors (if leaves are the sample, they may be plucked manually).
- Remove all the contaminations by gentle brushing. If any washing is required, it should be done prior to the drying of the plant material.
- Transfer plant samples to the laboratory as early as possible for further processing. During transport, the material should be lightly packed in loosely woven bags.
- Dry the sample in an oven or in air for storage.

### 4.1.1 For Mineral Constituents

- Remove thoroughly all the foreign matter from the plant material.
- Mix corresponding plant parts from more than one plant to obtain a representative or random sample for analysis.
- Weigh a portion of this sample for those analyses, which are to be performed on fresh weight basis.
- Dry the sample in a hot air oven at 100°C immediately to check decomposition, grind and store in airtight bottles.

### 4.1.2 For Carbohydrates

- Rapidly grind or chop the plant material into fine pieces (1-2 cms).
- Add weighed sample to enough hot re-distilled 80% alcohol (5-10 ml of ethanol per g sample), to which enough precipitated $CaCO_3$ has been added to neutralize acidity.
- Heat nearly to boiling point on steam or water bath for 30 minutes with stirring frequently.
- Cool the extract, crush the tissues with a mortar pestle thoroughly and then pass through double layered cheese cloth.
- Re-extract in hot 80% ethanol for 5 minutes, using 2-3 ml alcohol per g sample. The second extraction ensures complete removal of alcohol soluble substances.
- Cool the extract and pass through cheese cloth.
- Pool both the extracts and filter through Whatman No. 41 filter paper.
- Raise the volume with 80% alcohol or reduce the volume by evaporating it to represent 5-10 ml of the extract for every g of the tissue used. This extract (alcohol soluble) contains reducing and non-reducing sugars, some glycosides, amides and amino N, other forms of soluble nitrogen, phenols, flavones, tannins, oils, lipids, chlorophylls, carotene and xanthophylls.
- The residue may be used for determination of macromolecules like starch, pectin, hemicellulose, cellulose, lignin, proteins, insoluble ash and minerals.

### 4.1.3 For Proteins

Samples for protein estimation are prepared as per the Standard A.O.A.C. method (1950).

- Macerate the residue (obtained after alcohol extraction in the above extraction) in a mortar pestle and transfer the homogenate into centrifuge tubes.
- Centrifuge at 2000 rpm for 20 minutes and then discard the supernatant.
- Suspend the pellets in suitable volumes of 5% TCA at 0-2°C for 15 minutes in an ice bath. (While processing enzyme extracts for estimation of buffer soluble proteins, pipette 1 ml of aliquots in centrifuge tubes and add 1 ml of 10% TCA or perchloric acid at 0°C to precipitate the proteins).
- Allow the pellets to stand for 15 minutes in ice bath and then centrifuge. Discard the supernatant. Repeat the process twice.
- Re-extract the pellets once with absolute alcohol and twice with hot ethanol-ether mixture, every time discarding the supernatants after centrifugation. The pellets contain proteins and nucleic acids.
- The pellets are suspended in 5 ml 1N NaOH solution at 100°C for 4-5 minutes and the extract is used for protein estimation.

### 4.1.4 For Nucleic Acids

- Suspend the pellets (obtained earlier in protein sample preparation) in suitable volumes of 0.3N KOH and incubate for 16-18 hours at 37°C.
- Centrifuge the solution at 2000 rpm for 15-20 minutes and collect the supernatant.
- Wash the residue twice with distilled water, centrifuge and collect the supernatants.
- Pool all the supernatants and make up the volume to 100 ml.
- Wash the residue twice with distilled water, centrifuge and collect the supernatants.
- Pool all the supernatants and make up the volume to 100 ml.

## 4.2 PROXIMATE ANALYSIS OF FEEDS

Proximate system of analysis was developed by Henneberg & Stohmann in 1860, which is also called as Weende system of analysis. It is a system for approximating the chemical composition and nutritive value of biological materials especially a feed. The principle of the analysis is to separate feedstuffs into various fractions e.g. water or moisture, crude protein, ether extract, crude fibre, nitrogen free extract and mineral matter or ash. The total mineral matter represents the inorganic part of the food whereas crude protein, ether extract, nitrogen-free extract and crude fibre are referred as organic matters.

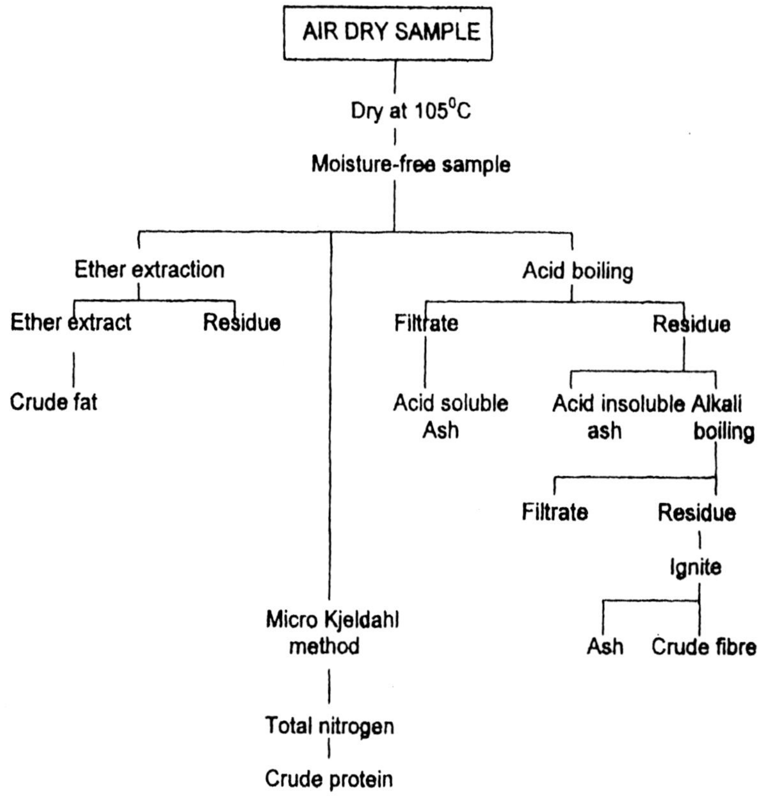


### 4.2.1 Determination of Moisture

**The constant weight of a sample after complete removal of moisture or water is generally called dry matter (DM % = 100 - % moisture).** The dry matter consists of ether extract, crude protein, crude fibre, ash etc, devoid of moisture. The moisture of the sample represents the most abundant constituent ranging from 50-90% of total chemical composition. It can be determined by several methods viz., oven drying, toluene distillation, freeze drying, infra-red moisture meter etc. However, oven drying method is most common. The moisture is lost by volatilization caused by heat. The amount of material left after removal of the moisture is the dry matter.

*Principle:* The moisture content of a sample is determined by heating it in an oven to a constant weight at 100-105°C under atmospheric pressure. The moisture is removed as vapors.

*Procedure:* Dry the moisture cup in an oven at 100°C temperature, cool it in desiccator and record its weight ($W_1$). Take about 10g fresh leaves in moisture cup and weigh the cup with leaves ($W_2$). Dry the sample in a hot air oven at 100 - 105°C for 5-6 hours till the constant weight is obtained. Remove the moisture cup from the oven and cool it in a desiccator. Repeat the process of heating and cooling till a constant weight is obtained. Record the constant weight of the cup with sample ($W_3$).

*Calculation :*

$$\text{Moisture \%} = \frac{W_3 - W_2}{W_2 - W_1} \times 100$$

$$\text{Dry Matter \%} = \frac{\text{Weight of dried sample}}{\text{Weight of fresh sample}} \times 100$$

$$\text{or Dry Matter \%} = 100 - \text{Moisture \%}$$

### 4.2.2 Determination of Total Ash (Mineral content)

*Principle:* Ash is the inorganic or mineral component of the sample left after complete ignition of the sample at 600°C temperature in muffle furnace. It is determined by AOAC method (1965).

*Procedure:* Dry the crucible in an oven, cool it in a desiccator and record its weight. Take weighed oven dry sample (1-2g) in this crucible. Ignite the sample in a burner or on an electric heater till the smoke ends up. Transfer the crucible to a muffle furnace with the help of a pair of long tongs. Keep it at 600°C temperature and left until white ash is obtained to a constant weight (approximately

2-3 hours). Remove the crucible from the furnace, cool it in the desiccator and weigh it. Substrate this weight from the weight of crucible plus sample to get the weight of ash. The ash should be preserved for extraction of acid soluble minerals and for estimation of acid insoluble ash.

*Calculation:*

$$\text{Ash or mineral matter \%} = \frac{\text{Weight of ash}}{\text{Weight of sample}} \times 100$$

### 4.2.3 Determination of Acid Soluble and Insoluble Ash

*Principle:* When the ash of sample is boiled in dilute acid, filtered and washed, the filtrate contains acid-soluble ash while the residue after ignition gives the acid insoluble ash.

*Procedure:* Use the residue obtained from the ash determination. Add 25 ml HCl (1:1.25; v/v) in the crucible containing ash, cover with watch glass and digest for 20-30 minutes on water bath. Remove the watch glass and rinse with water into the contents of the crucible. Filter through ash-less Whatman no. 42 filter paper into a 100 ml volumetric flask, wash and transfer the residue from the crucible repeatedly with small volumes of hot water until acid free. Mix the contents thoroughly and preserve this solution for estimation of different inorganic ingredients. Place the filter paper and residue in a porcelain dish/crucible and heat in muffle furnace at 600°C for 2 hours until carbon-free. Cool the crucible in a desiccator. Weigh the dried residue, which gives acid insoluble ash.

$$\text{Acid insoluble ash} = \frac{\text{Weight of dried residue}}{\text{Weight of sample}} \times 100$$

### 4.2.4 Determination of Crude Fat (Ether extract)

Crude fat is a combination of simple fat, fatty acid esters, compound fat, neutral fat, sterols, waxes, vitamins (A, $D_2$, $D_3$, E, K), carotene, chlorophyll etc.

*Principle:* Ether extract or crude fat includes all the portions of a feed, which are soluble in ether. The ether extract content of the sample is estimated by extracting it with a fat solvent like petroleum ether, benzene, chloroform, diethyl ether etc. Ether is continuously volatilized at 55-60°C, condensed and allowed to pass through the sample in a Soxhlet's apparatus, extracting ether-soluble materials. When the process is completed, the ether is distilled and collected in another container and the remaining crude fat is dried and weighed.

*Procedure:* Weigh 2g dry leaf sample and place it in a completely dried and weighed extraction thimble. Weigh an empty fat extraction flask, introduce the thimble containing sample into the Soxhlet extractor in a straight direction so that the condensed ether may drop on it. Connect the flask with it. Pour sufficient anhydrous ether (boiling point 40 – 60°C) through the condenser mouth to cause comfortably sufficient flow of ether through the siphon to the flask. Pour a further amount of ether so that its level in the extractor remains just below the siphon level. The flask is kept in warm water bath at 55 – 60°C and cold water is circulated through all the condensers. The ether will boil and condense into the extractor till the collected ether again siphon off to the flask. The repeated extraction is carried out for 4 hours in Soxhlet apparatus at a condensation rate of 5-6 drops per second. Remove the condenser, take out the thimble and collect the ether in the extractor and keep it in a separate bottle for further use. The flask containing the crude fat may still contain some ether, which is removed by drying the flask in an oven at 100°C for 10 minutes. Cool and weigh the flask. Increase in weight of flask gives the amount of crude fat present in the sample. Keep the thimble at room temperature for evaporation of ether and dried in an oven at 100-105°C overnight. Remove the thimble from the oven, cool it in a desiccator and weigh.

Weight of sample = (Wt. of thimble + sample) – Wt. of thimble

Weight of fat = [Wt. Of thimble + sample] - [Wt. Of thimble + sample after extraction]

$$\% \text{ Crude fibre} = \frac{\text{Weight of dry fat material}}{\text{Weight of dry sample}} \times 100$$

### 4.2.5 Determination of Crude Fibre

Crude fibre consists of celluloses, variable proportion of hemicelluloses and highly variable proportion of lignins along with some minerals.

*Principle:* The estimation of crude fibre is done by treating the moisture- and fat-free samples successively with dilute acid and alkali. The organic residue is collected and the loss of weight on ignition is called crude fibre.

*Reagents:*

1. 2.5% Sulphuric Acid - 15 ml of concentrated sulphuric acid (Sp. Gr. 1.62) are diluted to 1000 ml with distilled water.
2. 2.5% NaOH - 25g sodium hydroxide pallets are dissolved in distilled water and the volume is made up to 1000 ml with water.
3. 1% HCl
4. 95% ethyl alcohol.

*Procedure :*

**Method - 1:** Transfer the residue after ether extraction to a 400 ml beaker. Add 100 ml distilled water (boiling) and 100 ml of 2.5% sulphuric acid. Mark the level of the mixture in the beaker by a pencil. Reflux the mixture for 30 minutes from onset of boiling. Keep the volume of the liquid the same during boiling by adding hot water. Filter through a linen spread over a Buckner funnel, which is fitted to a filter pump. Wash with hot water till free from acid. Transfer back the entire residue to the original beaker with a jet of hot water. Add 100 ml of 2.5% sodium hydroxide solution and then make up the volume to 200 ml with boiling water. Boil gently for 30 minutes. The mixture is filtered through same linen, washed with hot water several times, then with dilute HCl and again with hot water till free from acid. Squeeze out the excess moisture. Finally, wash twice with 95% alcohol. Transfer the residue to a weighed platinum basin and dry in an oven at 105°C until constant weight. Cool in a desiccator and weigh. Ignite the material in a muffle furnace at 550-600°C for 2-3 hours to get white ash, cool and then take weight again. The difference in two weights represents the weight of crude fibre.

$$\text{Crude fibre \%} = \frac{\text{Wt. of crude fibre}}{\text{Weight of dry sample}} \times 100$$

**Method - 2:** Extract 1g leaf sample with ether or petroleum ether. Transfer the residue to a 500 ml flask to which 100 ml of distilled water and 100 ml of 1N sulphuric acid are added. Immediately, the digestion flask is connected to condenser and heated. It is necessary to rotate the flask until the sample is wetted. The boiling is continued briskly for 30 minutes and afterwards the contents are filtered through Whatman No. 1 filter paper. The residue is washed with water until free from acid. The filter paper is then pierced with a glass rod and 100 ml of 1N NaOH are added so that the residue is transferred to the digestion flask. Next, it is refluxed for exactly 30 minutes and then filtered through filter paper, washed repeatedly to remove traces of sodium hydroxide and dried at 110°C for 4-5 hours to constant weight. It is cooled and final weight is taken.

$$\% \text{ Crude fibre} = \frac{\text{Weight of dried residue}}{\text{Weight of dry sample}} \times 100$$

### 4.2.6 Determination of Total Nitrogen (Crude Protein)

Total nitrogen or crude protein comprises of nitrogen value from all organic and inorganic sources of the material. It describes the energy of the nitrogenous fraction of food. It can be determined by the following methods.

- Macro-Kjeldahl method
- Micro-Kjeldahl method
- Modified Micro-Kjeldahl method

Crude protein content in plant material is calculated by multiplying total nitrogen content estimated by Micro-Kjeldahl method by 6.25.

#### *4.2.6.1 Macro-Kieldahl method*

*Principle*: Organic nitrogen when digested with concentrated sulphuric acid in the presence of a catalyst (selenium oxide, potassium sulphate or copper sulphate) is converted into ammonium sulphate. Ammonia liberated by making the solution alkaline is distilled into a known volume of standard acid, which is then back titrated.

$$\text{Organic nitrogen} + \text{conc. } H_2SO_4 = (NH_4)_2SO_4$$

$$(NH_4)_2SO_4 + 2\ NaOH = 2\ NH_3 + Na_2SO_4 + 2\ H_2O$$

$$2\ NH_3 + H_2SO_4 = (NH_4)_2SO_4$$

*Reagents:*

1. 30% NaOH - 30g sodium hydroxide pallets are dissolved in water and the volume is made up to 100 ml with water.
2. 0.1N NaOH - 0.4g sodium hydroxide pallets are dissolved in water and the volume is made up to 100 ml with water.
3. 0.1N Sulphuric acid - 2.78 ml of conc, sulphuric acid is diluted to 100 ml with distilled water.
4. Methyl red indicator - 0.1g methyl red is dissolved in 100 ml of 95% ethyl alcohol.

*Procedure:* 0.5-1g sample is taken in a Kjeldahl flask with 10 ml of water, 20 ml conc. $H_2SO_4$, 1g $CuSO_4$ and 10g $K_2SO_4$ or anhydrous sodium sulphate. Digest till the mixture is clear and assumes a pale green or yellowish green color. The flask is cooled. Dilute the contents of the flask to about 50 ml with water and transfer to a distillation flask. About 300 ml of water are added with 100-150 ml of 30% NaOH till alkaline and then distilled. The distillate is collected in excess of 0.1N $H_2SO_4$ with methyl red as indicator. Discontinue distillation when about 200 ml of distillate are collected. The excess acid is titrated back with 0.1N NaOH. The net amount of 0.1 N acid used refers to the amount of nitrogen in the material.

### 4.2.6.2 Micro-Kieldahl method

*Reagents:*

1. Concentrated sulphuric acid (Sp. Gr. 1.62)
2. Anhydrous Potassium sulphate
3. 2% boric acid
4. 30% NaOH
5. Methyl red indicator
6. Bromophenol blue indicator - 40 mg bromophenol blue is dissolved in 95% ethanol and the volume is made up to 100 ml.
7. 0.1N NaOH
8. 0.1N Oxalic acid - 0.63g oxalic acid is dissolved in water and diluted to 100 ml with water.

*Procedure:* Take 1g oven dry sample into a Kjeldahl flask, add 1-2g potassium sulphate, a pinch of copper sulphate to facilitate digestion and 5 ml of conc, sulphuric acid. The mixture is first digested at low flame and then strongly until a clear mixture is obtained. It is allowed to cool down and diluted with distilled water to 50 ml. Suitable aliquots of the digested mixture (5 ml) are transferred to a distillation flask. Add 4-5 ml of 30% NaOH solution and connect to the Kjeldahl distillation unit. In a 100 ml conical flask, take 10 ml of 2% boric acid solution and add few drops of mixed indicator (Methyl red and Bromophenol blue). The conical flask is connected to the receiving tube and the digested mixture is distilled till all the ammonia is liberated (10-15 minutes) and the color of boric acid solution changes from pink to greenish blue. Ammonia containing boric acid is back titrated with 0.1N NaOH and end point is noted when color changes to pink again.

1 ml of N/10 $H_2SO_4$ = 0.0014 g Nitrogen

*Calculation:*

$$\text{Nitrogen content of sample} = \frac{V_1 \times 0.0014 \times D \times 100}{W \times A}$$

where,

$V_1$ = Volume of N/10 NaOH required for neutralization

D = Dilution factor (50 ml)

W = Weight of sample (1g)

A = Aliquot taken (5 ml)

### 4.2.6.3 Modified Micro-Kieldahl method

*Principle:* In Micro-Kjeldahl method as described above, most of the nitrogen of nitrates is lost during digestion and hence don't represent the whole of the nitrogen of the sample. Accordingly, nitrate nitrogen is fixed by salicylic acid. Sodium thiosulphate is added with sulphuric acid to reduce the nitrates.

*Digestion*

$$2\ NaNO_3 + H_2SO_4 \longrightarrow Na_2SO_4 + 2\ HNO_3$$

$$\underset{\text{Salicylic acid}}{C_6H_6OHCOOH} + HNO_3 \longrightarrow \underset{\text{Nitrosalicylic acid}}{C_6H_3NO_2(OH)COOH} + H_2O$$

$$Na_2S_2O_3 + H_2SO_4 \longrightarrow H_2S_2O_3 + Na_2SO_4$$

$$H_2S_2O_3 \longrightarrow H_2SO_3 + S$$

$$C_6H_3NO_2(OH)COOH + 3H_2SO_3 + H_2O \longrightarrow C_6H_3(NH_2)(OH)COOH$$
Aminosalicylic acid $+ H_2SO_4$

$$2\,C_6H_3(NH_2)(OH)COOH + 27\,H_2SO_4 \longrightarrow (NH_4)_2SO_4 + 26\,SO_2 + 14\,CO_2 + 30\,H_2O$$

*Distillation*

$$(NH_4)_2SO_4 + 2\ NaOH \longrightarrow Na_2SO_4 + 2\ NH_3 + 2\ H_2O$$

*Absorption*

$$2\ NH_3 + H_2SO_4 \longrightarrow (NH_4)_2SO_4 \text{ (conical flask)}$$

*Titration*

$$H_2SO_4 + 2\ NaOH \longrightarrow Na_2SO_4 + H_2O$$

*Procedure:* Weigh 1g of the sample and place it into Kjeldahl flask. Dissolve 1.0 g salicylic acid in 30 ml of conc. Sulphuric acid and add the mixture to the flask. Shake well until thoroughly mixed, allow to stand at least 30 minutes with frequent shaking and then add 5-7 g sodium thiosulphate. The whole mixture is kept and shaken occasionally for another 30 minutes. The mixture is then digested first at low flame and then strongly for 5 minutes. Then add 10 g potassium sulphate and a pinch of copper sulphate and heated more strongly until a clear mixture is obtained. Allow to cool, dilute with water and transfer it to a distillate flask. Add about 120 ml of 30% NaOH solution and connect up to a condenser dipping into conical flask containing N/10 $H_2SO_3$ and suitable indicator. Distil until 150 ml passes over. Back titrate the excess of the acid in the receiver with N/10 NaOH. Then calculate the percentage of nitrogen.

1 ml of N/10 $H_2SO_4$ = 0.0014 g Nitrogen

### 4.2.7 Determination of Nitrogen-Free Extract

The nitrogen-free extract (NFE) represents the difference in dry weight of the fraction obtained by subtracting the total dry weight of the sample from the sum of the values of ash, crude fibre, crude fat and crude protein on dry weight basis. The main components of NFE are the sugars and starches. The digestible or the metabolizable energy of the body ultimately obtains from the NFE, from the crude fibre, or from total carbohydrates of a food. The celluloses and hemicelluloses yield less energy to the body than do the carbohydrates of the NFE. It is calculated as follows.

NFE (%) = 100 - [CP % + CF % + EE % + Total ash %]

Where,

CP – Crude protein

CF – Crude Fibre

EE – Ether Extract

### 4.2.8 Estimation of Sugars

*Total Sugars*

Take 5g dry leaf powder in a thimble and introduce the thimble in the middle part of the soxhlet apparatus. Add about 130-150 ml 80% alcohol and heat for 22-24 hours till the alcohol in the tube-containing thimble become colorless. After it, transfer the whole amount to a measuring flask and wash the thimble with distilled water. Make up the volume to the mark (say 250 ml). From this extract, take 50 ml in a beaker and evaporate till there is no smell of alcohol. The experiment may be conducted in duplicate. Add about 2g lead acetate to it to remove the impurities and filter. To remove excess of lead acetate, add sodium oxalate solution till there is no precipitation. Filter it again and volume is made up to the mark (say 100 ml).

Take 5 ml of this solution in a 500 ml conical flask and add 5 ml of conc, hydrochloric acid. Heat on a water bath at a constant temperature of 75°C for exactly 10 minutes. Cool the solution under tap water and neutralize with solid sodium carbonate (till there is no effervescence). Add 20 ml of Benedict's solution to this and boil for 3 minutes. Cool under running tap water and then add 100 ml of 5N acetic acid, 20 ml of N/25 iodine and 25 ml of dilute hydrochloric acid. Keep in dark for 5 minutes and then titrate against N/25 sodium thiosulphate solution. When the color changes to pale yellow (near end point), add 2 ml of starch solution and then immediately titrate. Note down the end point.

Let the end point be 12.3 ml

Iodine reacted = (20 – 12.3) = 7.7 ml

But 1 ml iodine = 0.00112g sugar

7.7 ml iodine = 0.00112 × 7.7 g sugar

Hence,

sugar % = (7.7 × 0.00112 × 100 × 250 × 100) ÷ (5 × 50 × 5)

***Reducing sugars***

The solution obtained after the precipitation with lead acetate and sodium oxalate is kept in a conical flask. Take 20 ml of the above solution and add 20 ml of Benedict's solution to it. Boil for 3 minutes and then cool under running tap water. Add 100 ml of 5N acetic acid, 20 ml of N/25 iodine and 25 ml of dilute hydrochloric acid (2-3N). Keep in dark for 5 minutes and then titrate against N/25 sodium thiosulphate solution.

Let the end point be 11.3 ml

Then iodine reacted = (20 – 11.3) = 8.7 ml

Reducing sugar % = (8.7 × 0.00112 × 100 × 250 × 100) ÷ (20 × 50 × 5)

***Carbohydrates***

Take 1 g powdered leaf in a 500 ml conical flask and add 150 ml distilled water and 20 ml conc. hydrochloric acid. Reflex the solution for 2.5 hours with Ledbig Condenser. The time is noted down when it starts boiling. Filter it and wash. Collect the filtrate in a 250 ml measuring flask and then volume is made up to the mark. Take 5 ml of this solution in a 500 ml conical flask and add a piece of litmus paper. Neutralize with 40% sodium hydroxide drop-wise until the litmus turns blue. To this solution, add 20 ml of Benedict's solution. Boil for 3 minutes and then cool. To this, add 100 ml of 5N acetic acid, 20 ml of N/25 iodine and 25 ml of dilute hydrochloric acid. Keep in dark for 5 minutes and then titrate against N/25 sodium thiosulphate solution when pale yellow color appears. 2 ml of starch solution is added and again titrated. The end point is noted down.

Let the end point be 11.7 ml

Then iodine used = (20 – 11.7) = 8.3 ml

Carbohydrate % = (8.3 × 0.00112 × 250 × 100) ÷ (5 × 2)

### *Starch*

After extraction of total sugars, the material left in the thimble is taken out, dried to remove alcohol and transferred in a 500ml conical flask. Add 150 ml distilled water and 20 ml of conc, hydrochloric acid. Reflex for 2.5 hours. Filter, wash with distilled water and make up the volume to 250 ml. Take 5 ml of this solution in a 500 ml conical flask and add a piece of litmus paper. Neutralize with 40% sodium hydroxide drop-wise until the litmus turns blue. To this solution, add 20 ml of Benedict's solution. Boil for 3 minutes and then cool. To this, add 100 ml of 5N acetic acid, 20 ml of N/25 iodine and 25 ml of dilute hydrochloric acid. Keep in dark for 5 minutes and then titrate against N/25 sodium thiosulphate solution when pale yellow color appears, 2 ml of starch solution is added and again titrated. The end point is noted down.

$$\text{Starch \%} = (\text{Iodine used} \times 0.00112 \times 250 \times 100) \div (5 \times 5)$$

## 4.2.9 Estimation of Total Phenolics

Total phenolics in plant materials are measured quantitatively using Folin method (Swain & Goldstein, 1964).

*Reagents:*

1. 2N HCl - 16.7 ml of concentrated hydrochloric acid (Sp. Gr. 1.18) are diluted to 100 ml with distilled water.
2. 6% ethanol - 60 ml of dehydrated ethanol are diluted to 1000 ml with distilled water.
3. Folin's reagent

*Procedure:* Hydrolyze air-dry powdered leaf sample (100 mg) with 2N HCl for 30 minutes at 100°C. After cooling and filtering, extract the filtrate with diethyl ether (2 x 5 ml) and then concentrate the ether layers to dryness. Dissolve the residue in 6% ethanol (500μl) and make 50 μl of samples to 3.5 ml with 6% ethanol. Add 0.5 ml of Folin reagent and mix the solutions thoroughly. Keep the mixture at 20°C for 1 hour.

Take absorbance at 710 nm with colorimeter using ethanol blank. Phenolic content is expressed as a percentage of dry weight by reference to standard curves.

Qualitative analysis of phenolics is done on plastic backed microcrystalline cellulose TLC sheets (Merck). The above ether extracts of the hydrolyzed leaf are chromatographed two-dimensionally in 1-butanol-acetic acid-water (6:1:2; v/v) and 20% acetic acid and also in toluene-acetic acid-water (6:7:3; upper layer) and 15% acetic acid and in one dimension in Forested solvent (acetic acid-conc, hydrochloric acid-water; 10:3:30, v/v). Phenolics are visualized with UV light before and after exposure to ammonia fumes and by spraying chromatograms with Folin's reagent followed by ammonia fume exposure. Compounds are identified by comparison of Rg values in the above solvent and color reactions with above reagents.

### 4.2.10 Determination of Tannins

50g fresh leaves of each food plants are taken separately and homogenized in 100 ml of acetone-water (70:30) containing a little of ascorbic acid. The suspension is stored in dark at room temperature for 2 days. A portion of the suspension is filtered and filtrate is transferred to a large conical flask. To each flask, add an excess of sodium chloride, which causes filtrate to separate in two layers. The upper layer (acetone) is separated, dried over anhydrous sodium sulphate and evaporated to dryness. The residue is extracted with absolute alcohol and alcohol soluble material is evaporated to dryness to yield tannins.

### 4.2.11 Determination of Essential Oils

Take 10g fresh leaves in a distilling flask, half-fill with water and connect with a condenser and receiver. Distil off about one third of water on a heating mantle. The distillate contains the essential oils, which are volatile in steam. The process is repeated till there is no odour in the distillate. The distillate is repeatedly extracted with small quantities of ether. The ether extract is dried over anhydrous sodium sulphate and the ether is distilled off. The residue possesses the characteristic odor of the essential oils of the plant.

### 4.2.12 Determination of Crude Glycosides

Extract 30g leaves repeatedly with ethanol (70%). Combine the alcohol extracts and then ethanol is distilled off. The aqueous extract is re-extracted with petroleum ether, ether and ethyl acetate. All the extracts, which are collected separately, are dried over anhydrous sodium sulphate. The ethyl acetate extracts are evaporated to dryness to yield glycosides. The traces of the solvent are removed by keeping the material in vacuum and then weighed.

### 4.2.13 Extraction of Lipids

Homogenize the leaf powder with chloroform-methanol (2:1). Add another portion of chloroform-methanol to the homogenate and extract the lipids at room temperature. The remaining lipids of leaves are extracted 4-5 times with the same solvent. The extracts thus obtained are pooled together and evaporated under reduced pressure. Re-extract 4-5 times with ethyl ether. The ethereal extracts are washed with 0.1% sodium chloride solution, twice with distilled water and finally dried with anhydrous sodium sulphate. The ethereal extracts are evaporated to dryness and the quantity of lipids present in leaves is determined by weighing the residue.

#### 4.2.14 Saponification and Extraction of Fatty Acids

The lipids of leaves are saponified for 3 hours in 95% (v/v) methanol containing 6% (w/v) potassium hydroxide at 80°C. The unsaponified residue is recovered by repeated extractions with ethyl ether from the sodium salts of lipids. The ether extracts are combined and dried over anhydrous sodium sulphate. Finally ether solvent is evaporated to yield sterols.

Fatty acids are repeatedly extracted with ethyl ether after acidifying the remaining sodium salts of lipids with 1N sulphuric acid. The ether extracts are washed with distilled water, dried with anhydrous sodium sulphate and finally ether is evaporated to yield fatty acids.

## 4.3 CHEMICAL & BIOCHEMICAL ANALYSIS OF FEEDS

Proximate analysis is the starting point of chemical analysis. It accounts about 50-60% of the chemical composition of the material. However, there are certain limitations in this analysis. The detailed chemical and biochemical analysis is desirable to know the exact and true nutritive value of a feedstuff or plant material.

### 4.3.1 Estimation of Carbohydrates

Carbohydrates are normally classified as mono-, oligo- and polysaccharides based on the number of monomer units present in the molecule. Carbohydrates are major constituents of plant materials.

They are of fundamental importance as a source pf metabolic energy, structural components and reserve or storage food. They exist as free sugars and polysaccharides in plant materials, which are estimated hydrolyzing them into mono-saccharides or simple sugars.

### *4.3.1.1 Estimation of Reducing Sugars*

Sugars with free aldehyde or ketone groups are able to reduce metal ions under alkaline conditions. Such sugars are called as reducing sugars. Glucose, galactose, lactose and maltose are reducing sugars. Reducing sugars are estimated following Nelson-Somogyi method (Somogyi, 1952) or the dinitrosalicylic acid (DNS) method (Miller, 1972).

*Nelson-Somogyi method*

*Principle:* The reducing sugars when heated with alkaline solution of copper tartarate reduce copper from cupric to cuprous state and thus cuprous oxide is formed. When cuprous oxide reacts with arsenomolybdic acid, the reduction of molybdic acid to molybdenum blue takes place and the blue color is developed, which is measured with a set of standards in a colorimeter at 620 nm. Sodium sulphate minimizes the chances of atmospheric oxygen causing re-oxidation of cuprous oxide in the solution.

```
    CHO                                        COONa        COONa
     |                                           |            |
H -  C - OH     COONa                       H - C - OH    H - C - OH
     |            |                              |            |
HO - C - H  + H - C - O - Cu  ——► Cu2O +  H - C - OH  + HO - C - H
     |            |      /       Heat            |            |
H -  C - OH   H - C - O                        COOK       H - C - OH
     |            |                                           |
H -  C - OH      COOK                                     H - C - OH
     |                                                        |
    CH2OH                                                   CH2OH
```

| D-Glucose | Sodium-Potassium Tartrate-copper Complex | Sodium potassium tartarate | Sodium D-Gluconate |
|---|---|---|---|

$Cu_2O$ + Arsenomolybdic acid ——► Molybdenum blue (Blue color complex)

The proteins must be removed before proceeding with determination of reducing sugars, which is carried out using zinc hydroxide as the protein precipitant to give a neutral protein-free solution.

*Reagents:*

1. Alkaline copper tartarate (Copper Reagent A) - Dissolve 25g anhydrous sodium carbonate, 25g sodium potassium tartarate (Rochelle salt), 200 g anhydrous sodium sulphate and 20g sodium bicarbonate in 800 ml distilled water and the volume is made up to 1 liter.
2. Copper Reagent B - Dissolve 15g copper sulphate in a small volume of distilled water. Add 1-2 drops of concentrated sulphuric acid and make up the volume to 100 ml.
3. Arsenomolybdate reagent - Dissolve 25g ammonium molybdate in 450 ml distilled water. Add 25 ml conc, sulphuric acid and mix well. Then add 3g di-sodium hydrogen arsenate dissolved in 25 ml water. Mix well and incubate at 37°C for 24 to 48 hours (If the reagent is needed urgently, heat to 55°C for about 25 minutes with stirring during heating).
4. Standard stock glucose solution - 100 mg glucose in 100 ml water.
5. Working standard - Dilute 10 ml of stock solution to 100 ml with distilled water (100ug/ml).

*Procedure:* Pipette suitable aliquots (1 ml) of alcohol-free extract into narrow test tubes graduated at 25 ml. Pipette out 0.2, 0.4, 0.6, 0.8 and 1.0 ml of working standard solution into a series of test tubes. Make up the volume in both sample and standard tubes to 2 ml with distilled water. Take 2 ml of distilled water in a separate test tube to serve as a blank. Add 1 ml of a mixture of 24 parts of Copper Reagent A and 1 part of Copper Reagent B and mix the solution. Heat the tubes in boiling water bath for 20 minutes. Cool the tubes and add 1 ml of arsenomolybdate reagent to all the tubes. Make up the volume to 25 ml with distilled water. Read the absorbance of blue color at 620 nm after 25 minutes. From the standard graph drawn with standard glucose solution, calculate the amount of reducing sugars present in the sample.

*Calculation:*

Reducing sugars % =

$$\frac{\text{Sugar Value } (\mu g)}{\text{Aliquot of sample (1 ml)}} \times \frac{\text{Total vol. of extract (ml)}}{\text{Wt. of sample (mg)}} \times \frac{1}{100}$$

*DNS Method*

*Principle:* Several reagents have been employed, which assay sugars by virtue of their reducing properties. One such compound is 3,5-dinitrosalicylic acid (DNS), which is reduced to 3-amino-5-nitrosalicylic acid, an orange coloured compound.

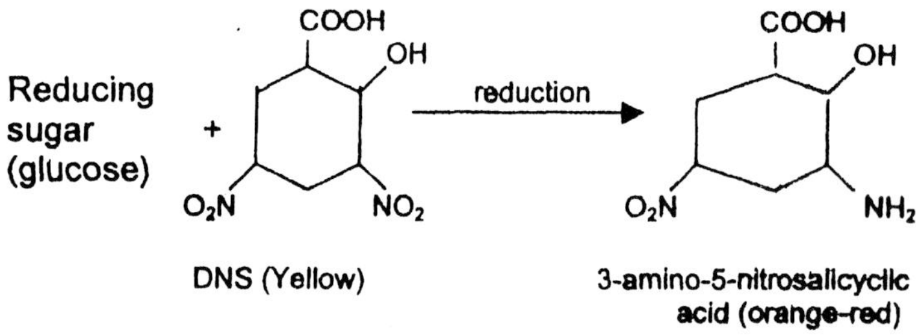

Reducing sugar (glucose) + COOH, OH, $O_2N$, $NO_2$ —reduction→ COOH, OH, $O_2N$, $NH_2$

DNS (Yellow) 3-amino-5-nitrosalicyclic acid (orange-red)

Reagents:

1. Dinitrosalicyclic acid (DNS) reagent: Dissolve 1 g dinitrosalicyclic acid, 200 mg crystalline phenol and 50 mg sodium sulphite in 100 ml of 1% sodium hydroxide solution by stirring. Store the reagent in a stoppered bottle at 4°C.
2. 40% Rochelle salt (sodium-potassium tartarate) solution.
3. Standard sugar solution (as prepared under Nelson-Somogyi's method).

*Procedure:* Pipette out 0.5 to 3 ml of alcohol-free extract into test tubes and make up the volume to 3 ml with water in all the tubes. Add 3 ml DNS reagent and mix. Heat for 5 minutes in a boiling water bath. After the color has developed, add 1 ml of 40% Rochelle salt solution when the contents are still warm and mix. Cool the tubes under running tap water and measure the absorbance at 510 nm using reagent blank adjusted to zero absorbance. Calculate the amount of reducing sugar in the sample using a standard graph prepared with the working stock glucose solution in the same manner.

*Calculation:*

Reducing sugars =

$$\frac{\text{Sugar valu e from graph } (\mu g)}{\text{Aliquot sample used (ml)}} \times$$

$$\frac{\text{Total vol of alcohol – free extract (ml)}}{\text{Wt of sample } (100\ \mu g)} \times \frac{1}{100}$$

### 4.3.1.2 Estimation of Non-Reducing Sugars

Non-reducing sugars, present in plant extracts, are first hydrolyzed with either sulphuric acid or formic acid to reducing sugars, which are estimated by Nelson-Somogyi method (Somogyi 1952) or the dinitrosalicylic acid (DNS) method (Miller, 1972).

*Hydrolysis of the extract:* 1 ml of alcohol extract is placed in a test tube and evaporated to dryness. Add 1 ml glass distilled water and 1 ml 1N $H_2SO_4$ to the residue and hydrolyze by heating at 49°C for 30 minutes. Cool the test tubes, add 1-2 drops of methyl red indicator and neutralize the contents by adding 1N NaOH drop-wise from a pipette. Maintain appropriate reagent blank.

*Estimation of reducing sugars:* The sugars are estimated as described under Nelson's method or DNS method.

### 4.3.1.3 Estimation of Total Soluble Sugars

Total soluble sugars can be estimated following Anthrone method (Seiffer et. al., 1950) or the phenol-sulphuric acid method colorimetrically.

*Anthrone Method*

*Principle:* Carbohydrates are dehydrated by conc, sulphuric acid to form furfural. Furfural condenses with anthrone (10-keto-9,10-dihydro-anthracene) to form a blue-green colored complex, which is measured colorimetrically at 620 nm though some carbohydrates may give other colours. The reaction is not suitable when proteins containing a large quantity of tryptophan are present. This method is suitable for estimation of hexoses, aldopentoses and hexuronic acids either free or present in polysaccharides.

$$\begin{array}{c} CHO \\ | \\ H-C-OH \\ | \\ HO-C-H \\ | \\ H-C-OH \\ | \\ H-C-OH \\ | \\ CH_2OH \end{array} + H_2SO_4 \longrightarrow \text{Furfural (CHO)} \xrightarrow{\text{Anthrone}} \text{Blue color Complex}$$

D-Glucose

Furfural

*Reagents:*

1. 5N HCl
2. Anthrone reagent: Dissolve 200 mg anthrone in 100 ml of ice cold 95% sulphuric acid just before use.
3. Standard glucose solution: Dissolve 100 mg glucose in 100 ml distilled water (stock solution). Dilute 10 ml of stock solution to 100 ml with distilled water before use as working solution.

*Procedure:* Pipette out 1ml aliquots of the protein-free extract into test tubes. Prepare the standards by taking 0, 0.2, 0.4, 0.6, 0.8 and 1.0 ml of the working solution. Make up the volume to 1 ml in all the tubes by adding distilled water. Add 4 ml of anthrone reagent. Place a glass marble on the top of each tube to prevent loss of water by evaporation. Put the tubes in a boiling water bath for 10 minutes. Remove the tubes and cool to room temperature in a water bath. A reagent blank is treated similarly. Read the green to dark green color at 620 nm. Draw a standard graph by plotting concentration of the standard against absorbance. Calculate the amount of carbohydrates present in the sample from the graph.

*Calculation:*

Carbohydrate (% mg) =

$$\frac{\text{Sugar value from graph (mg)}}{\text{Aliquot sample used (ml)}} \times \frac{\text{Total vol. of extract (ml)}}{\text{Wt. of sample (mg)}} \times \frac{1}{100}$$

*Phenol-sulphuric acid Method:*

*Principle:* Carbohydrates like simple sugars, oligosaccharides, polysaccharides and their derivatives give an yellow-orange color when treated with phenol and concentrated sulphuric acid. In hot acidic medium, glucose is dehydrated to form hydroxymethyl furfural. This produces an orange-yellow color compound with phenol, which is measured colorimetrically at 490 nm.

In this reaction, furfural is produced when pentoses are heated with dilute acids and hydroxyl-methyl furfural when hexoses are heated with dilute acids. These products condense with phenol and an orange to yellow product is formed.

$C_5H_{10}O_5$ (Pentose) — Dilute HCl or Dilute $H_2SO_4$ (Heat) → Furfural + $3H_2O$

$C_6H_{12}O_6$ (Hexose) — Dilute HCl or Dilute $H_2SO_4$ (Heat) → 5 - Hydroxymethyl furfural + $H_2O$

Furfural / 5 - Hydroxymethyl furfural + phenol (OH) → Orange - yellow complex

*Reagents:*

1. 5% phenol: Dissolve 50g redistilled (reagent grade) phenol in water and dilute to one liter.
2. 96% sulphuric acid (reagent grade).
3. Standard glucose solution: Dissolve 100 mg glucose in 100 ml distilled water (stock solution). Dilute 10 ml of stock solution to 100 ml with distilled water before use as working solution.

*Procedure:* Pipette out 0.1 ml and 0.2 ml aliquots in two separate test tubes. Prepare the standards by taking 0, 0.2, 0.4, 0.6, 0.8 and 1.0 ml of the working solution. Make up the volume to 1 ml in all the tubes by adding distilled water. Set up a blank with 1 ml distilled water. Add 1 ml of phenol solution followed by 5 ml of sulphuric acid to each tube and shake well. Read the color at 490 nm. Calculate the amount of carbohydrates present in the sample from the graph.

### 4.3.1.5 Estimation of Cellulose

*Principle*: Cellulose undergoes acetolysis with acetic / nitric reagent to form acetylated cellodextrins, which get dissolved and hydrolyzed to form glucose units on treatment with sulphuric acid. On dehydration with sulphuric acid, glucose forms 5-hydroxymethyl furfural, which on reaction with anthrone gives a green colored product. The color intensity is measured at 620 nm.

*Reagents:*

1. *Acetic/nitric reagent* - Mix 15 ml of 80% acetic acid with 15 ml of concentrated nitric acid.
2. *Anthrone reagent* - 200 mg anthrone / 100 ml of concentrated sulphuric acid (95%) (prepare fresh and chill before use).
3. *Standard cellulose solution* - To 100 mg cellulose, add 10 ml of 67% sulphuric acid and leave for 1 hour. Dilute 1 ml of this solution to 100 ml (100 ug/ml).

*Procedure:* To 0.5 - 1g sample, add 3 ml acetic-nitric acid reagent and mix using a vortex mixer. Place it in a water bath at 100°C for 30 minutes. Cool and centrifuge for 15-20 minutes. Discard the supernatant. Wash the residue with distilled water, add 10 ml of 67% sulphuric acid and leave it for 1 hour. Dilute 1 ml of this solution to 100 ml. To 1 ml of the diluted solution, add 10 ml of anthrone reagent and mix well. Heat the tubes in a boiling water bath for 10 minutes, cool and measure the absorbance at 620 nm. Prepare a reagent blank with anthrone reagent and distilled water. Prepare a standard curve by taking 0.4, 0.6, 0.8, 1, 1.2, 1.4, 1 ,.6, 1.8 and 2 ml of standard solution, making the volume to 2 ml in each tube, adding anthrone reagent and developing color as discussed above. Draw a standard graph plotting concentration of cellulose against absorbance. Calculate the amount of cellulose in the sample from the standard curve.

### 4.3.1.6 Estimation of Neutral Detergent Fibre (NDF)

*Principle:* The sample is refluxed with neutral detergent solution to remove the water-solubles and minerals other than fibrous components. The residue is weighed after filtration and expressed as neutral detergent fibre (NDF).

*Reagent:*

1. *Neutral Detergent solution* - Put 18.61g disodium ethylene diamine tetra-acetate (EDTA) dehydrate crystals and 6.81g sodium borate decahydrate in a large beaker (2 liter). Add about 200 ml of distilled water, shake and heat until these chemicals are dissolved. Place 4.36g anhydrous disodium hydrogen phosphate in a separate beaker, add some distilled water and heat until it is dissolved. Mix both the solution properly and check pH to range between 6.9 and 7.1. If made properly, pH adjustment will rarely be required.
2. Decahydronaphthalene (reagent grade)
3. Acetone
4. Anhydrous sodium sulphite

*Procedure:* Weigh 0.5 - 1g dry sample (20-30 mesh size) into a beaker of the refluxing apparatus. Add 100 ml of neutral detergent solution, 2 ml of decahydronaphthelene and 0.5g sodium sulphite. Heat to boiling in 5-10 minutes.

Reduce heat as boiling begins, in order to avoid foaming. Reflux for 60 minutes, timed from onset of boiling. Filter through a weighed glass crucible on filter manifold. Rinse the sample into the crucible with minimum of hot water (90-100°C). Filter liquid and repeat the washing procedure. Wash twice with acetone in the same manner and suck dry. Dry the crucible in hot air oven at 100°C for 8 hours or overnight and weigh. Ash the residue in the crucible at 500-550°C for 8 hours and weigh. Record ash content as ash insoluble in neutral detergent solution.

*Calculation :*

Cell Wall Constituents =

$$\frac{(\text{Wt. of crucible} + \text{Cell wall contents}) - \text{Wt. of crucible}}{\text{Wt. of sample}} \times 100$$

### 4.3.1.7 Estimation of Acid Detergent Fibre (ADF)

*Principle:* Refluxing the sample with acid detergent solution removes the water-solubles and minerals other than fibrous components. The residue is weighed after filtration. This gives the acid detergent fibre (ADF).

*Reagents:*

1. *Acid detergent solution* - Add 20g of Cetyl trimethyl ammonium bromide (CTAB) to 1N sulphuric acid and make volume to 1 liter.
2. Decahydronaphthalene (reagent grade)
3. Acetone
4. n-hexane

*Procedure:* Weigh 1g air dry sample (20-30 mesh size) into a beaker of the refluxing apparatus. Add 100 ml of acid detergent solution and 2 ml of deca-hydro-naphthelene. Heat to boiling in 5-10 minutes. Reduce heat as boiling begins, in order to avoid foaming. Reflux for 60 minutes, timed from onset of boiling. Filter the contents through a weighed sintered glass crucible (G-2) on filter manifold by suction. Rinse the sample into the crucible with minimum of hot distilled water (90-100°C). Filter the liquid and repeat the washing procedure. Wash twice with acetone in the same manner. Break up all lumps so that the solvent may come in contact with all particles of fibre. Wash with hexane, if required. Hexane should be added while crucible still contains some acetone. Suck dry the acid detergent fibre free from hexane. Dry the crucible in hot air oven at 100°C for 8 hours or overnight and weigh.

*Calculation:*

Acid Detergent Fibre (%) =

$$\frac{\text{(Wt. of crucible + Cell wall contents)} - \text{Wt. of crucible}}{\text{Wt. of sample}} \times 100$$

### 4.3.1.8 Estimation of Hemicellulose

Hemicelluloses are non-cellulosic, non-pectic cell wall polysaccharides, which are rich in xylans, mannans, glucomannans, galactans and arabino-galactans. They are classified under unavailable carbohydrates since they are not digested by the digestive enzymes.

*Principle:* The sample is refluxed with neutral detergent solution and acid detergent solution separately to remove the water-solubles and minerals other than the fibrous components. The left out material is weighed after filtration and expressed as neutral detergent fibers (NDF) and acid detergent fibers (ADF). The difference between NDF and ADF is hemicellulose content.

Hemicelluloses = NDF - ADF

*Procedure:* NDF and ADF are estimated following the procedure given in 4.3.1.6 and 4.3.1.7 respectively and hemicellulose content is calculated by difference.

### *4.3.1.9 Determination of Acid Detergent Lignin (ADL)*

*Principle:* Refluxing the sample with acid detergent solution removes the water-solubles and minerals other than fibrous components. The residue is dried and weighed after filtration. The residue is then treated with sulphuric acid, filtered, dried and ashed. The loss of weight on ignition gives the acid detergent lignin.

*Reagents:*

1. 72% sulphuric acid (w/v) - Take 417 ml of distilled water in a volumetric flask and add 583 ml of pure concentrated sulphuric acid with occasional stirring (Caution: The flask should be cooled in a water bath while mixing sulphuric acid).

*Procedure:* Transfer ADF to the crucible (sintered) and add 25-50 ml of 72% sulphuric acid (15°C). Stir with glass rod to smooth the paste and break the lumps. Let the glass rod remain in the crucible. Refill it with sulphuric acid and stir at hourly intervals as acid drains away. Crucible does not need to be kept full at all the times. Three additions are sufficient. Keep crucible at 20-23°C. After three hours, filter off as much acid as possible. Wash the contents with hot water until it is free from acid. Rinse and remove glass rod. Dry the crucible in hot air oven at 100°C for 8 hours or overnight and weigh. Take the residue in a crucible and place it in muffle furnace at 500-550°C for 3 hours. Cool it and weigh.

*Calculation:*

$$\frac{(\text{Wt. of crucible} + \text{Lignin}) - \text{Wt. of crucible Acid detergent Lignin (\%)}}{\text{Weight of sample}} \times 100$$

### *4.3.1.10 Determination of Lignin*

*Gravimetric method:*

*Procedure:* Determine NDF and ADF present in the sample following the above methods. Calculate the lignin content as follows.

NDF = Hemicelluloses + Cellulose + Lignin + Minerals

ADF = Cellulose + Lignin + Minerals

Cellulose = ADF - Residue after extraction with 72% sulphuric acid

Lignin = Residue after extraction with 72% sulphuric acid -Ash

*Spectrophometric method*
*Method 1*

*Principle:* The sample is extracted in sodium hydroxide solution and the aliquot samples are adjusted to pH 7.0 and 12.3. The amount of lignin is calculated by a difference between $A_{245}$ (pH 7.0) and $A_{350}$ (pH 12.3).

*Reagents:*

1. Diethyl ether
2. 0.1M sodium phosphate buffer, pH 7.0
3. 0.1 and 0.5N sodium hydroxide
4. 2N HCl

*Procedure:* Moisten 100 mg of oven-dried material in a mortar with distilled water. Grind with ether until it is free from chlorophyll pigment. Centrifuge at 2000 r.p.m. for 5 minutes and decant the supernatant. Wash the sediment with distilled water, recentrifuge and discard the supernatant. Repeat washings twice. Add 2 ml of NaOH to the residue and extract at 70-80°C for 12-16 hours. Cool, add 0.45 ml of 2N HCl and adjust the pH 7 or 8 with NaOH. Make up the volume to 3 ml with water. Centrifuge at 2000 r.p.m. for 5 minutes. Collect the supernatants. To 0.8 ml of supernatant, add 0.8 ml of 0.1 M sodium phosphate buffer, pH 7. To another aliquot of 0.8 ml supernatant, add 0.8 ml of 0.1 N NaOH (pH 12.3). Measure the absorbance at 245 and 350 nm, respectively. Derive the lignin content from the difference between A245 and A350 on pH 7 and 12.3 samples diluted with buffer and NaOH respectively. Express the amount of lignin as $A_{350}$ /sample.

*Method 2*

*Procedure:* Pipette out 0.55 ml of hot alkali extract in water containing 1-3 μg of phenol into test tubes. Add 0.4 ml of 0.5M

Tris HCl-buffer pH 9 and 0.05 ml of freshly prepared alcohol solution containing 25 μg of 2,6-dichloro-quinone chlorimide. After incubating at room temperature for 1 hour, measure the absorbance at 610 nm using quaiacol as a standard. Use a conversion factor 32 to calculate the lignin content (32 × mg phenol = mg lignin, calculated from bigasse sample).

### 4.3.1.11 Estimation of Tannins

*Folin-Denis Method*

*Principle:* Tannin-like compounds reduce phosphotungstomolybdic acid in alkaline solution to produce a blue-coloured solution. The intensity of color is measured at 700 nm in a spectrophotometer.

*Reagents:*

1. *Folin-Denis reagent* - Dissolve 100g sodium tungstate and 20g phosphomolybdic acid in 750 ml distilled water and made up the volume to 1 litre with distilled water.
2. *Sodium carbonate solution* - Dissolve 350g sodium carbonate in 1 litre of distilled water at 70-80°C. Filter through glasswool after allowing it to stand overnight.
3. *Standard tannic acid solution* - Dissolve 100 mg tannic acid in 100 ml distilled water.
4. *Working standard solution* - Dilute 5 ml of stock solution to 100 ml with distilled water. One ml of working solution contains 50 μg tannic acid.

*Procedure:* Weigh 0.5g powdered plant material and transfer it to a 250-ml conical flask. Add 75 ml water and heat the flask gently. Boil the contents for 30 minutes. Centrifuge at 2000 rpm for 20 minutes and collect the supernatants in 100-ml volumetric flask. Make up the volume with distilled water. Transfer 1 ml of the supernatant to a 100-ml volumetric flask containing 75 ml distilled water. Add 5 ml of Folin-Denis reagent, 10 ml of sodium carbonate solution and dilute to 100 ml with distilled water. Shake well and read the absorbance at 700 nm after 30 minutes. Prepare a standard graph by using 0-100 μg tannic acid. Calculate the tannin content of the sample as tannic acid equivalent from the standard graph.

*Vanillin-Hydrochloride Method*

*Principle:* The vanillin reagent reacts with any phenol that has an unsaturated resorcinol or pholoroglucinol nucleus and forms a coloured product, which is measured at 500 nm.

*Reagents:*

1. Vanillin-hydrochloride reagent - Mix equal volumes of 8% HCl in methanol and 4% vanillin in methanol. The solutions must be mixed just before use.
2. Catechin stock solution - Prepare a standard stock solution containing 1 mg catechin per ml of methanol.
3. Working standard solution - Dilute 1 ml of stock solution to 10 ml with methanol (100 μg / ml).

*Procedure:*

*Extraction* - Extract 1g grounded plant material in methanol (50 ml). Mix occasionally by swirling. After 20-28 hours, centrifuge and collect the supernatant.

*Estimation* - Pipette out 1 ml of supernatant into test tube. Quickly add 5 ml of vanillin hydrochloride reagent and mix. Read in spectrophotometer at 500 nm after 20 minutes. Prepare a blank and also draw a standard graph with 20 - 100 μg catechin using working solution. Calculate the amount of catechin i.e. tannin in sample as per the absorbance value and express the results as catechin equivalent.

### 4.3.2 Estimation of Proteins

Total protein is estimated by the method of Lowry et. al. (1951) using Folin-Ciocalteau reagent (Folin & Ciocalteau, 1927), Biuret method and Bradford's method.

*Folin-Ciocalteau Method:*

*Principle:* Protein reacts with the Folin-Ciocalteau reagent to give a blue-colored complex. The color is formed due to the reaction of the copper ions with the peptide bonds (-CO-NH- groups) of protein in alkaline solution and reduction of phosphomolybdate-phosphotungstatic components of the reagent by amino acids like tyrosine and tryptophan present in protein. The intensity of blue color

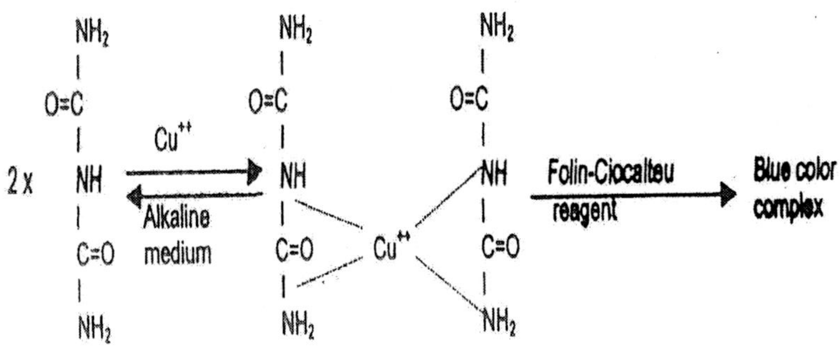


is measured colorimetrically at 660 nm. The intensity of color will depend on the quantity of these aromatic amino acids present.

*Reagents*:

1. Folin-Ciocalteau reagent - Reflux 20g sodium tungstate, 5g sodium molybdate, 140 ml distilled water, 10 ml of 85% orthophosphoric acid and 20 ml conc. hydrochloric acid for 10 hours in dark and then add 30g lithium sulphate, 10 ml water and 5 drops of liquid bromine. Boil the mixture again for 15 minutes, cool and filter. The clear solution is diluted to 200 ml with distilled water and neutralized with 5N NaOH. Then add 60 ml of conc. HCl and the solution is diluted to 600 ml with distilled water. This reagent is then stored in a brown bottle.
2. Solution A - Dissolve 2g sodium hydroxide and 10g sodium carbonate in water and make the volume to 500 ml.
3. Solution B - Mix 5 ml of 1.1% Rochelle salt solution and 0.5 ml of 5.5% copper sulphate.
4. Solution C - Mix 50 ml of solution A and 1 ml of solution B.

*Procedure:* Take 1 ml of the extract prepared for protein estimation as per 4.1.3 and add 5 ml of solution C to it. Stir thoroughly. After allowing the mixture to stand at room temperature for 10 minutes, add 0.5 ml of Folin-Ciocalteau reagent with constant stirring. Allow the solution to stand for 30 minutes and measure the intensity of the resultant color against blank (containing 1 ml water, 5 ml of solution C and 0.5 ml Folin-Ciocalteau reagent) at 660 nm (light path = 1 cm) on spectrophotometer. A calibration curve is prepared with a standard solution of crystalline bovine serum albumin. Calculate the amount of protein in the sample with the help of calibration curve.

### *Biuret Method*

*Principle:* The peptide bonds of the protein form a purple color complex with copper ions in an alkaline solution. The intensity of purple complex is measured at 520 nm colorimetrically.

*Reagents:*

1. Biuret reagent - Dissolve 3g copper sulphate and 9g sodium potassium tartrate in 500 ml of 0.2 N NaOH solution with constant stirring. To this, add 5g potassium iodide and make up to 1 liter with 0.2 N NaOH. Mix thoroughly.
2. Standard protein solution - Dissolve 50 mg bovine serum albumin in 10 ml of distilled water. Prepare fresh.

*Procedure:* Turn on the colorimeter or spectrophotometer to be used and allow to warm. Number ten test tubes (16 x 150 mm) serially and place them in a test tube rack. In each tube, carefully pipette 0, 0.1, 0.2, 0.3, 0.4, 0.5, 0.6, 0.7, 0.8 and 1 ml of standard protein solution (10 mg/ml) and make up the volume in each tube to 1 ml with distilled water. Take 1 ml of the extract in another test tube. Add 4 ml of Biuret reagent in each tube and vortex the mixture for a few seconds to effect thorough mixing of solutions. Incubate the tubes at 37°C temperature for 20 minutes. Cool the content and read the absorbance at 520 nm (green filter) against a reagent blank. Calculate the protein content in the sample using a standard curve prepared with bovine serum albumin protein.

### *Bradford's Method*

*Reagents:*

1. Standard protein solution - Dissolve 25 mg bovine serum albumin in 0.25M NaCl and make up the volume to 25 ml with NaCl solution (1 mg / ml).
2. 0.01% Protein reagent - Dissolve 100 mg Coomassie Brilliant Blue G-250 in 50 ml of 95% alcohol and add 100 ml of 85% (w/v) phosphoric acid and dilute to 1 liter with water. Prepare fresh before use.

*Procedure:* Number ten test tubes (16 x 150 mm) serially and place them in a test tube rack. In each tube, carefully pipette 0, 0.01, 0.02, 0.03, 0.04, 0.05, 0.06, 0.07, 0.08 and 0.1 ml of standard protein solution (1 mg/ml) and make up the volume in each tube to 0.1 ml with an appropriate buffer. Take 0.1 ml of the extract in another test tube. Add 5 ml of protein reagent in each tube and mix thoroughly by inverting or vortexing for a few seconds. Read the absorbance at 595 nm after 2 minutes and before 1 hour against a reagent blank. Calculate the protein content in the sample using a standard curve prepared with bovine serum albumin protein.

### 4.3.3 Estimation of Nucleic Acids

#### *4.3.3.1 Separation of RNA and DNA*

RNA and DNA are separated by the Schmidt-Thannhausen-Schneider procedure. Acidify an aliquot of nucleic acid extract with $HClO_4$ to pH 1.5 to 2.0. This will precipitate DNA and excess of $KClO_4$. Centrifuge it at 2000 rpm for 15-20 minutes, collect the supernatant containing RNA and the residue containing DNA.

The RNA present in the supernatant is estimated directly while DNA is estimated after solubilizing it in $HClO_4$ and removing the excess perchlorate.

#### *4.3.3.2 Estimation of RNA*

RNA is estimated directly with spectrophotometrically or by Orcinol method (Merchant et. al., 1969).

**Orcinol Method**

*Reagents:*

1. Orcinol reagent - Dissolve 100 mg orcinol and 100 mg ferric chloride in conc. HCl (Sp. Gr. 1.19) and make up the volume to 100ml with conc. HCl.
2. Standard Ribose solution - Dissolve 100 mg Ribose in distilled water and make up the volume to 100 ml. A working standard containing 100 µg Ribose per ml is prepared by diluting 1 ml of sock solution to 100 ml with distilled water.

*Procedure:* Pipette 0.2 ml of the extract in a test tube and add 2 ml of orcinol reagent. Cover the test tubes with glass stoppers and heat for 8 minutes on a water bath maintained at 100°C. Cool the test tubes under running tap water and measure the absorbance against reagent blank at 665 or 670 nm. If the color intensity is strong, dilute the contents to 10 ml with n-butanol. Calculate the concentration of pentose present from a standard curve prepared with ribose (10 μg/ml) and express the results as ribose equivalents.

**Direct Method**

*Procedure:* Measure the absorbance of RNA extract at 260 nm and calculate the concentration of RNA from a standard curve prepared with yeast RNA.

### 4.3.3.3 Estimation of DNA

DNA is estimated by the Method of Burton (1968) [Diphenylamine Reaction Method] and UV spectrophotometric method.

**Diphenylamine Reaction Method**

*Reagents:*

1. Diphenylamine reagent - Dissolve 1.5g pure white crystalline diphenylamine in 100 ml glacial acetic acid followed by addition of 1.5 ml of conc, sulphuric acid (Sp. Gr. 1.84). Just before use, 0.1 ml of 1.6% aqueous acetaldehyde is mixed with 20 ml of the reagent.
2. Standard DNA solution - Dissolve 3 mg Calf Thymus DNA in 10 ml of 5mM NaOH (0.3 mg / ml). The working solution is prepared by mixing equal volumes of stock solution with 1N perchloric acid and heating it for 15 minutes at 70°C. The solution is stable for three weeks.

*Procedure:* Dilute the extract with 0.5N perchloric acid so that the sample for analysis contains 0.02 - 0.25 μ mole of DNA-phosphate / ml. Pipette 2 ml of the sample in test tubes and add 4 ml of diphenylamine reagent. Maintain appropriate standards and blank tubes containing the same amount of perchloric acid. Incubate the tubes for 15-17 hours at 25-30°C and then measure the absorbance at 600 nm. Draw a standard graph plotting absorbance against quantity of DNA and then calculate the concentration of DNA in the sample from the graph.

**UV Spectrophotometric Method**

*Principle:* The nucleic acids absorb light strongly in the ultraviolet region due to the conjugated double bond systems of the constituent purines and pyrimidines. They show characteristic maxima at 260 nm and minima at 230 nm. The concentration of DMA is calculated using the following relationship.

Concentration = $A_{260}$ = 1.0 = 50 μg duplex DNA/ml

*Procedure:* Suspend the residue obtained earlier in suitable aliquots of 5%TCA at 0°C, centrifuge at 2000 rpm for 15 minutes and discard the supernatant. Repeat the process and wash the residue once with absolute alcohol and then with ethanol-ether mixture. The residue is suspended in 0.5N $HClO_4$ and incubated at 90°C for 7 minutes in a constant temperature water bath. Centrifuge and collect the supernatant. Combine the supernatants and make up the volume. Add equal volume of 1N KOH to a known aliquot to precipitate excess perchlorate as $KClO_4$. Centrifuge and collect the supernatants. Measure the absorbance of the DNA extract at 260 nm and calculate the concentration from a standard curve using the above relationship.

## 4.3.4 Estimation of Amino Acids

### *4.3.4.1 Estimation of Total Amino Acids*

#### *Method of Moore & Stein (1948)*

*Principle:* Ninhydrin, a powerful oxidizing agent, reacts with α-amino acids between pH 4 and 8 and decarboxylates to give an intensely bluish purple coloured compound, which is measured colorimetrically at 570 nm. The amino acids proline and hydroxyproline give a yellow colour.

*Reagents:*

1. *0.2M citrate buffer (pH 5.0)* - Dissolve 21g citric acid in 200 ml of 1N NaOH in a 500 ml volumetric flask and make up the volume to 500 ml with distilled water.
2. *Ninhydrin reagent* - Dissolve 800 mg hydrated stannous chloride in 500 ml of citrate buffer and 20g re-crystallized ninhydrin in 500 ml of methyl cellosolve (ethyl glycol mono methyl ether). Mix both the solutions. The reagent is prepared afresh daily.
3. *Diluent's solution* - Mix equal volumes of distilled water and n-propanol.
4. *Stock solution* - Dissolve 50 mg leucine in 50 ml distilled water (1 mg / ml).

5. *Working standard solution* - Dilute 10 ml of stock leucine solution to 100 ml with distilled water.

*Preparation of samples:* Samples are prepared as in protein estimation.

*Procedure:* Pipette out 1 ml of alcohol soluble extract in a test tube, add one drop of methyl red indicator and neutralize with 0.1N NaOH. Add 1 ml of ninhydrin reagent to the tube, mix thoroughly and place a glass marble on the top of the tube. Heat the contents of the tube for 20 minutes in a boiling water bath to which are added 5 ml of diluent's solution while still on water bath. The blank is prepared with 1 ml of distilled water instead of sample. Remove the tubes, cool and mix thoroughly. Measure the intensity of purple color at 570 nm. Calculate the amount of total free amino acids using standard curve prepared with leucine by pipetting 0.1 - 1.0 ml (10 - 100 µg) of the working solution and estimating the absorbance as above after developing the colour. Express the results as percent equivalent of leucibne.

**Method of Yemm & Cocking (1955)**

*Reagents:*

1. 0.2M citrate buffer (pH 5.0) - Dissolve 21.008 g citric acid in 250 ml of distilled water in a 500 ml volumetric flask and make up the volume to 500 ml with 1N NaOH. The reagent is stored in cold with a little thymol as a preservative.
2. Potassium cyanide solution - Dissolve 0.1628g potassium cyanide in distilled water and make up the volume to 250 ml. It is stable for 3 months at room temperature.
3. Potassium cyanide-methyl cellulose solution - Dilute 5 ml of stock potassium cyanide solution to 250 ml with methyl cellosolve. It is stable for one month.
4. Methyl cellosolve-ninhydrin solution - Dissolve 5g ninhydrin in methyl cellosolve and make the volume up to 100 ml. It is stable for six months.
5. Ninhydrin reagent (KCN-methyl cellosolve-ninhydrin solution) - Mix 50 ml of methyl cellosolve-ninhydrin solution with 200 ml of potassium cyanide-methyl cellosolve (ethyl glycol mono methyl ether) solution. The resulting solution is initially red but turns yellow subsequently. It is stored overnight before use to

ensure low blank readings. At room temperature, the solution is stable for one week.

6. Standard Glycine solution - Prepare stock solution by dissolving 150 mg glycine in 0.1N HCl and making the volume to 100 ml. Store in cold with a little thymol. The working solution is prepared by diluting 10 ml of stock solution to 100 ml with citrate buffer.

*Preparation of samples:* Samples are prepared as in protein estimation.

*Procedure:* Pipette out 1 ml of alcohol soluble extract in a test tube and add 0.5 ml of 0.2M citrate buffer. Add to this solution either 0.2 ml of methyl cellosolve-ninhydrin solution and 1 ml of KCN-methyl cellosolve solution or 1.2 ml of KCN-methyl cellosolve-ninhydrin solution. Cover the test tube with glass marble, heat for 15 minutes at 100°C in water bath and cool them under running tap water. Similarly maintain reagent blank with 1 ml of citrate buffer instead of sample. Make up the solution to a convenient volume with 60% ethanol and mix the contents well. Measure the absorbance at 570 nm for all amino acids except proline and hydroxyproline, which are measured at 440 nm. Determine the amino acid content with the help of standard curve prepared with glycine.

**Method of Frame, Russell & Wilhelmi (1943)**

*Reagents:*

1. *Glycine standard* - Dissolve exactly 0.268g glycine in water and transfer it to a 500 ml volumetric flask. Add 35 ml of 1N HCl and 1g sodium benzoate. Add water to dissolve and dilute to 500 ml with water. Mix well.
2. *Glutamic acid standard* - Dissolve exactly 0.525 g glutamic acid in water, transfer to a 500 ml volumetric flask, add 35 ml of 1N HCl and 1g of sodium benzoate as above, dilute to 500 ml and mix well. It contains 0.1 mg of amino acid nitrogen per ml.
3. *Mixed standard* - Transfer 3 ml each of glycine and glutamic acid standards to a 100 ml volumetric flask and dilute to 100 ml with distilled water. It contains 0.03 mg amino acid nitrogen per ml.

4. *Phenolphthalein solution* - Dissolve 0.25g phenolphthalein in 95% alcohol and dilute to 100 ml.
5. *Sodium hydroxide* - Dilute 10 ml of 1N NaOH to 100 ml with distilled water.
6. *Borax solution* - Dissolve 15g borax in distilled water and dilute to 1 liter with water.
7. *Naphthoquinone solution* - Dissolve 0.25g of β-naphthoquinon-4-sulfonic acid in water and dilute to 50 ml. Prepare daily before use.
8. *Acid-formaldehyde solution* - Dilute 11.3 ml of 40% formaldehyde to 1 liter with distilled water. Mix 4 volumes of this solution with 3 volumes of 1.5N HCl and 1 volume of glacial acetic acid. It Is stable indefinitely.
9. *Sodium thiosulphate* - Dissolve 25g of sodium thiosulphate in water, dilute to 1 litre with water and mix well. Usable indefinitely.

*Procedure:* Transfer 5 ml of protein-free extract (filtrate) to a test tube. Place 5 ml of mixed standard and 5 ml of distilled water in separate test tubes as a standard and reagent blank respectively. Add one drop of phenolphthalein solution to each tube followed by 0.1N NaOH drop by drop until a permanent pink color is obtained. Adjust the volume in each tube to be the same. To each tube, add 1 ml of borax solution, mix and then add 1 ml of freshly prepared naphthoquinone solution. Mix and place the tubes in a boiling water bath 10 minutes. Remove the tubes from the water bath for and place them in cold water for 5 minutes. To the cooled tubes, add 1 ml of acid-formaldehyde solution and mix. Add 1 ml of 0.1 N sodium thiosulphate, dilute immediately to 15 ml with distilled water and mix well. Allow to stand for 10-30 minutes and measure the absorbance at 490 nm. Calculate the amount of total amino acids with the help of the following formula.

$$\frac{\text{Density of Unknown sample}}{\text{Density of Standard}} \times 0.03 \times \frac{100}{0.5} = \text{mg amino acid N/100 ml}$$

### 4.3.5 Estimation of Plant Pigments

#### *4.3.5.1 Chlorophyll*

***Method 1:*** Weigh 1-5g leaf sample into mortar and pestle and add small quantity (0.1g) of calcium carbonate or sodium carbonate.

Macerate the tissue with pestle, add quartz sand and grind for short time. Then, add 85% acetone, little at a time and continue grinding until the tissue is finely ground. Transfer the mixture to a funnel, filter with suction and wash the residue with 85% acetone. Place the residue into mortar with more acetone and grind again. Filter and wash as before. Repeat the procedure until the tissue is devoid of any green color and washings are colorless. When extraction is complete, pool all the filtered extracts to a volumetric flask of appropriate size and dilute it with acetone to the mark.

Measure the T of this solution at 660 nm with photoelectric colorimeter and read quantity of chlorophyll from the standard curve prepared with standard solution of chlorophyll.

Extract samples are diluted in different concentrations and T of original and diluted solutions is measured. Transfer aliquot of original extract to ether and evaluate total chlorophyll spectrophotometrically.

A. Pipette out aliquot of 25-50 ml into separator containing 50 ml ether. Add distilled water carefully until it is apparent that all fat soluble pigments have entered into ether layer. Drain and discard water layer. Place separator containing ether layer in upper rack of support. Add 100 ml water to second separator. Continue water washing of ether solution until all acetone is removed (5-10 washings). Then, transfer ether solution to 100 ml volumetric flask, dilute to volume and mix well.

B. Add teaspoonful anhydrous sodium sulphate to 60 ml reagent bottle and fill it with ether solution of pigments. When this solution is optically clear, pipette out aliquots into another dry bottle and dilute with enough dry ether to cause logio $\log_{10}$ I/I value (A) to fall between 0.2 and 0.8 at 660 nm wavelength.

$C = (\log_{10} I/I)/a\ l$

where,

| | |
|---|---|
| $I_0$ | = Intensity of light transmitted by solvent filled cell |
| I | = Intensity of light transmitted by solution filled cell |
| C | = Concentration of chlorophyll |
| a | = Absorptivity |
| l | = Thickness of solution layer |

Total chlorophyll (mg/C) = 7.12 $\log_{10}$ I/I (660 nm) + 16.8 $\log_{10}$ $I_0$/I (642.5 nm)

Chlorophyll a (mg/C) = 9.93 $\log_{10}$ I/I (660 nm) - 0.777 $\log_{10}$ $I_0$/I (642.5 nm)

Chlorophyll b (mg/C) = 17.6 $\log_{10}$ Io/I (642.5 nm) - 2.81 $\log_{10}$ $I_0$/I (660 nm)

***Method 2:*** Cut 1g fresh leaves into small pieces and homogenize with excess 80% acetone in a mortar pestle. Decant and filter the supernatant through a Buckner funnel using Whatman No. 42 filter paper (or centrifuge at 5000 r.p.m. for 5 minutes). Add sufficient acetone and repeat the extraction till colourless supernatant is obtained. Pool the supernatants and make up the volume to 100 ml with 80% acetone. Transfer 5 ml of the extract into a 50 ml volumetric flask and dilute it upto the mark. Measure the absorbance of this extract at 645 nm and 663 nm for determination of chlorophyll a, b and total chlorophyll. For routine measurements of total chlorophyll, measure the absorbance at 652 nm.

$$\text{Total chlorophyll (mg/g)} = \frac{20.2\ A_{645} + 8.02\ A_{663}}{a \times 1000 \times W} \times V$$

$$\text{or} = \frac{27.8\ A_{663}}{a \times 1000 \times W} \times V$$

$$\text{Chlorophyll a (mg/g)} = \frac{12.7\ A_{663} - 2.69\ A_{645}}{a \times 1000 \times W} \times V$$

$$\text{Chlorophyll b (mg/g)} = \frac{22.9\ A_{645} - 4.68\ A_{663}}{a \times 1000 \times W} \times V$$

a = length of light path in the cell (1 cm)

V = volume of the extract in ml

W = fresh weight of the sample in g

### 4.3.5.2 Carotenes

*Method 1:* Place the weighed material (2-5g) in a high-speed blender, and add 40 ml acetone, 60 ml hexane and 0.1g magnesium

**PHYTOHORMONMES**

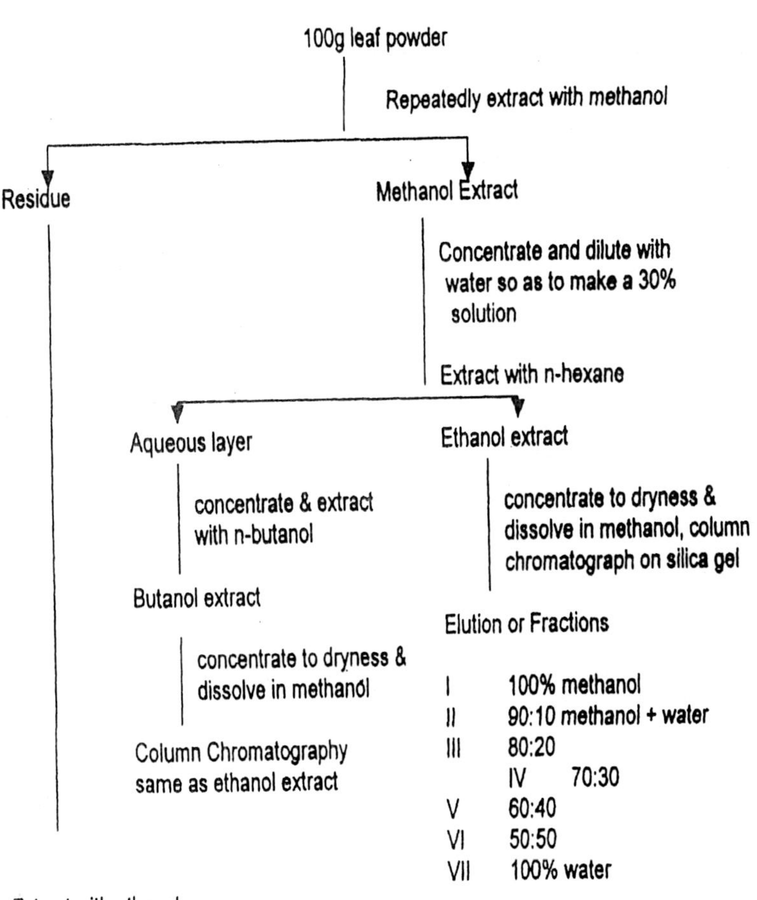


Extract with ethanol on water bath

↓

Alcohol extract

↓ concentrate to dryness & dissolve in methanol

Methanol extract
Repeat the cycle as above

carbonate. Blend for 5 minutes. Filter with suction or let residue settle down and decant into separator. Wash residue with 25 ml portions (two) of acetone, then with 25 ml hexane and combine the extracts. Wash acetone from extract with five 100 ml portions water, transfer upper layer to 100 ml volumetric flask containing 9 ml acetone and dilute to the mark with hexane.

Prepare chromatographic column with 1:1 mixture of activated magnesia and diatomaceous earth. To prepare column, place small glass wool or cotton plug inside tube, add loose adsorbent to 15 ml depth, attach tube to suction flask and apply full vacuum of water pump. Place 1 cm layer of anhydrous sodium sulphate above adsorbant.

With vacuum continuously applied to flask, pour extract into column. Use 50 ml of acetone-hexane (1:9) or slightly more to develop chromatograms and wash visible carotenes through adsorbant. Keep top of the column covered with layer of solvent during entire operation. Collect entire elute. Transfer elute, (which has been reduced in volume by loss of vapour through water pump) to 100 ml volumetric flask, dilute to volume with acetone-hexane (1:9) solvent and determine carotene content photometrically.

Determine absorbance (A) of solution at 436 nm by Klett photometer with No. 44 filter. Calibrate the instrument first with solution of high purity β-carotene.

$$C = (A \times 454) \div (196 \times L \times W)$$

where,

C = carotene content (mg/lb)

L = Length of cell in centimeter

W = Weight of sample in g / ml final dilution

***Method 2:*** Chop the freshly harvested leaves weighing about 2g and transfer to a mortar pestle along with 20 ml of distilled acetone or methanol. Homogenize the tissue thoroughly. Decant the extract and filter through Whatman No. 42 filter paper. Repeat the extraction twice or thrice until the tissue is free from pigments. Pool the filtrates and, partition with equal quantities of peroxide-free ether thrice using a separating funnel. Add water if necessary to produce two layers during the initial ether extraction. The ether phase contains carotenoids. Evaporate the combined ether extracts under reduced pressure at 35°C on a hot water bath or rotatory evaporator. Dissolve the residue in minimum quantity of ethanol. Add 60% KOH at the rate of 1 ml for every 10 ml of the ethanol extract to saponify it. Keep this mixture in dark in the presence of nitrogen or boil it for 5-10 minutes and leave it overnight at room temperature. Add equal amount of water and partition twice with peroxide-free ether. Evaporate the ether under reduced pressure and dissolve the residue in a minimum volume of ethanol. Measure the absorbance at 450 nm in a colorimeter. Calculate the carotenoid content using a calibration curve of β-carotene.

Make suitable dilution of the acetone or methanol extracts of plant tissue and measure its absorbance at 450 nm and 670 nm in a colorimeter. If $A_{450}$ is 10 times that of $A_{670}$, carotenoids present in the extract may be estimated from the $A_{450}$ values using the following formula.

$$C = \frac{D \times V \times f \times 10}{2500}$$

where,

C = total amount of carotenoids in mg

D = absorbance at 450 nm in 1 cm cell

f = dilution factor

V = volume of the original extract in ml.

## 4.3.6 Estimation of Plant Enzymes

### 4.3.6.1 Extraction of Plant Enzymes

*Extraction in acetone* - Weigh the tissue, cut into pieces of 1-2 cm size, transfer to a blender and add chilled acetone (-20°C) enough to cover the tissues (5 ml acetone for each g of tissue). Blend at high speed for 3-5 minutes, filter the extract through Buckner funnel using Whatman No. 1 filter paper and wash the residue (powder) with chilled acetone at least thrice. Wash again with cold diethyl ether. Dry the residue on the funnel with suction. Spread the residue on Whatman No. 1 filter paper and air dry for about 1 hour. Store the residue in containers with tight cap in a freezer.

Just before use, weigh 0.1g of the powder, grind in 5 ml of phosphate buffer (0.1M, pH 6.6) at 4°C for 10-15 minutes in a mortar and pestle. Centrifuge for 30 minutes at 2000 r.p.m. at 4°C temperature. Decant the supernatant and use the clear extract for enzyme assay. Determine the protein content of the extract by Lowry's method to express the results as specific activity per unit of protein.

*Extraction in buffer* - Weigh the tissue, cut into 1-2 cm pieces, transfer to a blender and add chilled (2-4°C) phosphate buffer (0.1M, pH 6.5) (5 ml buffer for each g of tissue). Grind at low speed for 2-3 minutes, squeeze the extract through cheese cloth to remove pulp and centrifuge the extract for 30 minutes at 4000 r.p.m. at 4°C temperature. Decant the supernatant and use the clear extract for assay of nuclease enzymes. Determine the protein content of the extract by Lowry's method to express the results as specific activity per unit of protein.

*Extraction in water* - Weigh the tissue, cut into 1-2 cm pieces, transfer to a blender/mortar-pestle and add distilled water enough to cover the tissues (5 ml water for each g of tissue). Blend for 7-10 minutes, filter the extract through cheese cloth and centrifuge at 2000 r.p.m. for 30 minutes at 4°C temperature. Decant the supernatant and dialyze it against several volumes of distilled water at 2-4°C for 24 hours. The enzyme may also be extracted in acetic acid-acetone buffer at pH 5.2. Determine the protein content of the extract by Lowry's method to express the results as specific activity per unit of protein.

### 4.3.6.2 Assay of Plant Enzymes

*Cellulase* - Pipette 4 ml of carboxy-methyl cellulose solution (Dissolve 0.5g carboxy-methyl cellulose in 100 ml of sodium acetate-acetic acid buffer at pH 5.2 at 50-60°C kept in a blender. Blend for 3-5 minutes at low speed, stir the contents and again blend at high speed for 3-5 minutes. Filter through Whatman No. 1 filter paper and add 2 ml of 1% aqueous merthiolate solution. Store at 4°C in a test tube followed by 1 ml of buffer and 2 ml of enzyme extract (water extract). Incubate the tube in water bath, withdraw the aliquots of 1 ml from each tube at pre-fixed intervals and determine the amount of reducing sugars released by Nelson Somogyi's method.

Express the enzyme as the amount of glucose released / ml of enzyme extract / unit time.

*Cellobiase* - Pipette 1.5 ml of the buffer (sodium acetate-acetic acid, pH 5.8), 2.5 ml of 5mM cellobiose (Dissolved in buffer) and 1 ml of enzyme extract (water extract) into a test tube. Incubate at 30°C for 2 hours. Terminate the reaction by placing the tube in a boiling water bath for 10 minutes. Measure the amount of glucose released as per Nelson's method.

*Aryl-β-glucosidase* - To 1 ml of 1mM p-nitrophenyl-β-D-glucoside solution in sodium acetate buffer (pH 5.0; 0.05 M), add 2 ml of enzyme extract (water extract) and 1 ml of the buffer. Incubate the reaction mixture at 37°C for 1 hour. Add 20 ml of 1M sodium carbonate and raise the volume to 30 ml with distilled water. Measure the amount of p-nitrophenol released by measuring the absorbance of the solution at 425 nm. Express the enzyme activity in units.

*Deoxyribonuclease* - Pipette aliquots of 2 ml of the enzyme extract (buffer extract) into test tubes. Add 0.5 ml of 0.05% calf thymus DNA solution in 0.05M phosphate buffer pH 6.5 followed by 0.5 ml of 0.2% magnesium chloride solution. After incubating the reaction mixture at 37°C for 4 hours in a water bath, add 0.5 ml of chilled perchloric acid to terminate the reaction and precipitate the unhydrolyzed nucleic acid. Maintain a zero time blank. Cool the contents of the tubes in tap water and centrifuge at 2000 r.p.m. for 15 minutes. Measure the absorbance of the supernatant at 260 nm in a spectrophotometer. Consider an increase of 1.0 $A_{260}$ / hour as one unit of the enzyme.

*Ribonuclease* - Pipette aliquots of 0.2 ml of the enzyme extract (buffer extract) into test tubes and add 1 ml of 0.15% yeast RNA solution in 0.05M phosphate buffer pH 6.5 to the tubes. Incubate the tubes containing reaction mixture in a water bath at 37°C for 2 hours. Add 0.5 ml of chilled 25% perchloric acid containing 0.75% uranyl acetate to terminate the reaction and precipitate the unhydrolyzed ribonucleic acid. Maintain a zero time blank. Similarly, cool the contents of the tubes in a running tap water and centrifuge at 2000 r.p.m. for 15 minutes. Measure the absorbance of the supernatant at 260 nm in a spectrophotometer against zero time blank. Consider an increase of 1.0 $A_{260}$ / hour as one unit of the enzyme.

Unit/ml =

$$\frac{A_{260} \times (\text{ml assay solution} + \text{ml precipitate agent}) \times \text{dilution factor}}{\text{ml enzyme extract solution} \times \text{minutes}}$$

### 4.3.7 Determination of Individual Minerals

The ash is digested with dilute HCl solution to dissolve inorganic salts and the digested aliquot is used for estimation of different inorganic constituents. The digested filtrate contains acid-soluble ash and residue after ignition is called the acid-insoluble ash.

*Procedure:* Moisten the ash obtained in the above method with 10 - 20 ml of dilute HCl (1:1), boil for 2 minutes, and evaporate to dryness (heat on steam bath for 3 hours to render silica $SiO_2$ insoluble). Moisten the residue with 5 ml HCl, boil for 2 minutes, add 50 ml $H_2O$, heat on water bath for few minutes (20-30 min) and filter through Whatman no. 42 filter paper into a 100 ml volumetric flask. Wash the residue thoroughly. The filtrate and the washings are diluted to 100 ml (this solution is designated as **solution A,** which can be used for determination of calcium, magnesium, sodium, potassium, copper, zinc, manganese, iron, aluminium and molybdenum). Transfer the residue into a crucible and ignite it in muffle furnace (600°C temperature). Cool the crucible in the desiccator and then weigh it.

*Calculation:*

$$\text{Acid insoluble ash \%} = \frac{\text{Weight of insoluble salt}}{\text{Weight of total ash}} \times 100$$

***Sand*:** Boil the residue for 5 minutes with 20 ml of saturated sodium carbonate solution, add few drops of 10% sodium hydroxide solution. Let the mixture settle and decant through ignited and weighed gooch. Boil the residue in dish with another 20 ml of sodium carbonate solution and decant as before. Repeat this process. Transfer the residue to the gooch and wash thoroughly first with hot water and then with little HCl and finally with hot water until chloride free. Dry filtered contents, ignite at 500-550°C temperature and weigh as sand.

***Alkali soluble $SiO_2$:*** Combine alkaline filtrate and washings from the above procedure, acidify with HCl and evaporate to dryness.

Add 5 ml of HCl and again evaporate. Dehydrate by heating for 2 hours at 110 - 120°C. Moisten the residue with 5-10 ml HCl and boil for 2 minutes. Add 50 ml water and heat on water bath for 10-15 min. Filter through ashless filter paper with hot water, ignite at 500 -550°C and weigh as $SiO_2$.

***Iron and alumunium:*** Take aliquot of **solution A**. Add few drops of conc. nitric acid, bromine water or hydrogen peroxide to oxidize iron. If solution does not already contain excess phosphate, add 0.5g ammonium hydrogen phosphate, stir until completely dissolved and dilute to 50 ml with water. Add few drops of thymol blue solution (0.1% - Dissolve 0.1g thymol blue in water, add enough 0.1N NaOH to change color blue and dilute to 100 ml) and then add ammonium hydroxide until the solution just turns yellow. Add 0.5 ml HCl and 25 ml of 25% ammonium acetate and stir. Let it stand at room temperature till the precipitate settles down (approximately 1 hour). Filter and wash ten times with hot 5% ammonium nitrate solution. Ignite the residue at 500 - 550°C and weigh as $FePO_4$ and $AlPO_4$.

Fuse ignited precipitate in platinium crucible with 4g mixture of equal parts of sodium carbonate and potassium carbonate. When the fusion is complete, let crucible cool down. Add 5 ml of sulphuric acid and heat until copious fumes of sulphite are evolved. Cool the contents and transfer to flask. Add water and digest until solution is clear. Reduce iron with zinc, cool and titrate with 0.1N $KMnO_4$. Correct for blank and calculate as % Fe or % $Fe_2O_3$. Calculate to $FePO_4$ and substrate from total Fe & $AlPO_4$ to obtain $AlPO_4$.

*Colorimetric Method for Iron*

*Reagents:*

1. *Acetic acid (2M):* Dilute 120g acetic acid to 1000 ml with water.
2. *Ammonium citrate (1%):* Dissolve 1g ammonium citrate in water and dilute to 100ml.
3. *Bromophenol Blue (0.04%);* Grind 0.1g bromophenol blue in mortar with 3 ml of 0.05N NaOH, transfer to volumetric flask and dilute to 250 ml with water.
4. *Buffer solution*: (a) pH 3.5 - Mix 6.4 ml of 2 M sodium acetate with 93.6 ml of 2M acetic acid and dilute to 1 liter, (b) pH

4.5 - Mix 43 ml of 2M sodium acetate with 57 ml of 2M acetic acid and dilute to 1 liter.

5. *Hydroquinone*: Dissolve 1g hydroquinone in 100 ml of buffer solution pH 4.5. Keep in refrigerator.
6. *Ortho-phenanthroline*: Dissolve 1g ortho-phenanthroline in water and dilute to 1 liter.
7. *Sodium acetate (2M)*: Dissolve 272g sodium acetate in water and dilute to 1 liter.
8. *Iron standard solution (1 mg/ml)*: Dissolve 1g electrolytic iron in 50 ml of 10% sulphuric acid, warm if necessary. Cool and dilute to 1 liter with water.

*Procedure:* Pipette identical volumes of **solution A** into 25 ml volumetric flask and test tube. Add 5 drops of bromophenol blue in the test tube and titrate with 2M sodium acetate solution until color matches with that of equal volumes of pH 3.5 buffer. Add 1 ml of hydroquinone solution and 2 ml of ortho-phenanthroline to aliquots in the volumetric flask and adjust pH to 3.5 by adding same volume, if found necessary, for aliquots in test tubes. If turbidity develops upon adjusting pH of aliquot in test tube, add 1 ml of ammonium citrate solution to volumetric fllask before adding sodium acetate solution. Dilute up to the volume, mix and let stand for one hour for complete color development. Determine iron colorimetrically at 470-520 nm.

### *Calcium*

*Reagents:*

1. EDTA 1M
2. Sodium hydroxide 1N (Dissolve 40g NaOH in distilled water to prepare 1 liter of solution)
3. Murexide indicator (grind 0.2g ammonium purpurate with 100 g NaCl (AR))

*Procedure:* Take 5 ml of **solution A**. Add about 100 ml of distilled water and 5 ml of NaOH solution. Add 100-200 mg of murexide and then the color turns to pink. Titrate with EDTA solution until the color changes to purple at the end point. The color change is not sharp, and, for better judgement of end point, compares the purple color with a distilled water end point.

*Calculation:*

$$\text{Ca \%} = \frac{A \times 400.8 \times V}{V \times 10000 \times S}$$

where,

A = Volume of EDTA used

V = Total volume of ash solution (100 ml)

v = Volume of ash solution titrated (5 ml)

S = Weight of plant material taken in g.

***Magnesium***

*Reagents:*

1. EDTA solution 0.1M (Dissolve 3.723g of di-sodium salt of EDTA in distilled water to prepare 1 liter)
2. Buffer solution (Dissolve 16.9g ammonium chloride in 1.43 ml of conc. Ammonium hydroxide. Dissolve separately 1.179g disodium salt of EDTA and 0.78g magnesium sulphate in 50 ml distilled water. Now, mix both the solutions and dilute to 250 ml with distilled water).
3. Eriochrome Black T indicator [Grind 0.4g Eriochrome Black T with 100g NaCl (AR)].
4. Sodium disulphide solution (Dissolve 5g of $Na_2S.5H_2O$ in 100 ml of distilled water and store in a tightly closed bottle.

*Procedure:* Take 5 ml of **solution A** in a conical flask and add 100 ml of distilled water. Add 15 ml of buffer solution and 10-20 mg of Eriochrome Black T indicator. Titrate with EDTA solution until, at the end point, the color changes to blue.

*Calculation:*

$$\text{Mg \%} = \frac{(B - A) \times 400.8 \times V}{V \times 1.645 \times 10000 \times S}$$

where,

A = Volume of EDTA used for Ca alone

B = Volume of EDTA for Ca + Mg

V = Total volume of ash solution (100 ml)

v = Volume of ash solution titrated (5 ml)

S = Weight of plant material taken in g.

***Flame Photometry***

This method is suitable for the determination of 0.01 - 1% sodium, 0.01 - 1% potassium, 0.01 - 5% calcium, 0.01 - 1% indium or 0.05 - 20% lithium. A sample is dissolved in suitable acid and the solution is analyzed by flame photometer.

*Reagents:*

1. Calcium chloride stock solution (1 ml = 1 mg Ca): Dissolve 2.497 g reagent grade Iceland spar ($CaCO_3$) in 300 ml of water and a minimum of HCI. Heat to boiling, cool and dilute to 1 liter with water.

2. Indium chloride (1 ml = 20 mg In): Dissolve 2g of pure Indium metal in 25 ml of water and a minimum of HCI. Mild heating and a few drops of nitric acid will help in dissolving the metal. Dilute to 1 liter with water.

3. Lithium chloride (1 ml = 10 mg Li): Dissolve 53.24g of GR grade dried lithium carbonate in 300 ml of water and a minimum of HCI. Heat to boiling on a hot plate to evolve carbon dioxide. Dilute to 1 liter with water.

4. Magnesium chloride (1 ml = 50 mg Mg): Dissolve 50g pure magnesium metal in 300 ml of water and a minimum of HCI. Dilute to 1 litre with water.

5. Potassium chloride (1 ml = 1 mg K): Dissolve 1.905g pure dried potassium chloride in water and dilute to 1 liter.

6. Sodium chloride (1 ml = 1 mg Na): Dissolve 2.543g of pure dried sodium chloride in water and dilute to 1 liter.

*Preparation of sample:* The sample weight should be taken depending upon the concentration of the elements and its emissivity intensity as per the followings.

For a sample weight of less than 1g, it is advisable to take a 1g sample and a suitable aliquot of the resulting solution. Dissolve the sample in 25 ml of water and a minimum of HCI and dilute to 100 ml with water.

| *Element* | *Wave length (mμ)* | *Percent* | *Sample weight (g)* |
|---|---|---|---|
| Sodium | 588.5 | 0.4-4.0 | 0.5 |
| | | 6.0-20 | 0.1 |
| Potassium | 767 | 0.4-4.0 | 0.5 |
| | | 6.0-20 | 0.1 |
| Calcium | 422.5 | 0.01-0.1 | 3.0 |
| | | 0.5-1.0 | 2.0 |
| Indium | 451.1 | 0.01-0.1 | 3.0 |
| | | 0.5-1.0 | 2.0 |
| Lithium | 671.0 | 0.4-4.0 | 0.5 |
| | | 6.0-27 | 0.1 |

*Preparation of Calibration Curve:* Prepare a series of 5 standard solutions of the element to be determined.

*Procedure:* After the instrument warms up on the red sensitive photo-tube, make the following settings: 50% emission, 0.03 mm slit width, 671 mμ, 0.1 selector switch, resistor switch at No. 2 position and sensitivity control at 1 turn-off of counter clockwise. Balance the dark current, ignite the burner and adjust the gas pressures. Position a sample cup containing standard solution No. 3 under the burner and re-balance the galvanometer needle with the sensitivity control. Now read the other standard solutions against this setting maintaining dark current and sensitivity balance at all the times. Flush the burner with water after each reading. The flame and solvent background should be determined. In the case of lithium, it should be zero.

Plot the readings obtained above on log-log graph paper as concentration of lithium verses percent emission.

*Calculation*: Convert the emission reading of the test solution to milligram of the element being determined by means of calibration curve.

$$\text{Element } \% = \frac{A}{B \times 10}$$

where,

A = mg of element being determined per 100 ml

B = Grams of sample represented in 100 ml of final solution.

# 5

# HISTOLOGY AND HISTOCHEMISTRY

## 5.1 MICROTECHNIQUES

For many biological studies, pure & applied histochemical studies and purely clinical & pathological investigations, a technique, which provides a disciplined channelized laboratory process for the preparation of material (slides) from the tissues so as to facilitate microscopic study, is necessary and this process of preparation is called Microtechnique.

Microtechnique: involves the use of a number of steps, like, collection and preparation of material, fixation, dehydration and clearing of material, embedding of material in wax and block-making, cutting of blocks into sections by microtome, downgrade hydration, of sections, staining and upgrade dehydration of sections, clearing of sections and finally mounting of sections in a mount on a slide.

### 5.1.1 Collection, Preparation of Material and Laboratory Procedure for Use of Collected Materials

The living organisms or products of their growth are the potential sources of histological studies. The material for microtomy from the living systems may be soft, hard, normal or pathological. Glands, uncalcified cartilage, skin, muscles, nerves and blood vessels are soft tissues, which do not need any special treatment; whereas bones, teeth, calcified cartilage, chitin and dermal derivatives are the hard materials, which need special treatment.

The specimens are cut open to take out the tissues (materials). Caution should be taken to avoid any damage to the tissue. Tissues for microtechnique should be removed from living anaesthetized animals by biopsy. The small portions of the tissue/organ removed from the killed or living animals/insects are put immediately in fixative-fluid. The worms can be fixed as a whole in a jar filled with fixative-fluid. Pathological tissues, which are infected with pathogens should be handled carefully. There are ;our techniques to prepare the histological specimens:

1. Whole mount
2. Sectioning
3. Teasing or masseration
4. Smearing

Insects are collected and maintained in separate boxes or cages. Cut their antennae, legs, wings, anal cerci; remove their mouth parts; clear these parts in xylene or benzene and mount them in DPX or Canada balsam on clean slides. Dissect these insects and take out different tissues like malpighian tubules, tracheae, fat bodies, ovaries, salivary glands, testes, etc. separately in cavity blocks; wash them with distilled water; dehydrate, stain with borax carmine, dehydrate further, clear and mount in DPX or Canada balsam.

### 5.1.2 Fixation

Literally, fixation means to immobilize and the main objective of fixation is to immobilize the cell structures while maintaining their morphological identity. The tissue is fixed after being taken out from the specimen. Fixatives are the agents that prevent autolysis. Fixation renders proteins of the tissue insoluble and converts living or fixed natural jelly condition of the cell into fine granular spongy mass. Fixed tissues can be washed for a long time without any change. The immobilization are done in two ways -

#### *5.1.2.1 Fixation by physical agents*

This process involves use of heat, drying at room temperature and drying at low temperature (cryodessication).

#### *5.1.2.2 Fixation by chemical agents*

The fixatives used for fixation of tissues chemically should possess the following properties,

- It should stop all the metabolic activities in a cell,
- It should coagulate the cell contents,
- It should conserve the cytological and histological architecture,
- It should conserve the original form of tissue,

- It should prevent putrification in the tissues, and
- It should impart selective staining by dyes to the different parts of the cell.

Some of the important fixative chemicals are methanol, ethanol, acetone, HCl, nitric acid, sulphuric acid, trichloroacetic acid, picric acid, chloroplatinic acid, mercuric acid, acetic acid, chromium trioxide, osmium tetraoxide, formaldehyde, potassium dichromate etc. The fixative chemicals used alone are called as **simple fixatives,** e.g., formalin, methanol, ethanol. The other fixatives are the mixtures of several fixing agents in a liquid form. Such fixatives are called **compound fixatives.** The chemical compatibility or incompatibility of the fixing agents that constitute a mixture is important while choosing the fixative for specific purpose.

### *5.1.2.3 Mode of Fixation*

1. Fixation by vapours
2. Fixation by liquids
   a. Fixation by immersion
   b. Fixation by injection into body cavity
   c. Fixation by perfusion

### *5.1.2.4 Classification of Fixatives*

1. Simple fixatives
2. Common fixatives
3. Other fixatives

*Simple Fixatives*

a. Formalin (formaldehyde): It is a well known non-coagulant fixative, obtained in an aqueous solution as formol (which is a formalin containing 35% to 40% of native formaldehyde). It is diluted generally to make 4% formaldehyde (1 volume formalin and 9 volume water). The tissue may remain in this fixative indefinitely.

b. Ethyl alcohol: This is a good fixative for enzyme histochemistry such as alkaline phosphatase, and also for glycogen, but does not fix lipids and carbohydrates. It is prepared taking 77 ml of absolute alcohol and 23 ml distilled water and the tissue may remain in this fixative for 12 to 14 hours only.

c. Acetone: Cold acetone at 5°C is useful as fixative for lipase and acid phosphatase. Dilute acetone is useful as a fixative for lipids. It has slow penetrative action and hence it should be used to fix small pieces of tissue. The period of fixation is 1 to 3 days.

d. Formalin-acetic acid: This is good fixative for parasitological material such as intestinal worms, and also insects and crustaceans. After the period of fixation, transfer the material to 90% alcohol directly. It is prepared by taking 90 ml of 70% alcohol, 2 ml glacial acetic acid and 8 ml commercial formalin. The period of fixation is 1 to 24 hours.

*Common Fixatives*

a. Aqueous Bouin's fluid

| | |
|---|---|
| Saturated picric acid solution | - 75 ml |
| Formaldehyde | - 20 ml |
| Acetic acid | - 5 ml |
| Period of fixation | - 12 to 24 hours |

After fixation, the material should be transferred directly to 90% alcohol.

b. Alcoholic Bouin's Fluid

| | |
|---|---|
| Picric acid | - 1 g |
| Formaldehyde | - 60 ml |
| 80% alcohol | - 150 ml |
| Acetic acid | - 15 ml |
| (to be added at the time of use) | |
| Period of fixation | - 1 to 3 days |

After fixation, the material should be transferred directly to 90% alcohol.

c. Allen's Bouin's Fluid

| | |
|---|---|
| Bouin solution | - 10 ml |
| Chromic acid crystals | - 1.5 g |
| Urea crystals | - 2 g |
| Period of fixation | - 12 to 24 hours |

This is commonly used for chromium fixation. After fixation, the material should be washed thoroughly in water or 50% or 70% alcohol. The yellow colour of Bouin's fluid must be removed before staining.

d. Zenker's Fluid

| | |
|---|---|
| Potassium dichromate | - 2.5 g |
| Mercuric chloride | - 5 g |
| Distilled water | - 100 ml (Stock solution) |
| Glacial acetic acid | - 5 ml |
| (to be added just before use) | |
| Period of fixation | - 8 to 24 hours |

After fixation, the material should be washed in running water overnight.

e. Helley's fixative

| | |
|---|---|
| Zenker's stock solution | - 100 ml |
| Commercial formalin 40% | - 5 ml (added just before use) |
| Period of fixation | - 24 hours |

After fixation, mercuric chloride of Helley's fluid should be removed by treating the cut sections with Lugol's iodine solution for 1 to 5 minutes and then washing with tap water followed by putting the sections in 5% sodium metabisulphite solution till the sections become colourless and final washing with tap water.

f. Formal sublimate acetate

| | |
|---|---|
| Mercuric chloride | - 5 g |
| Sodium acetate (anhydrous) | - 2 g |
| 10% formalin | - 100 ml |

g. Rossman's fluid

| | |
|---|---|
| Saturated picric acid solution in absolute alcohol | - 90 ml |
| Commercial formalin | - 10 ml |
| Period of fixation | - 6 to 48 hours |

h. Susa fixative

| | |
|---|---|
| Saturated mercuric chloride solution in 6% sodium chloride | - 50 ml |
| Trichloroacetic acid | - 2 ml |
| Glacial acetic acid | - 4 ml |
| Distilled water | - 30 ml |
| Period of fixation | - 24 hours |

After fixation, the tissue is directly put in 50% alcohol.

i. Regaud's fluid

| | |
|---|---|
| 3% potassium dichromate soln. | - 80 ml |
| Commercial formalin | - 20 ml |
| Period of fixation | - 12 hours |

This is good fixative for mitochondria and other cytoplasmic inclusions. The material is washed with running water overnight.

j. Glison's fluid

| | |
|---|---|
| Conc. Nitric acid | - 15 ml |
| Glacial acetic acid | - 4 ml |
| Mercuric chloride | - 20 g |
| Ethyl alcohol | - 100 ml |
| Distilled water | - 880 ml |
| Period of fixation | - 24 hours to several days |

This is good fixative for invertebrates because this does not harden the tissues. The material is washed directly in 50% alcohol after fixation.

k. Formol nitric acid

| | |
|---|---|
| Formalin (25 parts of 40% formalin and 75 parts of distilled water) | - 3 parts |
| Nitric acid conc. 10% | - 1 part |
| Period of fixative | - 10 min to 30 min |

This is a good fixative for embryos. Cut the egg shell and make a small window. Put the fixative through the window by a dropper.

*Other Fixatives*

*Preservation of lipids*

| | |
|---|---|
| Cadmium chloride | - 1 g |
| Commercial formalin | - 10 ml |
| Aqueous calcium chloride | - 10 ml |
| Distilled water | - 80 ml |
| Period of fixation | - 24 hours |

The material fixed in this fixative should be stained with Sudan Black.

Preservation of polysaccharides

| | |
|---|---|
| 90% alcohol saturated picric acid | - 85 ml |
| Commercial formalin | - 10 ml |
| Glacial acetic acid | - 5 ml |
| Period of fixation | - 1 to 4 hours |

After fixation, the material should be washed with 90% alcohol.

Preservation of enzymes

Cold absolute acetone at 5°C

| | |
|---|---|
| Period of fixation | - 24 hours |

After fixation, the tissue is washed with absolute alcohol directly.

Preservation of proteins

| | |
|---|---|
| Commercial formalin 40% | - 100 ml |
| Distilled water | - 900 ml |
| Sodium acid phosphate monohydrate | - 4 g |
| Sodium phosphate anhydrate | - 6.5 g |
| Period of fixation | - 24 to 72 hours |

### 5.1.3 Dehydration

Once fixation of tissues is over, the fixed tissues are dehydrated to remove water. **Dehydration is a procedure of step-wise removal and replacement of water from the tissue, small specimens and other biological materials by the dehydrating agents, like, ethyl alcohol, cellosolve (ethylene glycol monoethyl ether), dioxane**

**(diethylene dioxide) and isopropyl alcohol**. Generally, the progressively higher grades of alcohol are used for dehydrating the tissues. The optimum length of time, for keeping the tissue in a particular alcohol-grade, depends upon the thickness and nature of tissue. For step-wise dehydration, the tissue should be kept in the progressive alcohol-grades for the following durations - in 30% & 50% alcohol for 5 to 10 hours, 70% alcohol for 10 to 12 hours, in 90% alcohol for about 15 hours and in absolute alcohol for 6 to 8 hours. The following precautions should be taken during dehydration.

- Dehydration should be done in tightly closed specimen tubes.
- The specimen tubes should be completely drained before putting higher grade of alcohol.
- The specimen tube containing the material should be closed tightly with glass cork immediately after the change of next grade of alcohol.
- In case of high humidity in weather, the specimen tubes should be kept in desiccators.
- During the staining, there should be sufficient quantity of stain in the jars so that the slide is completely immersed.

### 5.1.4 Clearing

The process of dehydration leads to the saturation of tissue with alcohol. After dehydration, the tissue is impregnated with paraffin wax to make it firm so as to cut the sections easily. However, diffusion of paraffin into the tissue to replace alcohol is not possible because of its immiscibility in alcohol and hence alcohol present in the tissue needs to be replaced or removed before embedding is done. **After dehydration, the tissue is kept in a solution/fluid, which makes paraffin to penetrate into the tissue. This process is called clearing and the fluid used is known as clearing agent.** Important clearing agents are xylene (xylol), benzene, toluene, clove oil and cedar wood oil. Xylene is the most commonly used clearing agent, which brings about the quick removal of alcohol from the tissue.

### 5.1.5 Embedding and Block Making

**Embedding is the process of standardization of the tissue for microscopic examination by sectioning with microtome.** The following steps are involved in embedding.

a. Termination of Fixation
b. Embedding
c. Block Preparation
d. Labelling and storage of blocks

### *5.1.5.1 Termination of Fixation*

Generally, fixatives stop fixation of the material automatically. In case of the fixatives soluble in water and alcohol, desired results are obtained during dehydration or before embedding. However, washing in running water is required if the fixatives contain chromium chloride, potassium dichromate, osmium tetraoxide or chloroplatinic acid so that the fixative is completely removed.

### *5.1.5.2 Embedding*

Embedding of the tissues is done with paraffin, celloidin and gelatin. The tissue is soaked in molten wax at a standard temperature coinciding with the melting point of the embedding medium used. The duration of soaking depends upon the size & thickness and the texture of tissue. Insufficient or prolonged embedding may cause difficulties in section cutting.

### *5.1.5.3 Block Preparation*

After the tissue is embedded with wax, it is cast into a block of paraffin. This process is known as casting of block or block making. It involves the following steps,

- Fill the mould or L piece with molten paraffin wax.
- Place the impregnated tissue from the vial in the mould and orient it according to the plane of sections needed.
- Immediately move a warm needle in the molten wax on all sides of the tissue so as to remove any air bubble.
- Blow the air gently on the mould; the wax forms a thin layer on the surface.
- Gently immerse the mould in cold water so as to cool the wax rapidly. While immersing, it should be lightly tilted to avoid sudden contact with water.
- When the block is solid, remove it from the water and trim it according to need.

#### *5.1.5.4 Labelling and Storage of Blocks*

The blocks are labelled carrying all the details of the tissue, which is done by inserting the label on one side of the block during casting. The blocks are stored in small paper bags or clean wooden boxes in a cool place until they are sectioned.

### 5.1.6 Section Cutting

A section is a thin slice cut from the biological materials for studying either the cells themselves or their arrangement. The paraffin blocks containing the tissues are cut into sections with the help of microtome (rocking or rotatory microtome)

#### *5.1.6.1 Trimming the Block*

The paraffin block is trimmed to give it a correct shape. The paraffin block containing the material is held against the light so that the outlines of the material can be seen clearly and then trimmed until the material lies in the centre of a perfect rectangle with the major axis of the material exactly parallel to the long sides of the block. While trimming, care should be taken that the block is not cracked and the material is neither exposed nor remain covered with a very thin layer of wax.

#### *5.1.6.2 Mounting the Block*

The trimmed block is now attached to a holder, which can be inserted in the jaws of the microtome. The block holder consists of a disc with rough surface attached to a cylindrical rod. The disc is first covered by a thin layer of molten wax and allowed to cool. The paraffin block is now attached to the hard layer of the wax on the disc of the block holder. The block holder with the attached block is allowed to cool and attain room temperature.

#### *5.1.6.3 Clearance Angle and Other Adjustments*

Before inserting the block holder in the microtome, it is desirable to trim the cutting face of the block to its maximum until a very thin layer of wax is left in front of the material. The block holder is inserted in the microtome and all the set screws are thoroughly tightened so that the cutting face of the block is properly oriented in the desired plane with its upper and lower surfaces parallel to the cutting edge of the knife. The knife is now placed into the knife holder and adjusted to such an angle that the cutting face of the knife is not exactly parallel to the face of the block and there is some angle (called **clearance angle**). This angle should not be too large or too small. After adjusting an optimum angle, the feed screw handle is turned moving the feeding mechanism forward or backward as required until the face of the block barely touches the cutting edge of the knife. The micron indicator is now adjusted to the desired thickness of the sections and microtome wheel is rotated, whereby, the block starts cutting into sections. As soon as the entire width of the block begins to cut along with material, a dry soft brush is slipped under the ribbon with the left hand and lifted away from the knife. However, care should be taken not to give extra strain to the ribbon otherwise it will break at undesirable point and time. After a sufficiently long ribbon is obtained, it is conveniently spread over a wooden board. Continue these processes until the whole of the required portion of the block has been cut and the ribbons are laid on the board. Arrange the ribbons in such a way that the serial order of the sections is not disturbed. These ribbons are then divided into pieces of suitable length for mounting on a slide.

**Table 5.1: Some of the Defects in Section Cutting and Remedial Measures**

| *Defects* | *Possible causes* | *Remedies* |
|---|---|---|
| Ribbon gets curved | Edges of paraffin block are not parallel. | Trim paraffin block |
| | One side of the block may be warmer than other. | Allow the block to cool |
| Ribbon gets split | Nick in the cutting facet of the knife. | Change the cutting facet by moving the knife sideways. |
| | Any hard particle on the cutting facet of the knife. | Clean the cutting facet with the help of a soft cleansing cloth or cotton wool, moistened with xylol. |
| Ribbon shows thin and thick sections alternately | Block or the supporting wax of the holder still warm after mounting | Allow the block and holder to attain room temperature |

| | | |
|---|---|---|
| | Block holder loose in the jaws of microtome. | Check and tighten all the screws of the jaw of the microtome. |
| | Knife may be loose. | Tighten the holding screws of the knife carrier. |
| Ribbon of the compressed sections | Knife edge blunt | Try another portion of the knife edge or sharpen the knife |
| | Wax warmer than the room temperature | Cool block to room temperature |
| Sections fail to form ribbon | Paraffin wax of the block may be of high melting | Apply soft wax or bee hive wax on any two sides of the point paraffin block. |
| Tissues crumbles or fail out of paraffin | Improper dehydration and clearing | Nothing can be done |
| | Kept too long in the paraffin bath | Nothing can be done |
| Ribbon gets lifted up by the block | Clearance angle very small | Increase clearance angle slightly by tilting the knife |
| | Ribbon electrified due to continuous friction | Stop cutting for some time, cool the knife edge and the block. |
| | Wax fragments on the knife edge | Clean knife edge by soft cloth or cotton wool soaked in xylene. |

### 5.1.7 Mounting and Spreading of Ribbons

On cutting sections of the paraffin block, a continuous ribbon of paraffin containing the sections of the tissue is obtained. The ribbon can be placed in a section-box or section tray, if the mounting is not done immediately. Before mounting on a slide, the ribbon is cut into pieces of small lengths which could be conveniently accommodated on a slide with sufficient space on the one side of the slide for labelling. The slides, on which the ribbons are to be mounted, are first smeared with any adhesive, such as, Mayer's albumen so that the section remained fixed to the slide while staining.

Mayer's albumen is prepared by thoroughly mixing equal portions of egg albumen and glycerol, and adding a few crystals of thymol, which acts as a preservative. After smearing, small pieces of ribbon are placed on the slide and flood it with water by a dropper. Now, the slide is placed on hot plate to heat the water whereby the paraffin ribbon begins to stretch, so also the sections. Care should be taken that the wax does not begin to melt otherwise the sections of the tissue will lose their contour and become distorted. Soon after the ribbon is completely stretched, the slide is removed from the hot plate, drain off the water and allow it to dry.

After the slide is completely dried, it is labelled with the pencil. The label should include the name of the tissue, range of section numbers in the ribbon piece, the date, the number of the slide and the name of the special stain to be given. Such labelled slides are stored in the wooden slide boxes having grooves. The staining should be undertaken 48 hours after the drying of the sections.

### 5.1.8 Staining

Most of the tissues do not retain sufficient colour or become very transparent after processing, to be clearly visible, when observed under microscope. The process of staining renders them visible under microscope. In many tissues, an assortment of components is present, which is distinguishable by their ability to retain dyes of contrasting colours when stained in two or more stains during differential staining. In **mordant staining,** some other compound is applied followed by application of dye. In **indirect staining,** a single staining solution containing both the mordant and dye is used. In this process, whole tissue become coloured. On transferring the tissue in a differentiating solution, the colour is taken by the cytoplasm is removed while leaving it in nucleus.

There are two types of stains - natural stains and synthetic stains. **Natural stains are those dyes which are obtained from natural sources like plants and animals.** The most common stain is haematoxylin, which is extracted from the residue of an extract of Campeach wood. **The synthetic stains are usually artificial chemical compounds composed of an acid and a base** that are further classified as acid stains, basic stains and neutral stains. **Acid stain** like acid fuschin is usually used to stain the basic components of a cell such as nucleus. **Basic stains** like basic fuschin are used for staining acidic components of a cell. **Neutral stains** are formed due to interchange of ions when the sodium salt of an acid dye and the chloride of the basic dye are mixed.

There are different methods of staining. In **progressive staining,** the stain is added to the tissue until the correct depth of colour is reached. In the **regressive staining,** the sections are over-stained and subsequently excessive amounts of stains are removed by placing the slides in differentiating solution. When the correct intensity of colour is left in the tissue, the sections are thoroughly washed in running water.

Blood smears are commonly stained with a neutral stain, e.g., Wright's Stain (Dissolve 0.1 g Wright's stain powder in 60 ml methyl alcohol and grind it in a mortar until it is completely dissolved. Use after about 20 days). The smear is prepared by taking a fresh drop of blood on a clean slide and spreading it in the form of a thin film with the help of another slide. The smear is dried and stained by putting one or more drops of Wright's stain on the smear and allowing it to remain for a minute or so. Then equal quantity of distilled water is added on the slide and it is allowed to remain for about 2 to 3 minutes. The slide is then washed with water and gently blots with filter paper without disturbing the smear.

For routine histological studies, the sections of any tissue are commonly stained with haematoxylin and counter-stained with eosin. The Ehrlich's haematoxylin is prepared by dissolving 2 g haematoxylin in 100 ml ethyl alcohol and then adding 10 g potassium alum, 100 ml glycerine, 100 ml distilled water and 10 ml glacial acetic acid. The bottle is placed in sunlight for 6-8 weeks until the solution is deep red in colour. Eosin is prepared separately by dissolving 1 g eosin powder in 100 ml of 70% ethyl alcohol. For staining, the slides are first deparaffinized by keeping them in xylene for about 30 minutes to 1 hour. The deparaffinized slides are then passed through a down graded series of ethyl alcohol, a process termed as running down slide to water. During this procedure, the slides should not be allowed to get dried. The whole staining procedure is as follows.

- Hydrate deparaffinized slides by passing through a graded series of ethyl alcohol in descending order, i.e., 100%, 90%, 80%, 70%, 50%, 30% and water.

- After hydration, stain the slide in haematoxylin by keeping it for about 2-5 minutes in the stain.
- Wash the slide in water and observe it behind any white background. If over-stained, de-stain by giving one or two dips in ascending order up to 70% alcohol.
- Immediately transfer the slides under running tap water for about 3-5 minutes, when the sections turn blue in colour.
- Now dehydrate the slides by passing through a graded series of alcohol in ascending order up to 70% alcohol.
- Counter stain in eosin by giving one or two quick dips in the stain solution and then wash it in fresh grade of 70% alcohol.
- Dehydrate further by passing through ascending grades of ethyl alcohol i.e. 80%, 90% rectified spirit, absolute alcohol I and absolute alcohol II.
- Allow the slides to remain in absolute alcohol for about 5 to 10 minutes so that they are completely dehydrated.
- Now transfer the slides to Xylene I for cleaning where xylene penetrates the tissue and replaces the alcohol. Give one or two changes of xylene and mount in DPX mountant or Canada balsam.

To make a permanent slide of the sections of any tissue, all the traces of alcohol are replaced by a medium, preferably, toluene or xylene that maintains the tissue in a clear and transparent condition. A mounting medium is then applied. In **cover slip mounting**, the stained and clear slide is placed on a plane table on a piece of blotting paper and a thin streak of mounting medium is applied on the cover slip. The cover slip is turned over the slide and rested on one side besides the section. Gradually, the cover slide is lowered in such a way that all the air is escaped from under the cover dip. The cover slip is gently pressed from the centre outwards to distribute the medium evenly. Finally, the cover slip is pressed gently and allowed to dry.

## 5.2 HISTOCHEMISTRY

It is a branch of science that deals with the composition of the cells and tissues in terms of the chemical elements and their compounds in the material. Histochemical techniques are based on the advancements made in the organic chemistry, electron microscopy, fluorescence microscopy, biophysics and cell physiology. These techniques enable the identification and localization of specific substances within tissues. The methods depend on the chemical reactions between the substances to be identified and localized in a tissue section, and one or more reagents in which the tissue section is incubated. Histochemical methods are classified on basis of what elements, ions or radicals are demonstrated in cells or tissues; whereas morphological methods depend on such structures as nuclei, cytoplasm and the tissue as a whole for their classification.

*Inorganic constituents:*

a. Ionized material (anions such as Cl, $SO_4$, $PO_4$ etc. and cations like metals, $NH_4$ and positively charged organic ions )

b. Non-ionized material (soluble compounds in which a metal is bound as a part in the molecule 9 such as haemoglobin, and insoluble substances such as calcium compounds in bones and teeth)

***Organic constituents:***

a. Specific compounds (starch, glycogen, amino acids, sterols)

b. Reactive molecular sites, radicals or configurations (aldehydes, hydroxyl, carboxyl, amino -s-s- and -c=c- bonds).

### 5.2.1 Determination of Inorganic Substances

#### *5.2.1.1 Determination of Iron*

**Perl's Method for Ferric Iron (Lison & Bunting)**

**(Formalin-fixed, paraffin sections preferred)**

*Method:*

- Bring sections to distilled water.
- Expose sections to a freshly prepared mixture of equal parts of 2% potassium or sodium ferrocyanide and 2% HCl for 30 - 60 minutes. Change the mixture after 30 minutes.
- Wash in distilled water.
- Counter-stain nuclei with 1% aqueous neutral red solution for 3 minutes.

- Wash, dehydrate in alcohol, clean in xylene and mount in DPX.

  *Result*: Ferric iron - deep Prussian blue; nuclei - red.

**Turnbull Blue Method (Tirman & Schmettzer)**

(Formalin-fixed, paraffin sections)

*Method:*

- Bring sections to distilled water.
- Treat sections for 1 - 3 hours with a dilute solution of yellow ammonium sulfide.
- Rinse in distilled water.
- Treat with a freshly prepared mixture of equal parts of 2% potassium ferrocyanide and 1% HCl for 10 - 20 minutes.
- Wash in distilled water.
- Counter stain nuclei with 1% aqueous neutral red solution if required.
- Wash, dehydrate in alcohol, clean in xylene and mount in DPX.

  *Result:* Ferrous ion and ferric ion - deep blue; nuclei - red.

### *5.2.1.2 Determination of Calcium*

**Alizarin Red S Method (Dabl, 1952)**

(Alcohol fixed, paraffin sections)

*Preparation of solutions:*

1. Alizarin Red S: Stir 0.5 g dye in 45 ml distilled water and add 5 ml of 1:100 dilution of 28% ammonium hydroxide, with constant stirring. The pH should be between 6.3 – 6.5 and the solution is kept for at least one month.
2. Differentiator: $10^{-3}$ M HCl in 95% ethanol (conc. HCl 1 part; 95% ethanol 10,000 parts).

*Method:*

- Bring sections to distilled water.
- Stain with Alizarin red solution for 2 minutes
- Wash with distilled water for 5 – 10 seconds.
- Dehydrate, clear and mount in cedar wood oil.

*Result:* Calcium deposits appear orange red, inorganic iron (not hemosederin) purple.

**Calcium Red Method (McGee Russell, 1955)**
(Formalin or alcohol fixed, paraffin sections)

*Preparation of solution*

1. *Calcium Red Solution* - Wash 2g nuclear fast red (G.T. Gurer; Batch 3564 recommended) twice with about 100 ml distilled water. Take up the residue (about 0.25 g) with 100 ml distilled water.

*Method:*

- Bring sections to distilled water.
- Stain in Calcium Red solution for 1 - 10 minutes.
- Wash with distilled water.
- Dehydrate, clear and mount in synthetic resin media.

*Result:* Calcium deposits appear red; tissues attain various shades of pink colour.

***5.2.1.3 Determination of Phosphate and Carbonates***
**Von Kossa Method for Calcium deposits**
(Formalin fixed, paraffin sections etc.)

*Method:*

- Bring sections to distilled water.
- Immerse in 0.5 - 1.0% aqueous silver nitrate solution for 10-15 minutes in presence of sunlight or UV light.
- Rinse in: distilled water.
- Immerse in 5% aqueous sodium thiosulphate solution for 30 seconds.
- Wash in tap water.
- Counter stain with 1% neutral red solution for 30 seconds.
- Wash with distilled water.
- Dehydrate, clear and mount in Canada balsam.

*Result:* Phosphate and carbonate appear black; nuclei red.

### *5.2.1.4 Determination of Potassium*

**Gersh Method for Potassium**

(Freeze-dried material only)

*Method:*

- Cut paraffin sections at 15 μ and mount dry on cover slip by applications of finger pressure.
- Remove wax with light petroleum and dry section by burning of excess petroleum (condensation produced by evaporation is avoided).
- Cover the section with a drop of 12% aqueous cobalt nitrate.
- Allow to dry and mount directly in glycerine.

*Result:* Yellowish crystals and micro-crystals of potassium cobalt nitrate indicate the presence of potassium.

## 5.2.2 Determination of Organic Substances

### *5.2.2.1 Carbohydrates*

**Okamato Method for Glucose (Mulfer, 1955)**

*Method:*

- Fix the thin (2 mm) tissue slices in methanol saturated with barium hydroxide for 24 hours at 10°C.
- Dehydrate in three changes of absolute alcohol.
- Clear and embed in methyl benzoate (benzene paraffin series).
- Cut 10 μ sections and mount on albumenized slides without any contact with water.
- Place the paraffin sections over hot plate at 56 - 58°C until the wax melt.
- Remove wax with chloroform and bring to absolute alcohol.
- Immerse in alcoholic silver nitrate (add 90 ml ethanol to 10 ml of 20% silver nitrate aqueous solution) for 30 minutes.
- Wash repeatedly in 96% alcohol.
- Reduce in alcoholic formalin (90 ml ethanol and 10 ml of 40% formalin).
- Bring through descending strengths of alcohol to water.
- Fix for 30 seconds in 5% sodium thiosulphate.

- Wash with distilled water.
- Dehydrate, clear and mount in a suitable (non-reducing) resin media.

*Result:* Black silver deposits indicate sites containing glucose.

## Silver Method for Glycogen and Mucin

*Preparation of stain solution:*

10% aqueous $AgNO_3$ - 4 ml
Saturated aqueous $Li_2CO_3$ - 16 ml

Add conc. Ammonium hydroxide drop by drop until lithium silver complex is almost dissolved leaving a light turbidity. Add saturated lithium carbonate to make up to 100 ml. Filter and keep in well stoppered dark bottles at 40°C.

*Method:*

- Bring sections to alcohol, cover with 1% celloidin, dry and pass through alcohol to water.
- Immerse in 10% chromic acid for 20-30 minutes.
- Wash in running tap water and distilled water.
- Treat with dilute (1:10) lithium silver solution at 45°C for 15-60 minutes.
- After the first two silver solution, wash in water and third in formalin (neutral) for 30-60 seconds.
- Tone with 1:500 gold chloride, if necessary.
- Rinse in 5% sodium thiosulphate for 3-5 minutes.
- Wash in water, remove celloidin with acetone, dehydrate, clear and mount.

*Result:* Glycogen and mucin appear brown to black with colourless background.

## Best's Carmine Method (Best, 1905)

(Paraffin sections, freeze dried sections, frozen sections)

*Preparation of solutions:*

1. Carmine Stock solution: Add 2g carmine, 1 g potassium carbonate and 5 g potassium chloride to 60 ml distilled water. Boil gently for 5 minutes, cool and filter. To the filtrate, add 20 ml ammonia and keep at 0 - 4°C.

2. Best's staining solution: Dilute 15 ml of stock solution with 12.5 ml ammonia (Sp.Gr. 0.88) and 12.5 ml methyl alcohol.
3. Best's Differentiator: Mix 4 ml methyl alcohol, 8 ml absolute alcohol and 10 ml distilled water.

*Method:*

- Bring sections from xylene to absolute alcohol.
- Place sections in 1% celloidin in absolute alcohol and ether mixture (1:1) for 2 minutes.
- Dry in air. Pass through alcohol to water.
- Stain in Ehrlich's haematoxylin for 5 minutes.
- Rinse and differentiate in 1% acid alcohol.
- Rinse in water.
- Stain in Best's staining solution for 15-30 minutes.
- Differentiate in Best's differentiator for 5-60 seconds.
- Wash in 80% alcohol.
- Dehydrate in absolute alcohol, clear in xylene and mount in DPX.

*Results:* Nuclei dark blue and glycogen red.

## Lead Tetra Acetate Schiff Method

*Preparation of solution:*

- Lead Tetra acetate solution - Dissolve 1 g lead tetra acetate in 30 ml glacial acetic acid.
- Immediately before use, add 70 ml saturated sodium acetate solution.

*Method:*

- Bring sections to water.
- Immerse in 1M sodium acetate solution for 5 minutes.
- Treat with freshly prepared lead tetra acetate solution for 10 minutes.
- Immerse in 1M sodium acetate solution for 5 minutes.
- Wash in running tap water for 10 minutes.
- Place sections in Schiff's reagent for 15 minutes.

- Treat with bisulphite water (as in Feulgen reaction for nucleic acid).
- Wash well in running tap water for 10 minutes.
- Dehydrate, clear and mount in Canada balsam.

*Result:* Glycogen and various muco-substances are stained reddish purple.

## Silver Method for Ascorbic Acid

*Method:*

- 10 μ sections are mounted dry on albumenized slides and incubated at 37-40°C overnight.
- Without removing paraffin, place sections in 10% silver nitrate (in 3% acetic acid) for 8-14 hours.
- Wash in 95% alcohol and dehydrate in absolute alcohol.
- Remove paraffin wax with xylene.
- Mount in permount.

*Result:* Reduced silver granules indicate ascorbic acid.

## PAS Reaction (Mc Mannus)

*Preparation of solution:*

1. *Periodic acid solution* - Dissolve 1 g periodic acid in 100 ml distilled water.
2. *Schiffs reagent* - Dissolve 1 g basic fuschin in 400 ml boiling distilled water. Cool to 50°C and filter. To the filtrate, add 1 ml of thionyl chloride. Keep the solution in dark overnight. Add 2g activated charcoal, shale well and filter. Store the filtrate at 4°C in a dark coloured bottle. Allow this solution to reach room temperature before use. Discard the reagent after use.

*Method:*

- Bring all the sections to water.
- Place the sections in periodic acid solution for 5-8 minutes.
- Wash in tap water for 3 minutes followed by 1 min washing in distilled water.
- Treat with Schiff's reagent for 15 minutes.

- Wash in running tap water for 10 minutes.
- Counter-stain in haematoxylin or Celestin blue-haemalum sequence.
- Wash in running tap water for 5 minutes.
- Differentiate if necessary in 1% acid alcohol.
- Wash in running tap water for 5 minutes.
- Dehydrate through graded alcohols to xylene.
- Mount in DPX.

*Result:* PAS positive materials (hexose containing mucosubstances) are stained magenta, glycogen deeply red and nuclei blue.

**Periodic Acid-Schiff's technique (Hotch Kiss, 1948)**

*Preparation of solution:*

1. *Periodic acid solution* - Dissolve 0.4 g periodic acid in 35 ml reagent ethyl alcohol and add 5 ml 0.2 M sodium acetate (27.2 g of hydrated salt in 1000 ml) and 10 ml distilled water. This solution should be kept in dark at 17-22°C and used at this temperature.
2. *Reducing bath* - Dissolve 1 g potassium iodide and 1 g sodium thiosulphate in 30 ml reagent ethyl alcohol and 20 ml of distilled water. Add 0.5 ml 2 N HCl. A deposit of sulphur form can be ignored. Keep between 17 and 22°C.
3. *Schiffs reagent*-As prepared in PAS reaction.
4. *Celestin blue solution* - Dissolve 2.5 g iron alum in 50 ml distilled water and left overnight at room temperature. Add 0.25 g Celestin blue B and boil for 3 minutes. Filter when cool and then add 7 ml glycerol.
5. *Orange G solution* - Dissolve 2 g orange G (C.I. 16230 or 27) in 100 ml of 5% aqueous phosphotungstic acid. Stand for 24 hours and use the supernatant.

*Method:*

- Bring all the sections to water and remove mercury.
- Rinse in 70% alcohol.
- Place the sections in periodic acid solution for 5 minutes.

- Rinse in 70% alcohol.
- Immerse in reducing bath for 1 minute.
- Rinse in 70% alcohol.
- Treat with Schiff's reagent for 20 minutes.
- Wash in running tap water for 10 minutes.
- Stain nuclei lightly with Celestin blue for 2 minutes, followed by Mayer's haematoxylin for 2-3 minutes.
- Differentiate in 1% acid alcohol.
- Wash in running tap water for 5 minutes.
- Counter-stain with Orange G for 10 seconds.
- Wash in water till sections are pale yellow.
- Dehydrate in alcohol, clear in xylene and mount in DPX.

*Result:* Mucoproteins and hexose containing muco polysaccharides are stained deep purple red. Glycoproteins are usually pale red or pink. Nuclei are blue-black and RBC & acidophil proteins yellow.

## Bauer-Feulgen Method (Bauer, 1933)

(Frozen sections only)

*Preparation of solution:*

1. Oxidizing solution (4% chromic acid) - Dissolve 4 g chromium trioxide in 100 ml distilled water.
2. Schiff's reagent
3. Sulphuric acid rinse - Mix 5 ml of 10% sodium metabisulphite, 5 ml hydrochloric acid and 90 ml distilled water.

*Method:*

- Bring all the sections to water.
- Place the sections in oxidizing solution for 30 minutes.
- Wash in running tap water.
- Treat with Schiff's reagent for 15 minutes.
- Transfer to sulphuric acid rinse for 2 minutes.
- Transfer to fresh sulphuric acid rinse for 2 minutes.
- Wash in running tap water for 5 minutes.

- Counter stain with haematoxylin for 2-5 minutes.
- Wash in tap water.
- Dehydrate through graded alcohols to xylene and mount in DPX.

*Result:* Glycogen - red; nuclei - blue.

### *5.2.2.2 Proteins*

**Ninhydrin-Schiff Method (Yasuma and Ichikawa, 1953)**

(Freeze dried, paraffin sections, cryostat unifixed, cryostat prefixed)

*Preparation of solutions:*

1. Ninhydrin 0.5% solution: Dissolve 500 mg ninhydrin in 100 ml absolute alcohol.
2. Schiff's reagent: As prepared earlier.

*Method:*

- Bring sections to 70% alcohol.
- Treat with ninhydrin solution at 37°C overnight.
- Wash in running tap water.
- Place sections in Schiff's reagent for 45 minutes.
- Wash well in running tap water.
- Counter stain, if required, in haematoxylin.
- Wash in tap water.
- Dehydrate through graded alcohols to xylene and mount.

*Result:* Amino groups take pink to red colour.

**Millon Reaction (Miller, 1849; Baker, 1956)**

(Freeze dried, formalin, alcohol, paraffin sections)

*Preparation of solution:*

1. *Mercuric sulphate solution* - Add 10 g mercuric sulphate to a mixture of 10 ml conc. sulphuric acid and 90 ml distilled water and heat until dissolved. Cool to room temperature and add 100 ml distilled water.
2. *Sodium nitrite solution* - Dissolve 250 mg sodium nitrite in 10 ml distilled water.

3. *Staging solution* - Mix 10 ml of mercuric sulphate solution and 1 ml of sodium nitrite solution.

*Method*

- Bring all the sections to water.
- Place sections in a small beaker, add staining solution and boil gently.
- Allow to cool to room temperature.
- Wash sections in distilled water.
- Repeat wash in distilled water.
- Dehydrate through graded alcohols to xylene and mount in DPX.

*Result:* Tyrosine - red, pink or yellow red.

**Millon Reaction (Bensley & Gersh, 1933)**

(Freeze dried, formalin, alcohol, paraffin sections)

*Preparation of solution:*

1. *Miller reagent* - Add 400 ml of cone $HNO_3$ (Sp. Gr. 1.42) to 600 ml distilled water and stand for 48 hours. Dilute 1:9 with distilled water and saturate with crystalline mercuric nitrate. Filter and, to 400 ml of filtrate, add 3 ml of the original diluted nitric acid and 1 g sodium nitrate.

*Method:*

- Remove wax from paraffin or freeze-dried sections with light petroleum.
- Rinse in absolute acetone and dry in air.
- Cover the section with Millon reagent and either leave at room temperature until the maximum colour develops or place in a covered petridish in the incubator at 60°C.
- Rinse in cold or warm 2% nitric acid.
- Dehydrate rapidly in 70% and absolute alcohol.
- Repeat wash in distilled water.
- Clear in xylene and mount in Canada balsam.

*Result:* Proteins containing Tyrosine stain orange to rose red.

## Dinitrofluoro Benzene Method (Danielli 1950; Burstone 1955)

(for Tyrosine, SH and $NH_2$ groups)

*Preparation of solution:*

1. *Dinitrofluorobenzene solution* - 2,4-dinitrofluorobenzene (DNFB) saturated in 90% ethyl alcohol with sodium hydrogen carbonate.
2. *Nitrous acid solution* - Mix 10 ml of 5% sodium nitrite and 40 ml of 2N hydrochloric acid.
3. *'H' acid solution* - 'H'-acid saturated in 0.1 M veronal acetate buffer pH 9.4.

*Method:*

- Bring all the sections to absolute alcohol or acetone, remove and allow to dry in air.
- Place sections in saturated solution of DNFB for 2-16 hours at room temperature.
- Wash in three changes of 90% alcohol and finally in water.
- Treat with 5% sodium hydrosulphite solution for 30 minutes at 45°C.
- Wash sections in distilled water.
- Treat with cold nitrous solution at 4°C for 30 minutes.
- Wash in cold distilled water.
- Treat with cold 'H'-acid solution at 4°C for 15 minutes.
- Wash in tap water.
- Dehydrate through graded alcohols to xylene and mount in DPX/ Canada balsam.

*Result:* The sites of DNFB attached to the tissues appear in reddish purple.

## Acrolein-Schiff Method

*Method:*

- Bring sections to 95% ethanol.
- Incubate for 15-60 minutes in 5% acrolein in 95% ethanol.
- Pass through three changes of ethanol, 5 minutes to each and then to water.

- Immerse in Schiff's staining solution for 10-20 minutes.
- Wash in water, dehydrate, clean and mount in a synthetic resin medium.

*Result:* Magenta staining indicates reactive groups.

**Mercury-Bromophenol Blue Method (Bonhas)**

*Preparation of solution:*

1. Staining solution -There are two alternatives -
   (a) 1% alcoholic bromophenol blue saturated with $HgCl_2$ and
   (b) 1% $HgCl_2$ and 0.05% bromophenol blue in 2% aqueous acetic acid.

*Method:*

- Bring paraffin sections to water.
- Stain in one of the alternative staining solutions for 2 hours at room temperature.
- Rinse sections for 5 minutes in 0.5% acetic acid solution.
- Transfer sections directly to tertiary butyl alcohol.
- Immerse in Schiff's staining solution for 10-20 minutes.
- Clean in xylene and mount in a suitable synthetic resin medium.

*Result:* Proteins are stained in a deep clean blue.

**Sakaguchi Dichloronaphthol Reaction**

*Preparation of solution:*

1. 4% barium hydroxide
2. 1% sodium hypochlorite
3. 1.5% 2,4-dichloro-α-naphthol in ter-butanol

These solutions are to be prepared immediately before use.

*Method:*

- Bring slides to water using two changes of distilled water. Blot slides and place them in an empty staining jar.
- Pour 5 parts of barium hydroxide solution, 1 part of sodium hypochlorite and 1 part of dichloronaphthol into a flask in

succession. Stir after each addition. Pour contents of the flask into the staining jar.

- Allow reagent to act for 10 minutes at 22°C.
- Transfer slides to three 5-second changes of ter-butanol; stir vigorously in each change.
- Transfer to two-changes of xylene (30-60 seconds) containing 5% tri-N-butylamine.
- Drain and mount in Shillaber's oil containing 10% tri-N-butylamine.

*Result:* An orange red colour indicates sites containing arginine.

## Sakaguchi Method (Sakaguchi, 1925; modified by Baker 1947)

(Various paraffin sections, freeze-dried etc.)

*Preparation of solution:*

1. 1% sodium hydroxide
2. 1% naphthol (1 g α-naphthol in 100 ml of 70% alcohol).
3. Milton solution - 1 ml milton in 99 ml distilled water.
4. α-naphthol hypochlorite solution - Mix 2 ml sodium hydroxide, 2 drops of naphthol solution and 4 drops of Milton solution.
5. Pyridine-chloroform solution - Mix 30 ml pyridine in 10 ml chloroform.

These solutions are to be prepared immediately before use.

*Method:*

- Remove wax and cover with a thin film of 1 % celloidin.
- Bring all sections to water.
- Remove from water and flick off till nearly dry.
- Flood the sections with α-naphthol hypochlorite solution for 15 minutes.
- Drain and blot dry.
- Immerse in pyridine-chloroform solution.
- Mount in pyridine-chloroform or dry pyridine and ring cover-slip.

*Result:* Various shades of orange red colour indicate sites containing arginine.

### 5.2.2.3 Lipids

**Sudan Black B Method for Lipids (Lison & Dagnelie, 1935)**

(Free floating formalin-fixed frozen or cryostat)

*Preparation of solution:*

1. Staining solution - 40 ml of distilled water is added to 60 ml of triethyl phosphate followed by addition of 1 g Sudan Black B. The mixture is heated to 100°C for 5 minutes and stirred constantly. The mixture is filtered when hot and once again immediately before use. This is stock solution.

50% ethyl alcohol: take equal parts of ethyl alcohol and distilled water, but is filtered each time it is used.

*Method:*

- Bring sections to 70% alcohol.
- Stain for 30 minutes at room temperature in saturated solution of Sudan Black B.
- Remove excess dye by rinsing quickly in 70% alcohol.
- Wash in running tap water.
- Counter stain in Mayer's carmalum solution for 16 hours or in 1% aqueous neutral red for 1 minute.
- Wash in water and mount in glycerin jelly.

*Result:* Lipids stain black or blue if present in sufficient quantity. Even a brownish black stain may be an indicator of the presence of lipid or lipoprotein. Nuclei - red.

**Sudan Black B Method for Masked Lipids (AcKerman, 1952)**

(Blood smears)

*Method:*

- Fix in formalin vapour for 2-5 minutes.
- Immerse in 25% acetic acid for 2 minutes.

- Wash thoroughly in tap water, then in distilled water and allow films to dry.
- Stain in saturated solution of Sudan Black B in 70% alcohol.
- Differentiate in 70% alcohol.
- Wash in running tap water.
- Blot dry and mount in glycerin jelly.

*Result:* Bound lipids and lipids serving the staining procedure stain black.

**Acid Haematein Method (Baker, 1946)**

(Formal-calcium, post chromatin; frozen sections)

*Preparation of solution:*

1. *Formal-Calcium fixative* - 40% formalin (10 ml), 10% calcium chloride anhydrous (10 ml) and distilled water (80 ml).
2. *Dichromate Calcium solution* - Potassium dichromate (5 g), anhydrous calcium chloride (1 g) and distilled water (100 ml).
3. *Acid Haematein* - Place 0.5g haematoxylin in a flask, add 48 ml distilled water and exactly 1 ml of 1% $NaIO_4$. Heat until the water begins to boil, cool and add 1 ml of glacial acetic acid. Freshly prepare this solution for use.
4. *Borax-ferricyanide Differentiator* - Dissolve 0.25g potassium ferricyanide and 0.25g sodium tetraborate in 100 ml distilled water. This solution should be kept in dark.

*Method:*

- Fix small blocks of tissue in formal calcium for 6-18 hours.
- Transfer to dichromate-calcium solution for 18 hours at 22°C.
- Transfer to dichromate-calcium solution for 18 hours at 60°C.
- Wash well in distilled water.
- Cut frozen sections at 10 μ (embedded in gelatin).
- Mordant in dichromate-calcium for 1 hour at 60°C.
- Wash in distilled water.
- Transfer to acid haematein for 5 hours at 37°C.
- Rinse in distilled water.
- Transfer to borax-ferri-cyanide for 18 hours at 37°C.

- Wash in water.
- Mount in glycerin jelly.

*Result*: Nucleoproteins and phospholipids - dark blue or blue black. Mucin - dark blue. Fibrinogen - pale blue. Cytoplasm - pale yellow.

## Copper Phthatocyanin Method (Kluver & Barrera)

*Method:*

- Bring the sections to absolute alcohol.
- Stain in 0.1% Luxol Fast Blue MBS, Luxol Fast Blue G or Methanol Fast Blue 2G in 95% alcohol for 6-18 hours at 56-60°C.
- Rinse in 70% alcohol and wash in water.
- Differentiate in 0.05% aqueous lithium carbonate for 30-60 minutes.
- Rinse in water.
- Counter stain in 1% aqueous neutral red for 1-30 minutes.
- Rinse in water and dehydrate in 70 and 90% alcohol.
- Clear in xylene and mount in suitable media.

*Result:* Phospholipids except sphingomyelin stain blue.

## Fischler's Method for Fatty Acids

(Formalin; frozen sections)

*Preparation of solution:*

1. Weigert's lithium haematoxylin solution - **Solution A**: 10% haematoxylin in absolute alcohol. **Solution B**: Saturated aqueous lithium carbonate (10 ml) and distilled water (90 ml). Mix equal parts of each solution.
2. Weigert's borax-ferricyanide solution - Dissolve 20g borax and 25g potassium ferricyanide in 1000 ml distilled water.

*Method:*

- Mordant 10 μ frozen sections for 24 hours at 37°C in saturated aqueous cupric acetate.
- Wash the sections in distilled water.
- Stain for 20 minutes in Weigert's lithium haematoxylin solution.

- Differentiate in borax-ferricyanide until the red cells are no longer dark blue.
- Wash in distilled water and mount in glycerine jelly.
- Control sections should be stained, which have previously been extracted for 24 hours with three changes of alcohol / ether (equal parts).

*Result:* Tissue components staining dark blue in the unextracted and colourless in the extracted sections are presumed to contain fatty acids.

## Digitonin Method (Windaus, 1910)

(Formalin-fixed frozen sections free floating)

*Preparation of solution:*

1. *Ethyl alcohol solution (50%)* - Add 100 ml ethyl alcohol in 100 ml distilled water.
2. *Digitonin solution* - Dissolve 500 mg digitonin in 100 ml of 50% ethyl alcohol.

*Method:*

- Incubate sections for 3 hours in digitonin solution at room temperature.
- Rinse the sections in 50% ethyl alcohol.
- Float the sections onto slides.
- Mount in glycerin jelly.

*Result:* Free cholesterol birefringent.

## Copper-Rubeanic Acid Method (Holezinger, 1959)

(Cryostat unfixed, cryostat prefixed, formalin fixed frozen sections)

*Preparation of solution:*

1. Copper acetate (0.005%): Dissolve 5 mg copper acetate in 100 ml distilled water.
2. EDTA (0.1%): Dissolve 50 mg Ethylenediamine tetra acetic acid in 50 ml distilled water.
3. Rubeanic acid (0.1%): Dissolve 50 mg rubeanic acid in 35 ml absolute alcohol by warming slightly and then add 15 ml distilled water.

*Method:*

- Place sections in copper acetate solution for 3 to 5 hours.
- Wash sections in EDTA solution for 10 seconds.
- Wash sections again in EDTA solution for 10 seconds.
- Wash sections in distilled water for 10 min.
- Immerse sections in rubeanic acid solution for 30 min.
- Wash sections in 70% alcohol for 3 minutes.
- Wash sections in running tap water
- Mount sections in glycerin jelly or dehydrate through graded alcohols and mount in DPX.

*Results:* Fatty acids - greenish black.

### *5.2.2.4 Nucleic acids*

### Feulgen Reaction (Feulgen & Rossenbeck, 1924)

*Preparation of solution:*

1. 1 N HCl - Add 8.5 ml conc. HCl in 91.5 ml distilled water.
2. *Schiff's reagent (Barger & Delamater, 1948)* - Dissolve 1g basic fuschin in 400 ml boiling distilled water, cool to 50°C and filter. To the filtrate, add 1 ml of thionyl chloride. Stand in the dark for 12 hours. Clear by shaking for 1 minute with 2g activated charcoal. Filtrate store in the dark at 0-4°C. Use in dark at room temperature.
3. *Bisulphite solution* - Add 5 ml of 10% potassium metabisulphite and 5 ml of 1N HCl in 100 mi distilled water.

*Method:*

- Bring the sections to water.
- Treat with 5N HCl at 20-22°C for 20 minutes to 2 hours.
- Rinse in distilled water.
- Transfer to Schiff's reagent for 30-60 minutes.
- Drain and rinse in three changes of freshly prepared bisulphate solution.
- Rinse in water.

- Counter-stain with 1% aqueous light green for 1 minute and 0.5% alcoholic fast green for 30-60 seconds.
- Dehydrate in alcohol.
- Clear in xylene and mount in Canada balsam or DPX.

*Result:* DNA appears in shades of reddish purple. Cytoplasm - green.

## Methyl Green-Pyronin Method (Kurnick, 1955)

*Preparation of solution:*

1. *Methyl Green solution* - Dissolve 2g methyl green in 100 ml distilled water by stirring well.
2. *Pyronin Y solution* - Dissolve 2g Pyronin Y in 100 ml distilled water.
3. *Staining solution* - Extract methyl green solution with chloroform by shaking in a separating funnel until the chloroform layer is no longer violet colour. Similarly, extract Pyronin solution with chloroform as above until the chloroform layer becomes colourless. For use, mix 12.5 ml of Pyronin Y solution and 7.5 ml of methyl green solution with 30 ml distilled water.

*Method:*

- Bring the paraffin sections to water.
- Stain for 6 minutes in methyl green-pyronin staining solution.
- Blot dry with filter paper.
- Immerse in two changes of n-butyl alcohol for 5 minutes in each.
- Immerse in xylene for 5 minutes.
- Mount in permount.

*Results*: Chromatin - clear green; nuclei - bright red; cytoplasmic RNA - bright red; Eosinophili granules and osteoid - bright red.

## Modified Turchini Method for DNA and RNA (Blakler & Alexander, 1952)

(Formal-mercury, Zenker, Bown, paraffin sections)

*Method:*

- Bring the sections to water as usual.
- Hydrate in N HCl at 60°C for 6-12 minutes.

- Transfer in 80% ethanol for 15 seconds.
- Immerse in the fluorine solution for 4-14 hours (Dissolve 0.5g of the dye in 100 ml of 95% ethanol containing 1 ml sulphuric acid. Filter before use.)
- Transfer in 1% aqueous sodium carbonate solution for 2 minutes.
- Immerse in distilled water for 2 minutes.
- Dehydrate in 50% acetone and in 100% acetone.
- Clear in acetone-xylene and in xylene.
- Mount in suitable synthetic resin.

*Results*: DNA in chromatin appears bluish purple and RNA reddish orange.

### 5.2.2.5 Enzymes

**Alkaline Phosphatase**

**Calcium Cobalt Method (after Gomori, 1948)**

*Method:*

- Remove wax from the slides by light petroleum.
- Pass water via absolute acetone.
- Incubate for ½ - 16 hours in the incubating medium containing 10 ml of 3% sodium glycerophosphate, 10 ml of 2% sodium diethyl barbieturate, 5 ml distilled water, 20 ml of 2% calcium chloride and 1 ml of 5% magnesium sulphate.
- Rinse in running water.
- Treat with 2% cobalt nitrate or acetate for 3-5 minutes.
- Rinse in distilled water.
- Treat with a dilute solution of yellow ammonium sulfide for 1-2 minutes.
- Wash in water, counter-stain in 1% eosin for 5 minutes, if desired.
- Dehydrate in alcohol. clear and mount in Canada Balsam.

*Results*: Various structures are stained black or brownish black in tissues possessing alkaline phosphatase activity.

### *5'-Nucleotidase*

### Calcium Methed (after Pearse & Reis)

*Method:*

- Prepare acetone fixed paraffin sections.
- Incubate one section with glycerophosphate or phenyl phosphate as for alkaline phosphatase at pH 9.2, and another section similarly at pH 7.5, third section in the following medium at pH 7.5 and fourth section in the same medium containing water instead of substrate, all for 3/8 hours at 37°C. [Medium contains 5 volume of barbiturate buffer pH 7.5 (sodium diethyl barbiturate 0.1M 3 volume; 2 volume of 0.1M HCl and one volume of distilled water), 1 volume 12% (w/v) calcium nitrate, 1 volume 2% (w/v) magnesium chloride and 1 volume 0.04M adenylic acid].
- Wash with 2% calcium nitrate (pH 8) and then in distilled water.
- Immerse in 1% silver nitrate and expose to daylight for 1 hour.
- Rinse in distilled water and treat with 5% sodium hyposulphite ($Na_2S_2O_3$) for 10 minutes.
- Wash, dehydrate, clear and mount in Canada Balsam.

*Results*: Various structures in neighbourhood of weak or moderately strong sources of 5'-nucleotidase are stained black. In case of very strong sources of enzyme, localization to a group of cells may be possible. (The glycerophosphate control at pH 7.5 must be subtracted from the total result given by adenylic acid at pH 7.5 to obtain a true estimate of 5'-nucleotidase.)

### Glucose-6-Phosphatase

### Method for Glucoce-6-phosphatase (after Chiquoine)

(cold microtome or cold knife sections; free floating or mounted)

*Reagents:*

1. *Conversion of barium salt of Glucose-6-phosphatase to potassium salt:* Dissolve 250 mg barium salt in 10 ml distilled water containing 2 drops of 2N HCl. Add 120g potassium sulfate. Allow to stand for two hours with frequent stirring. Centrifuge and test the supernatant for barium with a pinch of potassium

sulfate. If no precipitate occurs, dilute to 30 ml with distilled water and adjust pH to 6.7 with 1N NaOH.

2. *Preparation of Incubating mediu:* Dilute one part of the above solution with two parts of 0.006M lead nitrate (0.2%). Filter before use.

*Method:*

- Incubate sections for 5-15 minutes at 32°C.
- Wash in distilled water.
- Develop with dilute yellow ammonium sulfide.

Wash in water, dehydrate, clear in xylene and mount in clarite.

*Results*: Brownish-black deposits indicate the sites of glucose-6-phosphatase activity.

## Method for Glucose-6-phosphatase (after Wachstein & Meisel)

(cold microtome sections; free floating)

*Method:*

- Incubate 10-15 μ sections for 5-15 minutes at 32°C in a substrate mixture containing 20 ml of 125 mg% solution of potassium glucose-6-phosphate, 20 ml of 0.2M Tris buffer (pH 6.7), 3 ml of 2% lead nitrate and 7 ml of distilled water.
- Wash in distilled water.
- Develop in dilute yellow ammonium sulfide.
- Wash in water. Post-fix in 6% neutral formaldehyde. Mount in glycerine and ring the cover slip with nail polish.

*Results:* Brownish-black deposits indicate the sites of glucose-6-phosphatase activity.

## Acid Phosphatase

## Lead Nitrate Method (after Gomon, 1950)

(Cold acetone, paraffin sections, cold formalin frozen sections, cold microtone-post fixed Wolman sections)

*Method:*

- Cut frozen sections, 10-15 μ thick and mount them on the slides.

- Incubate at 37°C for ½ - 16 hours (average 4 hours) in 0.07M sodium-6-glycerophosphate in 0.05M acetate buffer (pH 5.8) containing 0.004 M lead nitrate.
- Wash and counter-stain with 1% aqueous eosin for 5 minutes.
- Wash briefly and immerse in dilute yellow ammonium sulfide for 1-2 minutes.
- Wash well and mount in glycerine jelly.

*Results*: The presence of acid phosphatase activity in the sections is indicated by a black precipitate of lead sulfide.

### Modified Lead Nitrate Method (after Takenchi & Tanoue)

*Method:*

- Incubate at 37°C for ½ - 2 hours in the following medium.

  2 volume of 2% sodium-β-glycerophosphate

  1 volume of 0.1 M acetate buffer (pH 5.0-6.0)

  1 volume of 2% lead acetate

  0.3 volume of 1-5% magnesium chloride
- Rinse in distilled water.
- Develop in ammonical silver nitrate solution for 30 minutes (add 28% ammonia water drop by drop to 5% aqueous silver nitrate until the precipitate just dissolves).
- Rinse in 5% sodium thiosulphate for 5 minutes.
- Wash briefly and immerse in dilute yellow ammonium sulfide for 1-2 minutes.
- Dehydrate, clear and mount in a suitable medium or mount directly in glycerine jelly.

*Results*: A brownish precipitate indicates sites of acid phosphatase activity.

### Phosphamidase

### Lead Nitrate Method (after Gomori, 1948)

(Cold acetone, double-embedded sections)

*Reagents:*

1. *Stock solution* - Dissolve sufficient p-chloroanilidophosphoric acid to make 0.1M solution in an excess of 10% $NH_4OH$. Adjust

pH to about 8.0 by adding dilute acetic acid and make up the volume with distilled water. This solution is stable at 4°C.

2. *Maleate Buffer* - Dissolve 5.8g maleic acid in 500 ml distilled water. Add 62 ml 1N NaOH and make up to 1000 ml. The pH should be in and around 5.6.
3. 0.1M lead nitrate
4. 10% manganese chloride
5. *Incubating medium* - Add 2 ml of 0.1 M stock substrate solution to 50 ml maleate buffer with 1.5 ml of 0.1 M lead nitrate and a few drops of manganese chloride. Incubate at 60°C for 30 minutes and filter into a Coplin jar.

*Method:*

- Incubate mounted frozen sections for 2-4 hours keeping the jar at an angle and slide face downwards.
- Rinse in distilled water; wipe precipitate from around the section and from the backs of the slides.
- Rinse in 0.1M citrate or acetate buffer at pH 4.5 until the diffuse white precipitate covering the section disappears.
- Rinse in running water.
- Treat with dilute yellow ammonium sulfide solution for 1-2 minutes.
- Wash in running water.
- Counter-stain in 1% aqueous eosin for 3-5 minutes.
- Wash well and mount in glycerine jelly.

*Results*: A black precipitate is present to indicate phosphamidase activity.

## Modified Phosphamidase Technique (after Meyer & Weinmann)

(Cold acetone; double embedded sections)

*Reagents:*

1. *Stock solution A* - Dissolve 2.08g p-chloro-anilido-phosphoric acid in 15 ml 1N NaOH and make up to 100 ml with distilled water.

2. *Solution B* - Dissolve 534 mg maleic acid in 5 ml of 1N NaOH and make up the volume to 100 ml with distilled water. Add 175 mg sodium chloride and 94 mg lead nitrate. The mixture becomes turbid and should be heated gently until it is clear. Add 4.5 ml of stock solution A and heat to 44°C. Filter it.

*Method:*

- Mount to (serial) sections, one at either end of the slide.
- Immerse one of the sections in 10% nitric acid for 90 minutes without removing paraffin wax.
- Wash this section and blot dry.
- Remove wax from both the sections with xylene.
- Incubate slides for 1.5 to 4 hours at 42°C in a horizontal position with the sections facing downwards.
- Differentiation - sections incubated for over 3 hours require treatment with 0.1M citric acid until the inactivated section is clear of precipitate.
- Wash in water and treat with dilute yellow ammonium sulfide for 2 minutes.
- Wash in water, dehydrate, clear and mount in a suitable synthetic resin.

*Results:* A brownish precipitate indicates sites of phosphamidase activity.

## Esterase

### α-Naphthyl-acetate Method for Esterase

(Cold formalin frozen sections; cold acetone paraffin sections; cold microtone formalin post fixed sections)

*Method:*

- Cut frozen sections 10-15 μ thick and mount on clean slides. Dry in air to ensure adherence.
- Incubate for 1-15 minutes at room temperature in the medium (Dissolve 10 mg α-naphthyl acetate in 0.25 ml acetone and add 20 ml 0.1M phosphate buffer pH 7.4, alternatively, add 1 ml of 1% α-naphthyl acetate in acetone to 10 ml buffer. Shake thoroughly until most of the initial conditions of cloudiness

disappear. Add 50-100 mg Fast B salt (I.C.I. Ltd.). Shake and filter directly on the sections. The sections should be dry if frozen sections but brought to water and left wet if paraffin sections).

- Wash in running water for 2 minutes.
- Counter-stain in Mayer's haemaleum for 4-6 minutes.
- Wash in running water for at least 30 minutes.
- Mount in glycerine jelly.

*Results:* Esterase black, nuclei dark blue (Lipase-A chE & chE can also hydrolyze α-naphthyl acetate and appear black).

### Naphthol AS Acetate Method for Esterase (Gomori, 1948)

(Cold formalin frozen sections; cold acetone paraffin sections; cold microtone post fixed sections)

*Reagents:*

1. *Preparation of substrate solution* - Dissolve 5g 2-hydroxy-3-naphthoic anilide (Naphthol AS or Brenthol AS, ICI Ltd.) in 10 ml dry pyridine with 20 ml acetic anhydride. Heat under a reflex condenser for one hour. Pour into cold water, filter off the pasty product and dry. Recrystalize from ethanol (containing a little charcoal) to obtain the acetate as a cream-coloured powder (m.p. 160-161°C)

*Method:*

- Cut frozen sections 10 μ thick and mount on clear slides.
- Dry in air to ensure adherence.
- Add 0.1 ml of 1% naphthol buffer at pH 7, shake to form a slightly turbid medium and add 10 mg of a suitable stable diazotate, stir and filter on the dry section.
- Incubate at 17-22°C for 20-30 minutes.
- Wash in running water for one minute.
- Counter-stain with either Mayer's haemaleum for 4-6 minutes or with Carmalum for 1-6 hours.
- Mount in glycerine jelly.

*Results*: A particular azo dye is deposited at sites of esterase activity. Nuclei are blue or red. Gomori (1952) recommended the diazotate of α-aminoazotoluene; this gives a finely particulate orange-red precipitate.

## Lipase

### Tween Method for Lipase (after Gomori, 1948)

(Cold formalin frozen sections; cold acetone paraffin sections)

*Reagents:*

1. *Stock solution* - 5% Tween 60 solution; 0.5M tris (hydroxymethyl) amino methane buffer pH 7.2 - 7.4 (0.2 M veronal buffer or bicarbonate buffer); 10% calcium chloride. Substrate and buffer solutions are preserved with 0.25% chloroform or with a crystal of thymol.
2. *Incubating medium* - 5 ml buffer solution, 2 ml calcium chloride solution, 2 ml Tween solution and 40 ml distilled water are mixed together with a thymol crystal. Keep this working solution in the refrigerator.

*Method:*

- Cut frozen sections and mount on the slides.
- After drying, incubate for 3-12 hours in the incubating medium.
- Wash thoroughly with distilled water.
- Immerse in 1% lead nitrate for 15 minutes.
- Wash in running water for 5 minutes.
- Immerse in dilute yellow ammonium sulfide solution for 1-2 minutes.
- Wash in distilled water and counter-stain with 1% aqueous eosin for 5 minutes.
- Wash well and mount in glycerine jelly.

*Results:* A brownish-black deposit indicates the presence of lipase activity.

### Modified Tween Method for Lipase & Esterase (after Martin)

*(Cold formalin frozen sections; cold acetone paraffin sections)*

*Reagents:*

1. *Stock solution* - Dissolve 2% Tween 60 in 0.2% calcium chloride solution in veronal acetate buffer (pH 7.4). Add a

crystal of thymol. Incubate at 37°C for 2 days and filter through a Seitz filter.

2. *Incubating medium* - 30 ml 0.05M veronal acetate buffer (pH 7.4) solution, 2-3 ml of 2% calcium chloride solution, 10-20 ml glycerol and 2-4 ml of stock solution water are mixed together with a small thymol crystal. Keep this working solution in the refrigerator.

*Method:*

- Cut frozen sections and mount on the slides.
- After drying, incubate for 3-12 hours in the incubating medium.
- Wash thoroughly with distilled water.
- Immerse in 1% lead nitrate for 15 minutes.
- Wash in running water for 5 minutes.
- Immerse in dilute yellow ammonium sulfide solution for 1-2 minutes.
- Wash in distilled water and counter-stain with 1% aqueous eosin for 5 minutes.
- Wash well and mount in glycerine jelly.

*Results:* A brownish-black deposit indicates the presence of lipase activity.

## Aminopeptidase

### Chelating Method for Leucine Aminopeptidase

Nachlos et al. (1957) have chelated copper to the azo dye, formed in by coupling di-azo blue B into naphthylamine and reported that a suitable dye was produced that could be carried away to mounting materials through alcohols and xylene.

*Reagents:*

1. Substrate solution - Dissolve 0.8g 1-leucyl-β-naphthylamine in 100 ml distilled water.
2. Buffer solution 0.1 M acetate (pH 6.5)
3. 0.85% sodium chloride
4. 0.02M potassium cyanide
5. Diazo Blue B powder

6. Incubating solution - Mix 1 ml substrate solution, 10 ml acetate buffer, 8 ml sodium chloride, 1 ml potassium cyanide and 10 mg Diazo Blue B.
7. Copper Chelating solution - 0.1M copper sulphate

*Method:*

- Bring frozen sections to water.
- Incubate at 37°C in incubating solution for 15-45 minutes.
- Rinse in saline for 2 minutes and place in 0.1 M cupric sulphate solution for 2 minutes.
- Rinse in saline, mount in glycerine jelly or after graded alcohols and xylene in Canada Balsam.

*Results:* Sites of leucine aminopeptidase activity are blue to red blue.

## Succinate Dehydrogenase

*Reagents:*

1. *Buffer* - 0.1M phosphate buffer pH 7.4 [8 ml $KH_2PO_4$ or $NaH_2PO_4$ and 42 ml $Na_2HPO_4$ solution].
2. *Substrate solution* - Mix 0.3 ml of 0.5M sodium succinate, 1 ml of 0.1M potassium phosphate buffer, 0.3 ml of 0.004M calcium chloride, 0.15 ml of 0.6M $NaH_2CO_3$, 0.7 ml of Blue tetrazolium (1 mg / ml), 0.05 ml of Methylene Blue (1mg/ml) and 0.5 ml of distilled water.

*Method:*

- Cut frozen sections of unfixed material into 0.1M phosphate buffer pH 7.4.
- Place sections in incubating solution at 37°C with slow stirring for 5-20 minutes. Oxygen flushed into the atmosphere above the beakers.

*Results:* Human kidney, uterus and ovary show blue shinning after only 20 minutes.

# 6

# BIOCHEMICAL ANALYSIS OF SILKWORM BIOMOLECULES

## 6.1 CARBOHYDRATES

### 6.1.1 Extraction of Carbohydrates for Estimation of Sugars

***From Tissues***

*Method 1:*

- Grind a known quantity of material in boiling 80% alcohol (5-10 ml per g material) thoroughly in a mortar or in a blender for 5-10 minutes and cool with cold water.
- Filter the extract through two layers of cheese cloth.
- Repeat the step 1 & 2 with the residue.
- Pool the extracts together and filter through Whatman No. 41. Filter Paper Collect the filtrate and record the volume. Store at 0-4°C. It can be used to estimate reducing and non-reducing sugars, phenolics, tannins, fats & oils, chlorophyll, carotene and xanthophyll content.
- The residue may be used for determination of macromolecules like starch, pectins, hemicellulose, cellulose, lignin, proteins, insoluble ash and minerals.

*Method 2:*

- Homogenize a known quantity of material in 0.3N perchloric acid at 0°C for 1 minute. Keep the homogenate in ice for 10 minutes. Remove the insoluble material by centrifugation at 2000 r.p.m. for 3 minutes.
- Wash the residue twice with ice-cold perchloric acid.
- Combine the supernatants and store at 0°C.

- When necessary, neutralize the extract by addition of solid potassium bicarbonate The neutralized extract is kept in ice for 30 minutes. Centrifuge it for sedimentation of $KHCO_3$ precipitates.
- Store the extract at -15°C.

## From haemolymph

- Place 1ml of silkworm haemolymph in a 50-ml flask.
- Add 9.5 ml of barium hydroxide solution (Dissolve 90g barium hydroxide in distilled water and dilute to 2 litres) and mix with rotation.
- Add 9.5 ml of zinc sulphate solution (Dissolve 100g zinc sulphate in distilled water and dilute to 2 litres).
- Shake vigorously, centrifuge and collect the supernatant for estimation.
- Discard the residue.

## 6.1.2 Qualitative Estimation of Carbohydrates

Individual carbohydrates can be identified based on the following qualitative tests.

***6.1.2.1 Fehling's test*** - To 2 ml of Fehling's solution (Dissolve 7 g copper sulphate in water and make up the volume to 100 ml. Dissolve 24g potassium hydroxide and 34.6g sodium potassium tartarate in water in a separate flask and make the volume to 100 ml. Just prior to use, mix both the solutions), add few drops of test solution and boil. If there is rusty brown colour or red precipitate, the test solution contains reducing sugars.

***6.1.2.2 Benedict's test*** - To 2 ml of Benedict's solution (Dissolve 17.3g sodium citrate and 10g sodium carbonate in about 75 ml of water and filter if necessary. Dissolve 1.73g copper sulphate in about 20 ml water in a separate flask and this is slowly added to the alkaline citrate solution with stirring. Make up the volume to 100 ml.), add few drops of test solution and boil. If there is rusty brown colour or red precipitate, the test solution contains reducing sugars.

***6.1.2.3 Barfoed's test*** - Prepare the reagent by dissolving 13.3g copper acetate in 200 ml water and adding 1.8 ml of glacial acetic acid. To 2 ml of this reagent, add 1ml of test solution and boil exactly for 1 minute. Note the changes in colour. Note the changes in colour.

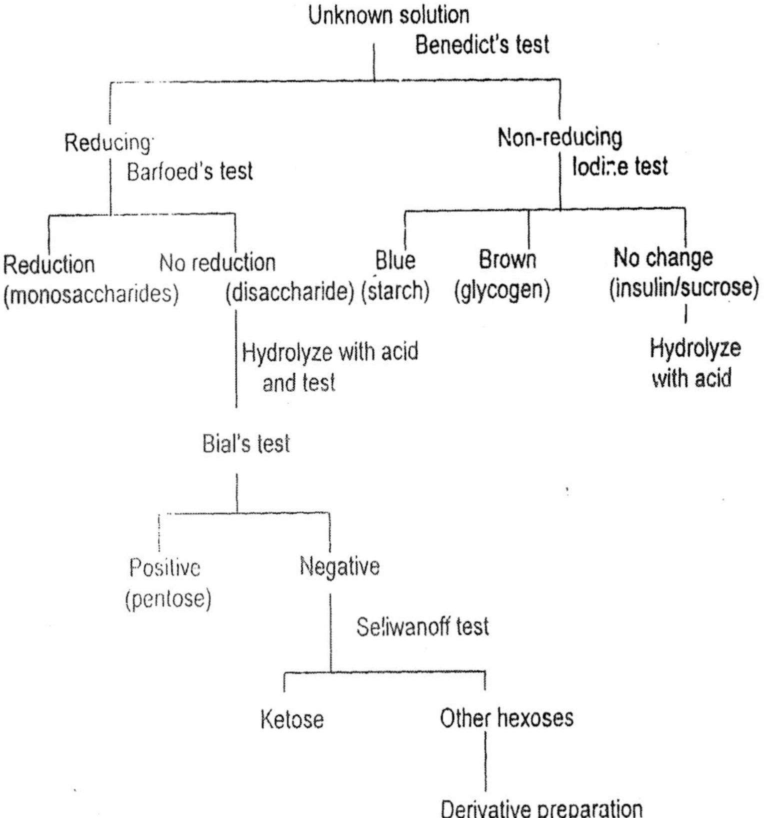


*6.1.2.4 Molisch's test* - Take 2 ml of test solution in a test tube and add 2 drops of an ethanolic solution (5%) of α-naphthol. Carefully run down the sides of the test tube, about 1 ml of concentrated sulphuric acid without mixing. A coloured ring at the junction of two liquids is positive test for carbohydrates.

*6.1.2.5 Anthrone test* - Dissolve 200 mg of the anthrone in 100 ml of concentrated sulphuric acid. Take 2 ml of anthrone reagent solution in a test tube, add 2 drops of the test solution and mix well. If there is no immediate colour change, boil in a water bath for 10 minutes.

*6.1.2.6 Bial's test* - Dissolve 150 mg orcinol in 50ml of concentrated hydrochloric acid. Mix 5 ml of this reagent with 2 ml of the test solution in a test tube and heat in a water bath. Note the time at which any colour change is observed; if it happens before 10 minutes, it is a pentose.

***6.1.2.7 Iodine test*** - Prepare 0.005 N iodine solution in 3% potassium iodide. Take a few drops of the test solution in a test tube add one or two drops of 0.1 N HCl followed by 2 drops of the iodine solutions. Iodine forms coloured complexes with polysaccharides.

***6.1.2.8 Seliwanoff test*** - Prepare the reagent dissolving 50 mg resorcinol in 100 ml HCl (3N or 1:2 diluted). To 5 ml of the reagent, add a few drops of the test solution and heat the mixture to boiling. This test is specific for ketoses.

## 6.1.3 Quantitative Estimation of Carbohydrates

### *6.1.3.1 Reducing Sugars*

*Nelson-Somogyi's Method*

*Principle*: The reducing sugars when heated with alkaline copper tartarate reduce the copper from the cupric to cuprous state and thus cuprous oxide is formed. When the cuprous oxide is treated with arsenomolybdic acid, the reduction of molybdic acid to molybdenum blue takes place. The blue colour developed is compared with a set of standards in a colorimeter at 620 nm.

| | | | | | | |
|---|---|---|---|---|---|---|
| CHO | | | | COONa | | COONa |
| H- C-OH | COONa | | | H-C-OH | | H-C-OH |
| HO-C-H | + H-C-O-Cu | → | $Cu_2O$ + | H-C-OH | + | HO-C-H |
| H-C-OH | H-C-O (/) | | | COOK | | H-C-OH |
| H-C-OH | COOK | | | | | H-C-OH |
| $CH_2OH$ | | | | | | $CH_2OH$ |
| D-Glucose | Sodium-Potassium Tartrate-copper Complex | | | sodium potassium tartarate | | Sodium D-Gluconate |

*Reagents:*

1. *Alkaline copper tartrate*: (a) Dissolve 2.5g anhydrous sodium carbonate, 2g sodium bicarbonate, 2.5g potassium sodium tartrate and 20g anhydrous sodium sulphate in 80 ml water and make up to 100 ml. (b) Dissolve 15g copper sulphate in a small volume of distilled water. Add one drop of sulphuric acid and make up to 100 ml. (c) Mix 4 ml of (b) and 96 ml of (a) before use.
2. *Arsenomolybdate reagent*: Dissolve 2.5g ammonium molybdate in 45 ml of water Add 2.5 ml sulphuric acid and mix well, then add 0.3g di-sodium hydrogen arsenate dissolved in 25 ml water. Mix well and incubate at 37°C temperature for 24 to 48 hours.
3. *Standard stock glucose solution*: 100 mg in 100 ml distilled water. Working standard is prepared by diluting 10 ml stock solution to 100 ml with distilled water (l00 ug per ml).

*Procedure:* Pipette out aliquots of 0.1 or 0.2 ml of alcohol-free extract to separate test tubes and estimate reducing sugars following the procedure given at 4.3.1.1 under Nelson Somoygi's method.

## Park and Johnson Method

*Reagents:*

1. Reagent A - Dissolve 0.5g potassium ferricyanide in 1000 ml distilled water. Store in brown bottle.
2. Reagent B - Dissolve 5.3g sodium carbonate and 0.65g potassium cyanide in 1000 ml distilled water.
3. Reagent C - Dissolve 1.5g ferric ammonium sulphate and 1g sodium dodecyl sulfate in 1000 ml 0.05N sulphuric acid (1.4 ml concentrated sulphuric acid per litre).
4. Working glucose solution - as prepared in Nelson-Somoygi's method.

*Procedure:*

- Pipette out aliquots of 0.1 or 0.2 ml of alcohol-free extract to separate test tubes.
- Pipette out 0, 0.1, 0.2, 0.3, 0.4, 0.5, 0.6, 0.7, 0.8, 0.9 and 1.0 ml of working standard glucose solution into a series of test tubes.

- Add appropriate amount of distilled water to make up the volume in each tube to 2 ml.
- Add 1 ml of solution B and then 1ml of solution A to each tube. Mix the contents of the tubes by vortexing. Cover each tube with a marble.
- Place the tubes in boiling water bath for 15 minutes.
- Cool to room temperature and add 5 ml of solution C to all the tubes. Mix thoroughly and left at room temperature for 10 minutes.
- Read the absorbance of blue colour at 620 nm after 10 minutes.
- From the graph, calculate the amount of reducing sugars present in the sample.

### *DNS Method*

*Principle*: Several reagents have been employed, which assay sugars by their reducing properties. One such compound is 3,5-dinitrosalicylic acid (DNS), which is reduced to 3-amino-5-nitrosalicylic acid.

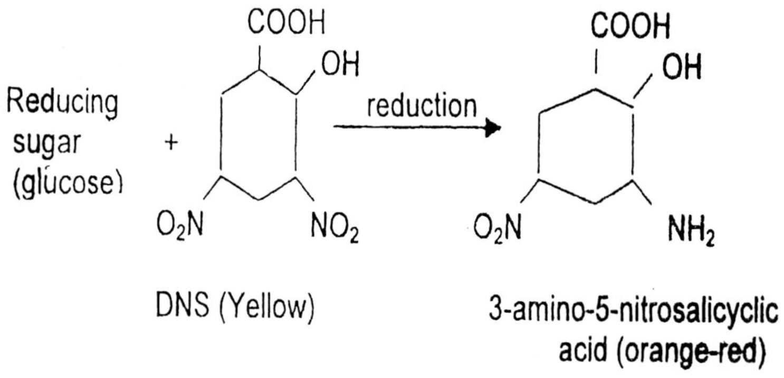

Reducing sugar (glucose) + DNS (Yellow) $\xrightarrow{\text{reduction}}$ 3-amino-5-nitrosalicyclic acid (orange-red)

COOH, OH, $O_2N$, $NO_2$ — DNS (Yellow)

COOH, OH, $O_2N$, $NH_2$ — 3-amino-5-nitrosalicyclic acid (orange-red)

*Reagents:*

1. Dinitrosalicyclic acid (DNS) reagent: Dissolve 1g dinitrosalicyclic acid, 200 mg crystalline phenol and 50 mg sodium sulphite in 100 ml of 1% sodium hydroxide solution by stirring. Store the reagent in a stoppered bottle at 4°C.
2. 40% Rochelle salt (sodium-potassium tartrate) solution.
3. Standard sugar solution (as prepared under Nelson-Somogyi's method).

*Procedure:* Estimate reducing sugars following the steps given under DNS method at 4.3.1.1.

### 6.1.3.2 Total Carbohydrates

#### Anthrone Method

*Principle:* Carbohydrates are dehydrated by cone sulphuric acid to form furfural. Furfural condenses with anthrone (10-keto- 9,10-dihydro-anthracene) to form a blue-green coloured complex, which is measured colorimetrically at 620 nm. This method is suitable for estimation of hexoses, aldopentoses and hexuronic acids either free or as polysaccharides.

*Reagents:*

1. 5N HCl
2. Anthrone reagent: Dissolve 200 mg anthrone in 100 ml of ice cold 95% sulphuric acid just before use.
3. Standard glucose solution: Dissolve 100 mg glucose in 100 ml distilled water (stock solution). Dilute 10 ml of stock solution to 100 ml with distilled water before use as working solution

*Procedure:*

- Weigh 100 mg of the sample into a boiling tube.
- Hydrolyze it with 5 ml of 2.5N HCl in a boiling water bath for 3 hours and then cool to room temperature.
- Neutralize it with solid sodium carbonate until the effervescence ceases.
- Make up the volume to 100 ml and centrifuge.
- Collect the supernatant and take 0.5 ml and 1 ml aliquots for analysis. Estimate the amount of total carbohydrates following the steps given under Anthrone method at 4.3.1.1.

### 6.1.3.3 Water-soluble Carbohydrates

#### Phenol-sulphuric acid Method (Dubois et al. 1956)

*Principle:* Carbohydrates like simple sugars, oligosaccharides, polysaccharides and their derivatives give a yellow-orange colour when treated with phenol and concentrated sulphuric acid. In hot acidic medium, glucose is dehydrated to form hydroxymethyl furfural.

This produces an orange-yellow colour compound with phenol and is measured colorimetrically at 490 nm.

*Reagents:*

1. 5% phenol: Dissolve 50g redistilled (reagent grade) phenol in water and dilute to one litre.
2. 96% sulphuric acid (reagent grade).
3. Standard glucose solution: Dissolve 100 mg glucose in 100 ml distilled water (stock solution). Dilute 10 ml of stock solution to 100 ml with distilled water before use as working solution.

*Procedure:*

- Follow the steps 1 to 4 as given in Anthrone method at 6.1.3.2.
- Pipette out 0.1 ml and 0.2 ml aliquots in two separate test tubes and estimate amount of water-soluble carbohydrates following the steps given under Phenol-Sulfuric Acid Method at 4.3.1.3.

## Benedict's Method

*Reagents:*

1. *Benedict's reagent* - Dissolve 200g sodium citrate, 75g anhydrous sodium carbonate and 125g potassium thiocyanate in about 600 ml water with gentle heating. Filter, cool and add 18g copper sulphate dissolved in about 100 ml water Pour this in slowly stirring continuously. Add 5 ml of 5% potassium ferrocyanide solution and make up to a litre with distilled water. If the solution is not clear, filter it.
2. Anhydrous sodium carbonate

*Procedure:*

- Pipette out 25 ml of the reagent in a conical flask and add 3g of anhydrous sodium carbonate.
- Heat the mixture to boiling. Meanwhile, take the standard sugar solution (0.5g/100 ml) in a burette and slowly run this solution into the boiling reagent. A bulky white precipitate is formed first, which is cuprous thiocyanate. At this stage, add the sugar solution slowly till the last traces of colour has disappeared. Note the volume of sugar solution required.

In this method, there is no standard curve and 25ml of Benedict's solution is equivalent to 50 mg of glucose, 53 mg of fructose, 68.8 mg of lactose and 74 mg of maltose.

### 6.1.3.4 Non-reducing Sugars

Principle: Non-reducing sugars present in the plant extract are first hydrolyzed with either sulphuric acid or formic acid to reducing sugars and then total reducing sugars are estimated by Nelson-Somogyi's or DNS method.

*Procedure:*

- Follow the steps in 1 to 3 as in Nelson-Somogyi's method to extract the reducing sugars from the sample, as grown at 4.3.1.1.
- Pipette out 1.0 ml of alcohol-free extract into test tubes and add 1 ml of 1N sulphuric acid.
- Hydrolyze the mixture by heating at 49°C for 30 minutes.
- Cool the tubes under running tap water and add 1 or 2 drops of methyl red indicator
- Neutralize the contents by adding 1N sodium hydroxide dropwise from a pipette.
- Estimate the reducing sugars by either Nelson-Somogyi's or DNS method.

*Calculation:*

$$\text{Non-Reducing sugars} = \frac{\text{Sugar Value from graph } (\mu g)}{\text{Aliquot sample used (1 ml)}} \times \frac{\text{(Total Volume extract (10 ml)}}{\text{Weight of sample (100 mg)}} \times 1000$$

### 6.1.3.5 Starch

#### Anthrone method

*Principle:* The starch is hydrolyzed into simple sugars by dilute acid in hot acidic medium and these simple sugar molecules are dehydrated to hydroxyl-methyl furfural, which forms a green coloured product with anthrone. The intensity of colour is measured colorimetrically.

*Reagents:*

1. 52% Perchloric acid
2. 80% ethanol
3. Anthrone reagent: Dissolve 200 mg anthrone in 100 ml of ice cold 95% sulphuric acid just before use.
4. Standard glucose solution: Dissolve 100 mg glucose in 100 ml distilled water (stock solution). Dilute 10 ml of stock solution to 100 ml with distilled water before use as working solution (100 ug/ml).

*Procedure*

- Homogenize 0.1 to 0.5g of the sample in hot 80% ethanol, centrifuge and the residue is repeatedly washed with hot 80% ethanol till the washings do not give colour with anthrone reagent. Dry the residue over a water bath.
- To the residue, add 5 ml water and 6.5 ml perchloric acid. Extract at 0°C for 20 minutes, centrifuge and keep the supernatant.
- Repeat the extraction with fresh perchloric acid. Centrifuge and pool the supernatants. Make up the volume to 100 ml.
- Pipette out 0.1 or 0.2 ml of the supernatant in separate test tube.
- Prepare standard test tubes taking 0.2, 0.4, 0.6, 0.8 and 1.0 ml of working standard and make up the volume to 1.0 ml in each tube with distilled water.
- Add 4 ml of anthrone reagent to each tube and heat for 8 minutes in a boiling water bath.
- Cool rapidly and read the intensity of green or dark green colour at 620/630 nm.

*Calculation:*

Find out the glucose content in the sample using the standard graph and multiply the value by a factor 0.9 to get the starch content.

### 6.1.3.5 Glycogen

*Principle:* Glycogen is released from the tissue by heating with strong alkali, precipitated by ethanol and separated by centrifugation. It is then hydrolyzed by dilute HCl, neutralized and the reducing sugars are estimated by any method described above.

*Reagents:*

1. 30% KOH
2. Saturated $Na_2SO_4$
3. 95% (v/v) ethanol
4. 1.2M HCl
5. 0.1% phenol red solution
6. 0.5M NaOH
7. Reagent for estimation of reducing sugars.

*Procedure:*

*Precipitation of Glycogen*

*Method 1:*

- Weigh about 1.5g sample into a calibrated centrifuge tube containing 2 ml KOH.
- Heat in a boiling water bath for 20 minutes with occasional shaking.
- Cool in ice, add 0.25 ml of saturated $Na_2SO_4$ and mix thoroughly.
- Precipitate glycogen by adding 5 ml ethanol, keep it on ice for 5 minutes and then centrifuge.
- Dissolve the precipitated glycogen in about 5 ml of distilled water with gentle warming.
- Dilute with distilled water to 10 ml mark and mix thoroughly.

*Method 2:*

- Add 0.25 ml of saturated $Na_2SO_4$ solution to 0.5 ml of acid extract obtained by the method given at 6.1.1 and then ethanol to 70%.
- Keep the mixture overnight at -15°C.
- Sediment the precipitate after centrifugation and then re-disperse in 4 ml ethanol.
- Centrifuge again and wash the precipitate.
- Dry the precipitate at 105°C and then disperse in distilled water.

*Estimation of Glycogen*

- Pipette out 1 ml aliquot sample into a test tube and add 1 ml of HCl.

- Heat in a boiling water bath for 2 hours and at the end add 1 drop of phenol red indicator. Neutralize with NaOH until the colour changes from pink to yellow through orange.
- Dilute to 5 ml with distilled water and determine the glucose content by either Nelson-Somogyi's method or DNS method.
- Calculate the amount of glycogen as g/100 g tissue

### *6.1.3.6 Protein-bound Hexoses*

**Lustig & Langer Method (1931) used by Weimer & Mashin (1952)**

*Reagents:*

1. 95% ethanol
2. Orcinol-sulphuric acid reagent - 7.5 volume of reagent A are mixed freshly with 1 volume of reagent B.
3. Reagent A - 60 ml of conc. Sulphuric acid are mixed with 40 ml of distilled water.
4. Reagent B - 1.6g orcinol are dissolved in 100 ml of distilled water.
5. Standard solution (0.2mg / ml) - 0.1 mg per ml of galactose and mannose.

*Procedure:* To 0.1 ml haemolymph, add 5 ml of 95% ethanol and mix. Centrifuge for 15 minutes and decant. Suspend the precipitate in 5 ml of 95% ethanol, centrifuge and decant. Dissolve the precipitate in 1 ml of 0.1N NaOH. Prepare a blank (1 ml of distilled water) and standard solution (1 ml of galactose-mannose standard solution). Add 8.5 ml of orcinol-sulphuric acid reagent. Cap the tubes with marble to minimize evaporation and place the tubes in a water bath at 80°C for exactly 15 minutes. Cool the tubes in tap water and take the reading in a photometer at 540 nm.

### *6.1.3.7 Fucose*

*Reagents:*

1. 95% ethanol.
2. $H_2SO_4$-$H_2O$ Mix - Mix 6 volume of conc. Sulphuric acid with 1 volume of distilled water.

3. Cysteine reagent- Dissolve 3g cysteine HCl in 100 ml of distilled water.
4. Standard solution - 20 µg fucose or rhamnose.

*Procedure:* To 0.1 ml haemolymph, add 5 ml of 95% ethanol and mix. Centrifuge for 15 minutes and decant. Suspend the precipitate in 5 ml of 95% ethanol, centrifuge and decant. Dissolve the precipitate in 1 ml of 0.1 N NaOH. Prepare a blank (1 ml of distilled water) and standard solution (1 ml of fucose or rhamnose solution). Add 4.5 ml of ice-cold sulphuric acid-water mix. Mix well while maintaining the solution cold in an ice bath. Heat for exactly 3 minutes in a boiling water bath and cool. After 60 to 90 minutes at room temperature, measure optical density in Beckman spectrophotometer at 396 & 430 nm with distilled water set at zero. Calculate the fucose content with the following formula

Fucose content (mg/100 ml) =

$$\frac{(O.D_{396} - O.D_{430}) - (O.D_{b396} - O.D_{b430})}{(O.D_{s396}\ O.D_{s430})}$$

where,

$O.D._{396}$ → Optical density of haemolymph extract at 396 nm

$O.D._{430}$ → Optical density of haemolymph extract at 430 nm

$O.D._{b396}$ → Optical density of blank at 396 nm

$O.D._{b430}$ → Optical density of blank at 430 nm

$O.D._{s396}$ → Optical density of standard soln. at 396 nm

$O.D._{s430}$ → Optical density of standard soln. at 430 nm

### 6.1.3.8 Hexosamine

*Reagents:*

1. 95% ethanol, 3N HCl, 3N NaOH
2. Acetyl-acetone reagent - Mix 1 ml of acetyl acetone with freshly prepared 50 ml of 0.5N $Na_2CO_3$ solution.
3. Erhrlich's reagent - Dissolve 0.8g p-dimethylamino-benzaldehyde in 30 ml of methanol and 30 ml of conc. HCl.
4. Glucosamine standard - 0.05 mg / ml glucosamine in distilled water.

*Procedure:* To 01 ml haemolymph, add 5 ml of 95% ethanol and mix. Centrifuge for 15 minutes, decant, suspend the precipitate in 5 ml of 95% ethanol, centrifuge and decant. To the precipitate, add 2 ml of 3N HCl and hydrolyze in a boiling water bath for 4 hours with air condenser. Neutralize the hydrolysate with 3N NaOH and dilute to 10 ml. To 1ml of aliquot (blank and standard also), add 1 ml of acetyl acetone reagent and mix. Cap the tubes and place them in a boiling water bath for 15 minutes. Cool and add 5 ml of ethanol and mix. Add 1 ml of Erhrlich's reagent, mix and dilute to 10 ml with enthanol. Take readings after 10 minutes at 530 nm.

### 6.1.4 Separation of Sugars

Individual sugars can be separated qualitative and quantitatively following Thin Layer Chromatography, Paper Chromatography and Gas Liquid Chromatography.

#### *6.1.4.1 Separation by Paper Chromatography*

*Principle:* In paper chromatography, the separation of sugars depends upon partition between two immiscible phases, the mobile phase and stationary phase. In this technique, the compounds are spotted on one end of a filter paper and the paper is kept dipped in a mixture comprised of hydrophobic and hydrophilic solvents. The solvent mixture starts wetting the paper by capillary action and when the solvent front reaches the spot, the compounds get partitioned. As the solvent front moves further, the counter current distribution of compounds in immiscible solvents gets achieved fully. The distance travelled by different compounds depends upon the differential solubility in two immiscible solvents and separation is measured in terms of a unit called $R_f$ (relative to front).

$$R_f = \frac{\text{Distance moved by the substance}}{\text{Distance moved by the solvent}}$$

Under given conditions of temperature, pH etc., the $R_f$ of a compound in a particular solvent system is constant and it is used to identify unknown compounds. If two compounds A and B give the same $R_f$ values in a solvent system, they are probably the same compound but they can separate in a different solvent system. There are different methods of paper chromatography -

- Ascending method - where the solvent is allowed to rise by capillary action.
- Descending method - where the solvent is fed to the paper from top.
- Horizontal or circular method - where the plane of the paper is kept horizontal.

Two-dimensional paper chromatography is an improved method in which the mixture of compounds is subjected to chromatography in two solvent systems in the same sheet of paper.

*Materials:*

1. *Paper*-Whatman No. 1 Filter paper
2. *Solvent systems* - (a) water-saturated phenol + 1% ammonia; (b) n-butanol-acetic acid-water (4:1:5; v/v); (c) Isopropanol-pyridine-water-acetic acid (8:8:4:1; v/v) or (a) n-butanol-acetic acid-water (5:1:4; upper layer); (b) ethylacetate-pyridine-water (2:1:2).
3. *Spray reagents* - (a) Ammoniacal silver nitrate [Add equal volumes of ammonium hydroxide to a saturated solution of silver nitrate and dilute with methanol to give a final concentration of 0.3M. After spraying the developed chromatograms, place them in an oven for 5-10 minutes, when the reducing sugars appear as brown to black spots.], (b) Alkaline permanganate [Prepare an aqueous solution of potassium permanganate (1%) containing 2% sodium carbonate. After spraying, the spots appear as yellow coloured in purple background.], (c) Aniline diphenylamine reagent [Mix 5 volumes of 1% aniline and 5 volumes of 1% diphenylamine in acetone with 1 volume of 85% phosphoric acid. After spraying the dried chromatograms with this solution, the spots appear by heating the paper at 100°C for a few minutes.], (d) Resorcinol reagent [Mix 1% ethanolic solution of resorcinol and 0.2N HCl (1:1; v/v). Spray the dried chromatograms and observe the spots by heating at 90°C] or Dissolve 930mg aniline and 0.6g phthalic acid in 100 ml of water saturated n-butanol. Store the reagent in brown bottles. Heat at 105°C for 5 minutes. Aldohexoses - brown, aldopentoses - red and uronic acid - pink.

**Table 6.1: Detection of sugars Sugar**

| *Sugar* | *Spray reagent* *a* | *b* | *c* | *d* |
|---|---|---|---|---|
| Aldohexoses | + | + | + | Pink |
| Ketohexoses | + | + | + | Red |
| Aldopentoses | + | + | + | Blue, green |
| Ketopentoses | + | + | + | – |
| Deoxy sugars | – | + | + | – |
| Glycosides | + | – | – | – |
| Amino sugars | + | + | + | – |

**Table 6.2: $R_f$ values of sugars**

| | *Solvent a* | *Solvent b* | *Solvent c* |
|---|---|---|---|
| Glucose | 0.39 | 0.18 | 0.64 |
| Galactose | 0.44 | 0.16 | 0.62 |
| Fructose | 0.51 | 0.25 | 0.68 |
| Ribose | 0.59 | 0.31 | 0.76 |
| Deoxyribose | 0.73 | - | - |
| Lactose | 0.38 | 0.09 | 0.46 |
| Maltose | 0.36 | 0.11 | 0.50 |
| Sucrose | 0.39 | 0.14 | 0.62 |

*Procedure:*

Cut a square sheet of filter paper (ABCD) and draw a thin line approximately 1 cm from two ends of the length (AD and AB) with the help of a pencil. Spot 5 μl of sample at the junction of two lines near one corner A, dry with dryer and repeat spotting once more. Place the solvent in the tray in the chromatography chamber. Develop the chromatogram in the direction AB in one solvent. The paper may be hung as such or can be rolled into a cylinder and hung if a smaller chamber is used. When the solvent reaches the top or the run is complete, take out the chromatogram and dry it in a stream of warm air. The separation of compounds is now along the axis AB. The chromatogram is rotated 90° and developed in a second solvent system in the direction AD. The separation of compounds will be along the axis AD. The chromatogram is taken out, dried and sprayed with appropriate spray reagents. Heat the chromatogram, measure the $R_f$ values of sugars from unknown mixture and identify using standards. The separation of sugars would be analogous to plotting on a graph sheet x-coordinate, the $R_f$ value in solvent 2 and y-coordinate, the $R_f$ value in solvent 1. The $R_f$ value of different sugars in solvent a, b and c is given in Table 6.2 and the response of the sugars on spraying of reagent a, b, c and d is given in Table 6.1

### 6.1.4.2 Separation by Thin Layer Chromatography

Separation of compounds on thin layer is based on the principle of differential adsorption where a substance gets attracted by electrostatic forces to the surface of a unit particle. If a substance is in solution and the particle insoluble in the solvent, then part of the substance is adsorbed and part remains in solution. The ratio between the amount adsorbed and amount in solution is constant, called as adsorption coefficient.

The separation of compounds is similar to paper chromatography (counter-current distribution) in many ways but it has the added advantages that a variety of supporting media (adsorbents) can be used. The molecules with least adsorptivity go along with the solvent and eluted first, followed by other components. In thin layer chromatography, the adsorbent is spread over a supporting media to form a thin layer of adsorbent. Once the sheet is coated and allowed to dry, the spotting of sample, developing in solvent and detection of spots by spraying etc. are done in the same way as in case of paper chromatography.

The adsorbents normally used are alumina, silicic acid, powdered sugar, activated charcoal, diatomaceous earth or magnesium silicate. Generally, adsorption is maximal in nonpolar solvents and decreases as the polarity is increased. Commonly used solvents can be arranged in the increasing order of polarity as follows.

Petroleum ether or hexane < benzene < carbon tetrachloride < ethyl ether < chloroform < acetone or alcohol < water < salt solutions < acid or base solutions

TLC combines the better resolutions of column chromatography with the easier manipulation of paper chromatography.

***Method 1:***

*Reagents:*

1. Slurry of silica gel G - Mix 25g of silica gel G powder with 50 ml of water or with 0.02M sodium borate buffer, pH 8 or 0.02M sodium acetate or boric acid.
2. Solvent systems
3. Spray reagents
4. Standard - 0.01M sugar in water

*Procedure:* Uniformly spread 250 μ thick, slurry of silica gel G (250 μ thick; prepared in 0.02M sodium acetate buffer) on glass plates (5 × 20 cm and 20 × 20cm and 0.4 cm thickness) using an applicator (spreader). Air dry the plates for 1-2 hours and allow the binder to set. Place the plates in a rack and activate them at 105°C for 30 minutes. If not used immediately, store the dried plates in a desiccator and activate just before use. Spot 5 μl of sample, dry with dryer and repeat spotting once more. Develop the plates in a solvent containing ethyl acetate-isopropanol-water-pyridine (26:14:7:2; v/v). After development i.e. when the solvent reaches the top, dry the plates in a stream of warm air and spray with appropriate developing reagents (Tab. 6.3). Heat the plate, measure the $R_f$ values of sugars from unknown mixture and identify using standards.

***Method 2:***

*Reagents:*

1. n-Butanol:pyridine:water (6:4:3, v/v)
2. Ethyl acetate:pyridine:acetic acid:water (5:5:1:3, v/v)
3. To saturate the tank under solvent (2) use pyridine:ethyl acetate:water (11:40:6, v/v).

*Procedure:* Prepare TLC plates coated with 250 m thick cellulose slurry. Spot 1μl of the sample at the origin and dry it in air. Place the plate in a thin layer tank containing the solvent 1 or solvent 2 and develop until the solvent front reaches the top of the plate. Dry the plates at room temperature. Spray the plates with developing reagents as given in Tab. 6.3. Calculate the $R_f$ or $R_g$ value and identify the sugars using standards.

### 6.1.4.3 Separation and Quantitative Estimation of Sugars and Amino Acids by Ion Exchange Chromatography

*Principle:* The charged resins hold compounds by electrostatic attraction. By suitable alteration of ionic environment, the compounds are released. Two types of resins, anionic and canonic are used.

*Procedure:*

- Remove pigments from the alcohol extract by portioning it twice with equal volumes of petroleum ether.
- Discard the petroleum ether fraction by using separating funnel.
- Evaporate the pigment-free alcohol extract to dryness on a boiling water bath.
- Dissolve the residue in 50 ml of water and centrifuge at 5000 rpm for 10 minutes. Preserve the supernatant.
- Wash the sediment with 10 ml of water, combine the supernatants and evaporate to dryness.
- Remove the phenolic substances by treating with purified insoluble polyvenylpyrollidone (PVP) and filter through 0.8 μ millipore filter under vacuum.
- Pass the above solution through a Dowex 50W × 8, $H^+$ form (20-50 mesh) cation exchange resin column. Adjust the flow rate to about 20 drops per minute.
- Fix the cation exchange resin column on the top of another column with anion exchange resin (Dowex 1 × 8, Cl- form, 20-50 mesh) so that every drop flowing from the former would fall into the latter.
- Collect the effluent through the anion column and analyse by chromatography and estimate colorimetrically.
- Wash the cation exchange column with water and then elute the amino acids with 50 ml of 2M ammonium hydroxide.
- Evaporate the elute till no ammonia is detected.
- Dissolve the residue in 5 ml of water and adjust the pH to 2.5 with 4M formic acid.
- Evaporate to dryness and dissolve the residue in 5 ml water. This solution contains amino acids.

- Separate the arnino acids by chromatography or estimate directly colorimetrically.

## 6.2 AMINO ACIDS

### 6.2.1 Qualitative Estimation of Amino Acids

*Xanthoproteic Reaction* - When a drop of concentrated nitric acid falls on skin, it turns yellow. This is because of the phenolic group in tyrosine unit of the protein reacts with nitric acid. Add a few drops of concentrated nitric acid to a test solution of amino acid. If yellow color is obtained, it shows the presence of amino acids. Subsequently, if yellow color turns orange on adding a few drops of sodium hydroxide, it confirms the presence of amino acids.

*Millon's Reaction* - Prepare a 15% solution of mercuric sulphate in 15% sulphuric acid. Add a few drops of this reagent to about 1 ml of the test solution and heat for 10 minutes. After cooling, add few drops of 1% sodium nitrite solution. Red color indicates the presence of tyrosine.

*Hopkin-Cole Test* - To 2 ml of glyoxylic acid (obtained by exposing glacial acetic acid to sunlight for a few minutes), add 2 ml of test solution, mix well and then carefully add through the sides of the tube 2 ml of concentrated sulphuric acid. A violet ring at the junction indicates the presence of tryphtophan.

*Pauly's Test-* Mix 1 ml of sulphanilic acid (1% solution in 10% HCl) with 2 ml of test solution and cool in ice bath. Add 1 ml of 5% sodium nitrite solution. After 5 minutes, add 2 ml of 1% sodium carbonate solution and note any change in colour. Tyrosine, histidine and tryptophan give strongly positive test.

*Ehrlich's Test* - The reagent consists of 10% p-dimethyl amino benzaldehyde in 10% HCl. To 1 ml of the reagent, add 1 ml of unknown solution. A red colour is developed, which indicates presence of tryptophan amino acid since this test is specific for indole and tryptophan contain indole nucleus.

*Sakaguchi Reaction* - Mix 3 ml of unkown solution with 1 ml of 40% sodium hydroxide solution. Add 2 drops of $\alpha$-naphthol (1% alcoholic solution) to the mixture followed by addition of a few drops of bromine water. Note the red colour, which is given by guanidine group of arginine amino acid.

## 6.2.2 Quantitative Estimation of Amino Acids

### *6.2.2.1 Estimation of Free Amino Acids*

*Extraction* - The tissue is first grinded in a mortar and pestle and then extracted 5-6 times with 70-80% ethanol or 0.01M Phosphate buffer, pH 7.0 (five times weight of sample). If the haemolymph is taken instead of tissue, it may be directly extracted with ethanol. If the tissue is tough, it may be necessary to heat the mixture to about 70-80°C during extraction. The pooled extracts are centrifuged and the clear supernatants are concentrated, preferably in vacuum. The extract may also be evaporated in a rotary evaporator to dryness and the residue is dissolved in 1-10 ml of 0.01 N HCl or in suitable sample dilution buffer.

*Estimation* - The content of free amino acids may be determined by any calorimetric methods viz. Moore & Stein (1948), Yemm & Cocking (1955) and Frame, Russell & Wilhelmi (1943) method, given at 4.3.4.1 or by chromatographic separation of amino acids or in an amino acid analyser.

### 6.2.2.2 Estimation of Protein-bound Amino Acids

*Extraction:*

*Acid hydrolysis of proteins* - Hydrolyze 1g dry sample powder with 6N HCl at 110°C for 22-24 hours (five times weight of the sample). The hydrolysate is repeatedly distilled off till all the hydrochloric acid is removed. Treat the dry residue with petroleum ether to remove lipids and then dissolve in an appropriate quantity of 10% iso-propanol solution or 1-10 ml of 0.01N HCl or 50 mM citrate buffer, pH 2.2 and store in cold

*Alkaline hydrolysis of proteins* - Weigh about 100 mg dry sample into a glass ampoule, add 3.2g barium hydroxide and 5 ml of water. Boil the solution over flame, seal at the constriction and keep at 110°C for 18-22 hours. Cool and transfer the contents into a centrifuge tube. Remove the excess basium sulphate salt by adding 8N sulphuric acid followed by centrifugation to remove the precipitate. Wash the precipitate once, centrifuge, pool the supernatants and make up to a known volume and store in cold.

*Estimation* - The content of protein-bound amino acids may be determined by any calorimetric methods viz., Moore & Stein

**Table 6.4: $R_f$ values of amino acids in Paper chromatography**

| *Amino acid* | *$R_f$ value* *Solvent a* | *Solvent b* |
|---|---|---|
| Histidine | 0.07 | 0.69 |
| Serine | 0.10 | 0.36 |
| Lysine | 0.10 | 0.48 |
| Arginine | 0.11 | 0.59 |
| Aspartic acid | 0.13 | 0.15 |
| Glutamic acid | 0.16 | 0.25 |
| Glycine | 0.17 | 0.40 |
| Alanine | 0.22 | 0.54 |
| Threonine | 0.22 | 0.50 |
| Proline | 0.30 | 0.91 |
| Tyrosine | 0.32 | 0.64 |
| Methionne | 0.40 | 0.80 |
| Valine | 0.47 | 0.77 |
| Tryptophan | 0.47 | 0.83 |
| Isoleucine | 0.55 | 0.86 |
| Phenylalanine | 0.58 | 0.89 |
| Leucine | 0.60 | 0.86 |

Solvent a: n-butanol-acetic acid-water 4:1:5 v/v

Solvent b: Phenol-water 4:1 v/v

(1948), Yemm & Cocking (1955) and Frame, Russell & Wilhelmi (1943) method, given at 4.3.4.1 or by chromatographic separation of amino acids or in an amino acid analyser.

### *6.2.2.3 Separation and Estimation of Individual amino acids*

#### Separation of Amino Acids by Two Dimensional Paper Chromatography

Amino acids are separated by two-dimensional technique of Dutta, Dent & Harris (1950) using n-butanol-acetic acid-water (4:1:5, v/v; upper layer) and phenol-water (4:1, w/v) as solvents and ninhydrin reagent (0.5%, w/v in acetone) as detection spray Chromatograms are developed by spraying ninhydrin solution and heated at 105°C for 30 minutes. Depending upon the concentration, pink spots may appear immediately. Individual amino acids are identified using specific spray reagents (Pant & Agrawal, 1965). The $R_f$ values of some standard amino acids are given in the Table 6.4.

*Procedure:* Take a large sheet of Whatman No. 1 filter paper. Cut the Whatman No. 1 filter paper and draw a line at one end of the paper (lengthwise). Using micro-pipette, spot a sample of amino acids (about 10 µ g) at one point on the line, dry the spot using a hot air dryer and repeat the application of sample. The paper is first developed in one direction with the solvent butanol-acetic acid-water by ascending or descending chrornatographic method. When the solvent front reaches the top, the filter paper is removed, marked the solvent front and dried thoroughly in a stream of warm air. The paper is now chromatographed in the second direction perpendicular to the first direction in a solvent containing phenol-water. The paper after development is dried and sprayed with ninhydrin reagent. The colour is developed by heating the filter paper at 105°C for 20-30 minutes. Plot the position of all the amino acids and construct a standard map, which will be used to compare unknown amino acids chromatographed under identical conditions. The amino acids will occupy a position almost worth x and y coordinates. Identify the individual amino acids using standard chromatograms with known amino acids. Calculate the $R_f$ value of different amino acids by the following formula and compare the $R_f$ values of amino acids for each solvent system (Tab. 6.4).

$$R_f = \frac{\text{Distance moved by the substance from the origin (cm)}}{\text{Distance moved by the solvent from the origin (cm)}}$$

For quantitative estimation, the amino acid spots are cut individually into small pieces and eluted with aqueous ethanol (50%; v/v; 4ml) in separate test tubes. Optical density is measured at 570 nm against a blank, which is obtained eluting the plain portion of the chromatogram with aqueous ethanol.

Alternately, two chromatograms are developed under identical conditions as described above. Spray one of the papers with ninhydrin reagent and identify the spots using standard chromatogram. On the other paper, mark the positions corresponding to these spots The amino acid spots are cut individually into small pieces and eluted with aqueous ethanol (50%; v/v; 4ml) in separate test tubes. The concentration of amino acid is determined calorimetrically by ninhydrin method as described earlier.

## Separation of Amino acids by Thin Layer Chromatography

Amino acids can also be separated by thin layer chromatography using silica gel G coated plates. The solvent generally used are (a) 96% ethanol-water (7:3, v/v) and (b) n-butanol-acetic acid-water (8:2:2, v/v). The spray reagent used is 0.3% solution of ninhydrin in butanol containing 3 ml acetic acid. The Thin layer plates coated with silica gel G (250 μ) are prepared as described earlier. 5 μl solution of standard amino acid solution and a known volume of unknown sample are spotted on separate plates. The chromatograms are developed in solvent a and solvent b. The plates are sprayed with ninhydrin solution and developed by heating the plates at 110°C for 10 minutes. Spraying of 1 ml solution of saturated copper sulphate mixed with 0.2 ml of 10% nitric acid and 100 ml of 96% ethanol on the developed plates can also be made since this helps in the development of maximum intensity of colour. Once the colour is developed, the plates are exposed to vapours of concentrated ammonium hydroxide, which helps in stabilization of colour.

Calculate the $R_f$ values and compare between standard and unknown samples. Some $R_f$ values of amino acids in Thin Layer Chromatography are given in Table 6.5.

**Tab. 6.5: $R_f$ values of Amino Acids in TLC**

| *Amino acid* | *$R_f$ value* | |
|---|---|---|
| | *Solvent a* | *Solvent b* |
| Serine | 0.48 | 0.22 |
| Aspartic acid | 0.55 | 0.21 |
| Glutanmc acid | 0.63 | 0.27 |
| Glycine | 0.43 | 0.22 |
| Alanirne | 0.47 | 0.27 |
| Threonine | 0.50 | 0.25 |

| | | |
|---|---|---|
| Proline | 0.35 | 0.19 |
| Tyro sine | 0.65 | 0.47 |
| Methionine | 0.59 | 0.40 |
| Valine | 0.55 | 0.35 |
| Tryptophan | 0.65 | 0.56 |
| Isoleucine | 0.60 | 0.46 |
| Phenylalanine | 0.63 | 0.49 |
| Leucine | 0.61 | 0.47 |
| Hydroxyproline | 0.44 | 0.20 |
| Cysteic acid | 0.69 | 0.14 |

Solvent a : 96% ethanol-water 7:3 v/v

Solvent b : n-butanol-acetic acid-water 8:2:2 v/v

*Ion Exchange Chromatography of Amino Acids* -

Ion exchange chromatography is used for separation of compounds on an insoluble matrix containing ionizable groups capable of exchanging with ions in the surrounding media. Ion exchange resins are best suited to separate small molecules like amino acids but are unsuitable for large molecules as proteins. The resins used in ion exchange chromatography may be anionic or cationic depending upon the nature of their affinity for either positive or negative ion. The concentration of ionizable groups present in a resin may be as much as 6 to 10 mol /l, so that these materials have a very high exchange capacity. Some of the widely used resins are Dowex 50, Dowex 1, Amberlite IR 120, Amberlite IRC 50 etc.

*Principle*: Amino acids exist as charged molecules in solution since they have ionizable groups on amino acid molecule. At the isoioninc point, the amino acid molecule has a net charge of zero and therefore its interaction with water is minimum. That is why most of the amino acids do not dissolve in water at neutral pH. This property of amino acids is used for separating them by ion exchange chromatography. The separation of amino acids by ion-exchange chromatography is based on acid-base or ion exchange principle wherein amino acids are made to bind to the oppositely charged ion-exchangers and can be eluted by employing buffers of slightly changed pH.

The method involves the use of a cationic exchanger Dowex 50 and the elution of the bound amino acids is done by using Tris-HCI buffer (0.2M, pH8.5). The amino acid eluted in order may be quantitatively estimated by ninhydrin method.

**Table 6.6: Ionization properties of Amino acids**

| *Amino acid* | *pK value* | | |
|---|---|---|---|
| | *Carboxyl* | *Amino* | *Others* |
| Serine | 2.21 | 9.15 | |
| Aspartate | 2.09 | 9.82 | 3.86 carboxyl |
| Glutamate | 2.19 | 9.67 | 4.25 carboxyl |
| Glycine | 2.34 | 9.60 | |
| Alanine | 2.35 | 9.69 | |
| Proline | 1.99 | 10.60 | |
| Isoleucine | 2.36 | 9.68 | |
| Phenylalanine | 1.83 | 9.13 | |
| Leucine | 2.36 | 9.60 | |
| Hydroxyproline | 1.92 | 9.73 | |
| Cysteine | 1.71 | 10.33 | 8.33 -SH |
| Cystine | 1.65 & 2.26 | 7.85 & 9.85 | |
| Arginine | 2.17 | 9.04 | 12.48 |
| Histidine | 1.82 | 9.17 | 6.0 |
| Lysine | 2.18 | 8.95 | 10. 83 amino |

*Procedure:*

*Step 1 (Preparation for column)* - Before use in chromatography, the cationic exchanger resin (Dowex 50) is suspended in 4N HCl (4-8 litres/100g dry resin) for about 15 minutes until fully swollen. It ensures that the resin is saturated with $H^+$ ions. Then, the suspension is filtered through Buchner funnel and repeatedly washed with distilled or deionized water, till the filtrate is of neutral pH.

*Step 2 (Setting the column)* - Suspend the Dowex 50 resin in water, stir and allow to settle. The supernatant containing any light particles is decanted. Repeat the washings with distilled water. Then, add citrate buffer (pH 3.4; prepared by mixing 40 ml of 0.1 M citric stir well and let it stand for one hour. Use the suspension to set up a glass column (about 30 × 0.9 cm). Run through the column, about two column volumes of the buffer. Adjust the flow rate to about 5 ml per 15 minutes.

*Step 3 (Loading the column)* - The amino acid mixture, to be separated, is adjusted to pH 1.0. At this pH, all amino acids will have positive charges. Place 1 ml of the sample on to the top of the column without disturbing the top layer (about 50-60% of the maximal capacity). Now, the tap is opened and the sample is allowed to percolate through the resin. When the level just reaches the top layer, stop the flow.

*Step 4 (Eluticn of amino acids)* - The amino acids are now bound to the column and these can be released and eluted by increasing the pH of the eluent buffer. As the pH increases, the protonated form of the amino acids will start ionizing and when the pi is reached, they will have no net charge and therefore will not bound to the resin any mere. They will be eluted out. Since the pi value of each amino acid is different, they will be eluted at different pH values.

The loaded column is connected with a pH gradient generating device. The vessel B contains buffer of pH 3.4 and vessel A of pH 10.0. The outlet of the vessel B is made air tight at the top of the column tube. The pinch cocks are now opened and the pH gradient solution is allowed to pass through the column. Eluant fractions are collected in 1 ml aliquots in serially numbered test tubes.

*Step 5 (Detection of amino acids)* - 1 ml aliquots from the collected fractions are pipetted out into different tubes and the ninhydrin reagent is added as described in Expt. 6.2.2.1. The tubes, which develop colour, indicate the presence of amino acids.

*Step 6* - Plot the amount of amino acid in each fraction against the volume eluted.

*Step 7* - From the tubes, which show the presence of amino acids, take suitable aliquots and subject them to two-dimensional paper chromatography. From the $R_f$ values as compared to standard samples, identify each amino acid.

### Separation or Amino Acids by Paper Electrophoresis

*Principle:* Many biological molecules carry an electric charge, the magnitude of which depends on the particular molecule and also the pH and composition of the surrounding medium. These charged molecules migrate in solution to the electrodes of opposite polarity when an electric field is applied, and this principle is used in electrophoresis to separate molecules of differing charges.

The charge carried by the amino acids depends upon the pH of the medium. At pH 7.6, histidine carries zero net charge, aspartic acid will be negatively charged and lysine will be positively charged. The charged amino acids are separated by paper electrophoresis and the spots are detected by dipping the paper in ninhydrin reagent. Electrophoresis is done at low voltage, which not only separates low molecular weight compounds but also illustrate the relationship between charge and pH.

*Procedure* - Cut a strip of Whatman No. 3 filter paper (10 cm x 2.5 cm) and draw a pencil line 3 cm away from one end. Wet the paper with Tris-acetate buffer (0.07M, pH 7.6)and blot off excess buffer between two sheets of filter paper. Fill both parts of each electrode compartment with buffer solution to the same level. Place the paper strip with the ends dipping in the buffer and 3 cm line towards cathode. Turn on the power supply and allow the paper to equilibrate for 15 minutes at 200V. Turn off the power supply, and carefully apply a streak of the amino acid mixture (25 μ l) along the line marked on the paper. Turn on the power supply and carry out horizontal electrophoresis for 3 hours at 8V / cm. Remove the paper strip and dry in an oven at 110°C. Dip the paper strip in freshly prepared ninhydrin solution, allow the acetone to evaporate in air and develop the colour by heating in oven at 105°C for 2 - 3 hours. Locate and identify the amino acid on the paper.

Repeat the experiment with 0.07mol/l cirate buffer, pH 3.0 and explain the result using the data given in Table 6.6.

## 6.3 PROTEINS

### 6.3.1 Extraction of Proteins

In biological systems, proteins exist in different forms - some are in free soluble form, some are bound to nucleic acids as nucleoproteins and some are bound to lipid-rich membranes. Each type of protein has to be extracted differently. The solvents used for extraction are selected based on the solubility of the proteins under investigation. The common solvents are distilled water, dilute acids and alkalis, salt solutions, 60-80% alcohol, buffers etc.

*From Tissues* - Grind 0.5g sample with a suitable solvent (buffer or distilled water) in a mestle and mortar. Centrifuge and use the supernatant for protein estimation.

*From Haemolymph* - Add cold TCA (10%) to the haemolymph to a final concentration of 5%, which will precipitate out almost all the proteins and the interfering materials like carbohydrates, pigments etc. remain in solution. The solution is centrifuged for 10-15 minutes at 5000 r.p.m. The precipitate is dissolved in suitable volumes of distilled water and used for protein estimation.

### 6.3.2 Estimation of Proteins

The proteins can be estimated by the indirect method of Micro-Kjeldahl and by the direct methods, such as, Lowry's method using Folin-Ciocalteau reagent (Folin & Ciocalteau, 1927), Biuret method and dye-binding method of Bradford's method.

#### *6.3.2.1 Micro-Kjeldahl's Method*

This method is generally used for estimating total or crude protein content of food materials, agricultural and clinical samples. This is also used to estimate the non-protein nitrogen of a sample after precipitating out the proteins. In this method, the sample is digested with concentrated sulphuric acid in the presence of a catalyst so that the nitrogenous compounds are converted to ammonium sulphate. By steam distillation, ammonia is liberated, which is collected in boric acid and then estimated by titration method. By finding out the nitrogen content of the sample, total or crude proteins can be determined by multiplying the nitrogen content by 6.25 since 1 mg nitrogen is equivalent to 6.25 mg protein. The method has already been described at 4.2.6.

#### *6.3.2.2 Biuret Method*

This is the most common method and is based on the principle that the peptide bonds of the protein (-CO-NH-) form a purple colour complex with copper ions in an alkaline solution. The intensity of purple complex is measured at 520 nm calorimetrically. The method has been described at 4.2.6.

#### 6.3.2.3 Lowry Method

In this method, the aromatic amino acids present in the protein reacts with the Folin-Ciocalteau reagent to give a blue-coloured complex. The colour is formed due to the reaction of the copper ions with the peptide bonds (-CO-NH- groups) of protein in alkaline solution and reduction of phosphomolybdate-phosphotungstatic components of the reagent by amino acids present in protein. The intensity of blue colour is measured calorimetrically at 660 nm. The method has been described at 4.2.6.

#### 6.3.2.4 Bradford's Method

In this method, the dye Coomassie Brilliant Blue G-250 is allowed to bind with the protein molecules and the absorbance of the solution is then measured at 595 nm. The method has been explained at 4.2.6.

#### 6.3.2.5 UV Absorption Method

*Principle:* Most proteins exhibit a distinct absorption peak in the UV range at 280 nm because of the presence of tyrosine and tryptophan amino acids in the proteins. The specific extinction coefficients of the proteins vary in the range of 6 to 62 depending upon their tyrosine and tryoptophan content. On an average, most proteins have a value close to 10, that is, 1 mg/ml of protein gives an extinction at 280 nm of about 1 when viewed through 1 cm path length. Since nucleic acids are present as contaminants in protein extracts and they mask the absorption by proteins by virtue of their strong absorption at 260 nm, the protein content may be calculated by Kalckar formula, which is as follows.

Protein concentration (mg/ml) = $1.55 \times O.D._{280} - 0.76 \times O.D._{260}$

*Procedure*: Pipette out 1, 2, 3, up to 10 ml of the standard protein solution into a series of test tubes and make up the volume in each tube with water to 10 ml. Mix well. Measure the absorbance of the standards and sample solution at 280 nm against blank in a spectrophometer. Draw the standard curve and calculate the amount of protein in the sample from the graph. Express the results as mg/g or mg/100g of sample or percentage.

Alternately, the optical density of the test solution is measured at 260 and 280 nm and using the Kalckar formula the concentration is calculated.

### 6.3.3 Isolation and Separation of Proteins

Biological materials contain a very large number of proteins and the detailed study of any single protein is possible only after it has been isolated up to a satisfactory level of purity The separation of protein molecules from the biological materials without affecting their biological activity or function or property is a complicated process requiring very sophisticated techniques, great skill and practice.

A number of physical techniques are employed for separation of proteins but all proteins cannot be separated by the same techniques. The best technique has to be found to separate a particular protein by practice. Some of approaches that can be adopted in isolation of proteins are indicated in Table 6.7.

**Table 6.7: Methods for fractionation of proteins**

| *Property on which the method is based* | *Method* |
|---|---|
| Solubility | Salting out (precipitation)<br>Salting in (extraction)<br>Alcohol fractionation (manipulation of dielectric constant of solvent) |
| Net charge | Adjustment to pH to the isoelectric point of proteins<br>Zone electrophoresis<br>Isoelectric focusing |
| Density | Ion exchange chromatography<br>Density gradient centrifugation (flotation) |
| Size | Gel filtration on molecular sieve |
| Adsorption | Adsorption-elution using celite, silicic acid, silica, alumina and calcium phosphate gel etc. |

### 6.3.3.1 Isolation of Proteins

*Preparation of homogenate* - A homogenate of biological tissue is prepared with distilled water or buffer as the case may be. First, dissect out the tissue from the body, dry it by folding between filter paper sheets and drop it in a beaker kept in ice. Using a scissor, mince it into small pieces. Add about 10 ml of cold distilled water to the mince, transfer it to a homogenizer and grind well. The homogenate so obtained is filtered through 2 to 3 layers of muslin cloth to remove any lump that may be present. The filtrate is diluted to a suitable volume and kept in cold.

*Differential extractability of proteins with salt* - Proteins show a variation in solubility that depends on the concentration of salts in the solution. This effect is due to interaction between charged side chains of the protein molecule and solution ions. Some neutral salts like sodium chloride have profound effect on the solubility of proteins. In low concentration, salt increases the solubility of many proteins, a phenomenon referred as '**salting-in**'. The ability of neutral salts to influence solubility of proteins is a function of their ionic strength. Salt decreases the ionic strength of the solution and hence the proteins get soluble. The salting-in effect is related to the nonspecific effect the salt has on increasing the ionic strength of the solution. The higher the ionic strength, the smaller are the interactions between the charged groups on the same protein or different proteins. In this method, tissue homogenate in water is centrifuged at 3000 r.p.m. for 15 minutes. Then, the residue is extracted with differential concentration of NaCl (0.2M to 1M) followed by 0.1N HCl solution and the residue is finally dissolved in 1N NaOH solution. The step-wise experiment is as follows.

Determine the protein content of various fractions (Supernatant A to E) by any suitable method, preferably by Modified Biuret method and tabulate the results.

**Differential extractability of proteins with buffers** - Prepare the tissue homogenate with the buffers of different pH values ranging from 4.0 to 9.0 (10 ml per g tissue) and centrifuge. Measure the protein concentration in the supernatants by suitable method.

### 6.3.3.2 Precipitation of Proteins from Solution

The tissue homogenate is a mixture of proteins, which are required to be precipitated and separated by the suitable methods.

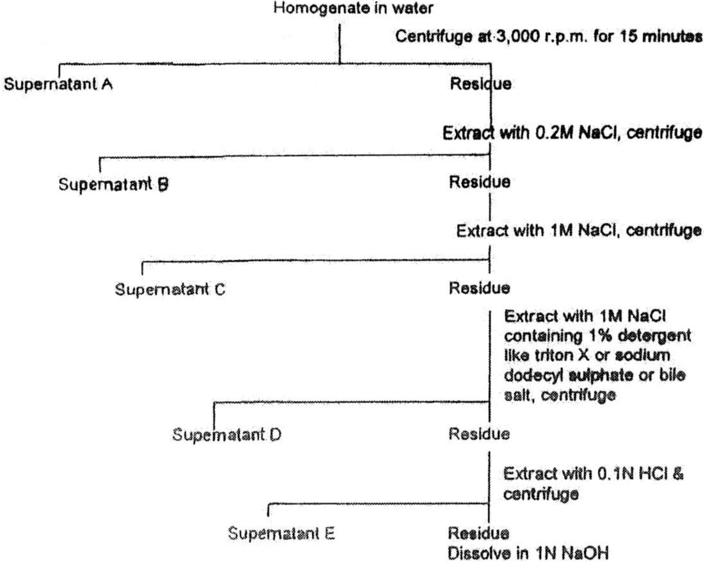


Some methods used for precipitation and separation of proteins are - isoionic precipitation, salt precipitation and solvent precipitation.

*Salt Precipitation or Salting out* - The proteins can be differentially precipitated by different concentration of neutral salts. The salts diminish the solubility of proteins by decreasing the activity of water in the solvent mixture, thereby causing dehydration of the hydrophilic groups of the protein molecules. The most commonly used salts are ammonium sulphate, magnesium sulphate, sodium sulphate, sodium sulphite and phosphatic mixtures at different pH values. By addition of ammonium sulphate in different amounts, the proteins can be fractioned into different component fractions. These agents do not denature the proteins. Fractionate the tissue homogenate into different protein. Fractions by adding different amounts of ammonium sulphate as follows and tabulate the results.

*Isoionic Precipitation* - Precipitation can also be brought about by diminishing net electric charge of the molecule. The solubility of proteins is minimal at its isoelectric pH. The pH at which a protein is least soluble is its isoelectric pH. **It can be defined as that pH at which the molecule has no net electric charge and fails to move in an electric field** Under these conditions, there is no

## Table 6.8: Table concerning ammonium sulphate solubility

Final concentration of ammonium sulphate (Percentage saturation)

initial concentration of ammonium sulphate (percentage of saturation)

| *10* | 20 | 25 | *30* | *33* | *35* | *40* | *45* | *50* | *55* | *60* | *65* | *70* | *75* | *80* | *90* | *100* |
|---|---|---|---|---|---|---|---|---|---|---|---|---|---|---|---|---|
| Grams solid ammonium sulphate to be added to 1 litre of solution | | | | | | | | | | | | | | | | |
| 56 | 114 | 144 | 176 | 196 | 209 | 243 | 277 | 313 | 351 | 390 | 430 | 472 | 516 | 561 | 662 | 767 |
| | 57 | 86 | 116 | 137 | 150 | 183 | 216 | 251 | 288 | 326 | 365 | 406 | 449 | 494 | 592 | 694 |
| | | 29 | 59 | 78 | 91 | 123 | 155 | 189 | 225 | 262 | 300 | 340 | 382 | 424 | 520 | 619 |
| | | | 30 | 49 | 61 | 93 | 125 | 158 | 193 | 230 | 267 | 307 | 348 | 390 | 485 | 583 |
| | | | | 19 | 30 | 62 | 94 | 127 | 162 | 198 | 235 | 273 | 314 | 356 | 449 | 546 |
| | | | | | 12 | 43 | 74 | 107 | 142 | 177 | 214 | 252 | 292 | 333 | 426 | 522 |
| | | | | | | 31 | 63 | 94 | 129 | 164 | 200 | 238 | 278 | 319 | 411 | 506 |
| | | | | | | | 31 | 63 | 97 | 132 | 168 | 205 | 245 | 285 | 375 | 469 |
| | | | | | | | | 31 | 65 | 99 | 134 | 171 | 210 | 250 | 339 | 431 |
| | | | | | | | | | 33 | 66 | 101 | 137 | 176 | 214 | 302 | 392 |
| | | | | | | | | | | 33 | 67 | 103 | 141 | 179 | 264 | 353 |
| | | | | | | | | | | | 34 | 69 | 105 | 143 | 227 | 314 |
| | | | | | | | | | | | | 34 | 70 | 107 | 190 | 275 |
| | | | | | | | | | | | | | 35 | 72 | 153 | 237 |
| | | | | | | | | | | | | | | 36 | 115 | 198 |
| | | | | | | | | | | | | | | | 77 | 157 |
| | | | | | | | | | | | | | | | | 79 |

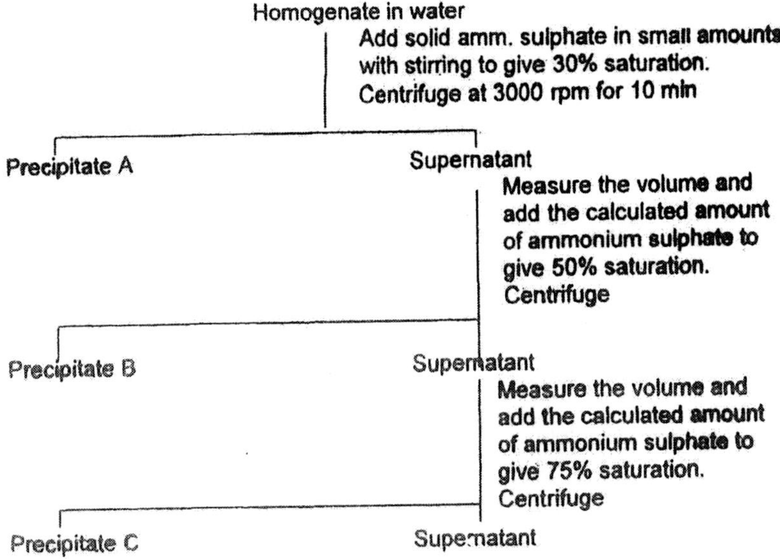


electrostatic repulsion between neighbouring protein molecules and they tend to coalesce and precipitate. Since different proteins have different isoelectric pH values, they can be separated by isoelectric precipitation. When the pH of the protein mixture is adjusted to the isoelectric pH of one of its components, much or all of that component will precipitate.

*Solvent Precipitation* - The addition of water-immiscible neutral organic solvents like acetone or alcohol decreases the solubility of most globular proteins in water to such an extent that they precipitate out of solution. At a fixed pH and ionic strength, the protein solubility is a function of the dielectric constant of the medium. Since ethanol has a low dielectric constant than water, its addition to an aqueous protein solution increases the attraction force between opposite charges, thus decreasing the degree of ionization of the R groups of the protein. As a result, the protein molecules tend to aggregate and precipitate. For example, Cohn's method of fractionation of plasma proteins is based on this principle.

*Desalting of Proteins by Dialysis* - Dialysis is commonly used for separating macromolecules from smaller molecules. Special semipermeable membranes called dialysis tubes have the property to allow compounds of small molecular weight to pass through them

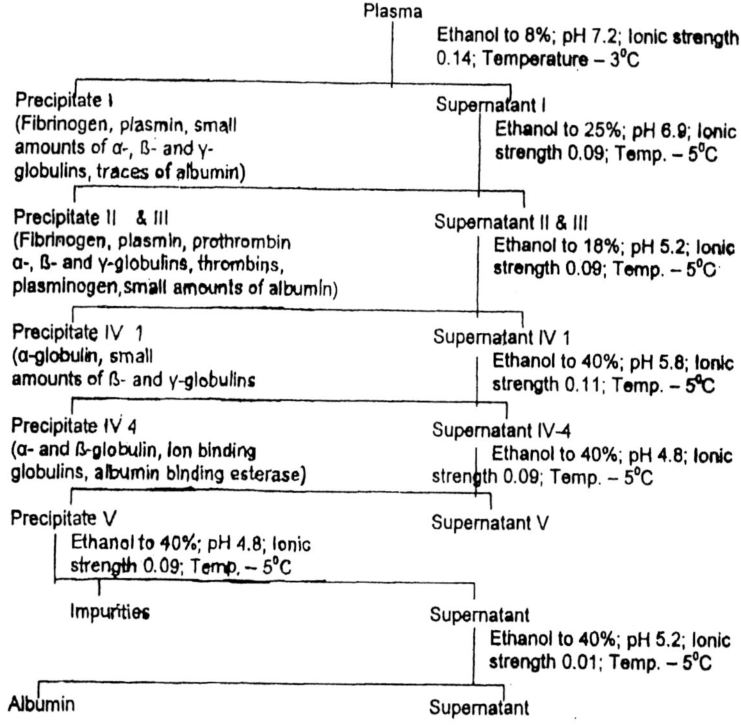


but not those with high molecular weight like proteins, which are held back. The solution is taken in a specially prepared dialysis tubes; which is then immersed in large volume of water or suitable buffer. There is a free exchange of small molecules across the membrane and if the volume of outer medium is large enough, their concentration inside the bag gets reduced. Repeated changes in the outer medium result in almost total removal of small molecular compounds from inside the bag. Dialysis can also be used to remove salt from the protein solution. The protein solution to be desalted is taken inside a dialysis bag and the two ends are tightly secured to prevent leakage. The bag should be about half full only. The bag is now suspended in a large vessel containing about 100 fold excess of water of preferably dilute buffer (0.001M phosphate buffer, pH 7) and the contents are stirred constantly in cold. Salt molecules pass freely and get diluted by the large volume of fluid in the external medium. Repeated changes of the dialysis fluid help in reducing the salt concentration inside the bag to negligible levels.

*Gel Filtration* - In gel filtration or molecular sieving, separation of biomolecules is done on the basis of their size and density. The compounds varying in size are passed through a column containing beads made on inert polysaccharide chains. These chains are made such that the space between them can allow only small molecules to enter and pass into the beads, whereas large molecules are excluded from entering into the structures. Because of this, the large molecules elute out of the column much earlier than :he small molecules. There are several types of beads with definite pore sizes. Since such polysaccharide sieves were first marketed under the trade name Sephadex, the technique itself is referred as Sephadex **chromatography**. Some of the properties of molecular sieves are given in Table. 6.9.

**Table 6.9: Properties of Molecular Sieves**

| *(Sephadex)* | *Exclusion limit (mw)* | *Water gain g/g dry gel* | *Bed volume ml/g dry gel* |
|---|---|---|---|
| G 10 | Up to 700 | 1.0 | 2-3 |
| G 15 | Up to 1 500 | 1.5 | 2.5-3.5 |
| G 25 | Up to 5,000 | 2.5 | 5 |
| G 50 | Up to 10,000 | 5.0 | 10 |
| G 75 | Up to 50,000 | 7.5 | 12-15 |
| G 100 | Up to 1,00,000 | 10.0 | 15-20 |
| G 150 | Up to 1,50,000 | 15.0 | 20-30 |
| G 200 | Up to 2,00,000 | 20.0 | 30-40 |

Suspend 5g Scphadex G-25 in water and leave it to swell for few hours. Stir after short intervals to prevent formation of lumps. Pour this into a column (about 1 cm dia) and allow the particles to settle down. Precautions should be taken to avoid trapping of air bubbles. This can be achieved by tapping gently with a rubber tube or pencil. Once the column is sat, about 100 ml of 0.1M tris-HCl buffer (pH 7.4) is passed through it to equilibrate the column. Now, the sample solution is loaded onto the column. The loaded volume should not be more than 10% of the column bed volume. After loading, the same buffer is continued to run through the column at

a flow rate of a bout 1 to 2 ml/minute. Collect 3 ml fractions of the eluant and check the absorbance at 280 nm. Aliquots of individual fractions can also be checked by suitable chemical tests for salts.

### 6.3.3.3 Separation of Proteins

Proteins contain many amino acids, which have ionizable side groups and hence they are electrically charged. This charge is subjected to variations in pH, which is the principle behind isoionic precipitation. The same principle is applied for separation of proteins by ion exchange chromatography and electrophoresis.

***Separation of Proteins by Ion Exchange Chromatography*** - In this method, the resins like DEAE-Cellulose (Diethylamino ethyl cellulose, weak anion exchanger) and CM-Cellulose (Carboxy methyl cellulose, weak cation exchanger) are used as ion exchangers instead of Dowex and Amberlite.

*Procedure:*

*Step 1 (washing of DEAE Cellulose)* - Take about 10g DEAE cellulose in a beaker and add 200 ml of 1M NaCl containing 0.1N NaOH. Sir vigorously for a few minutes and filter through a glass sintered funnel (or Whatman No. 1 paper in Buchner funnel). Continue washing till the filtrate becomes colourless. Then, wash the filter cake in the funnel with distilled water thoroughly till the pH of the filtrate becomes neutral. Stir the suspension in water well. Transfer the filter cake into a beaker and suspend in 200 ml of 0.1N HCl and stir well. This step charges the amino group of DEAE cellulose. The excess of acid is removed by washing with water as before till the filtrate is neutral. At this stage, the adsorbent is in the fully protonated stage. Suspend the material in 200 ml of buffer (pH 3.5) to be used in the experiment.

*Step 2 (Column packing)* - Mount a column on a stand vertically. The suspension of DEAE cellulose is poured into the column gently through the sides, avoiding any trapping of air bubbles. Simultaneously open the column outlet so that the adsorbent is settled down. Pass through the column about 500 ml of buffer (pH 3.5) to ensure equilibrium.

*Step 3 (Loading and elation)* - Load an aliquot of protein solution (1-2 ml; cell-free tissue homogenate or biological fluid like haemolymph) carefully onto the column without disturbing the top surface. Connect the column to a gradient generator capable of generating a salt gradient between 0 and 0.5M KCl. Vessel A contains 100 ml of 0.5m KCl and vessel B water. Open the stop cocks and adjust the flow rate to about 0.5 to 1 ml per minute. Collect different fractions of 3 ml volume.

*Step 4 (Detection)* - Measure the optical density of each of the eluant fractions at 280 nm detect for the presence of proteins. Plot a graph with the number of tubes versus optical density.

***Separation of Proteins by Electrophoresis*** - This is yet another technique based on the charges on biomolecules.

*Principle* - When the current is applied into a solution having charged substances, the positively charged substances move towards the cathode and negatively charged substances move towards anode. If a molecule of charge $q$ is present in an electric field of strength $x$, then the force on the particle causing it to accelerate is $qx$. This is balanced by frictional resistance to give a terminal velocity $v$:

$$qx = fv$$

Now, as per Stokes' law for spherical molecule of radius r moving through a medium of viscosity $\eta$,

$$f = 6\pi\eta r$$

$$\text{Or } qx = 6\pi\eta rv$$

The electrophoretic mobility v of a molecule is defined as the migration per unit strength of electric field,

$$v = v/x = q/6\pi\eta r$$

The rate of migration depends upon the size (r), molecular weight, total charge (q), viscosity of liquid ($\eta$) and other factors. When a liquid medium is used, the technique is referred as *zone elctrophoresis*. Recently solid mediums like paper, starch, agar gels, acrylamide gels etc. are extensively used and hence the techniques are referred as paper electrophoresis, disc gel electrophoresis, SDS gel electrophoresis etc. Uses of acrylamide gels, which have sieving properties, are commonly used.

*Zone Electrophoresis:* In this technique, electrophoresis is carried out on a supporting medium impregnated with buffer solution. The molecules are completed separated into distinct zones. However, this technique is not in much use because of other advanced techniques.

*Paper Electrophoresis:* The filter paper is quite cheap and easy to use. The procedure has already been discussed in this chapter earlier (6.2.2.3).

*Cellulose Acetate Electrophoresis:* This is a more convenient and rapid method for separation of proteins as compared with paper electrophoresis. In this technique, cellulose acetate is used in place of paper where adsorption of the material is minimal resulting in easy elution and good recovery. Moreover, the very low quantity of material is required and separation is effected in very short period of time.

*Disc Gel Electrophoresis:* Gel electrophoresis is widely used to separate and characterize proteins under an electric field. Electrophoresis using polyacrylamide is more convenient than other media like paper, cellulose, starch gel etc. because of the following advantages -

- It is transparent and can be scanned in the visible and ultraviolet range.
- The pore size can be controlled so that the separation is obtained based on the size and shape of the molecules as well as the charge.
- Finer resolution of complex mixtures can be obtained.

The gel is prepared by polymerizing acrylamide ($CH_2$=CH.CO.$NH_2$) and a small quantity of cross-linking reagent methylenebisacrylamide in the presence of a catalyst, ammonium persulphate. Tetramethylethylenediamine (TEMED) is also added to initiate and control the polymerization. The gel mixture is allowed to polymerize in small tubes sealed at the bottom with rubber cap. A layer of water is placed on the top of the gel to ensure a flat surface and also to exclude oxygen, which inhibits polymerization. The pore size of gel can be altered by varying the concentration of monomer in the gel solution. A gel containing 7-7.5% acrylamide can easily separate proteins having molecular weights between 104 and 106.

However, higher molecular weight molecules require gels with larger pore size (about 4% acrylamide). Thus, the proteins of high molecular weight are separated using gels of larger pore size whereas smaller proteins are separated with gels of smaller pore size. Different

Table 6.10 Selection of gel system for disc Gel Electrophoresis

| Anionic enzyme sample (upprer buffer compartment cathode; lowrer buffer compartment anc... tracing dye – methylene blue) | | | | | Cationic enzyme sample (upprer buffer compartment anode; lowrer buffer compartment cath... tracing dye – methyl green) | | | | | |
|---|---|---|---|---|---|---|---|---|---|---|
| | | System i | | | System II | | | System III | | |
| Range mol. Wt. of proteins | % acryl-amide | Separating gel run at pH 9.5[a] | Stacking gel run at pH 8.3[a] | Buffer system pH 8.3 | Separating gel run at pH 8.0 | Stacking gel run at pH 7.0 | Buffer system pH 7.0 | Separating gel run at pH 3.8 | Stacking gel run at pH 5.0 | Buffer system pH 4.5 |
| $10^6$ | 3.75 | 4 | 2 | 3 | - | - | - | - | - | - |
| | 5.00 | 5 | 2 | 3 | | | | | | |
| | 7.00 | 1 | 2 | 3 | | | | | | |
| $10^4$ - $10^6$ | 7.5 | 6 | 2 | 3 | 11 | 12 | 13 | 14 | 15 | 16 |
| | 10 | 7 | 2 | 3 | - | - | - | - | - | - |
| | 15 | 8 | 2 | 3 | - | - | - | 17 | 15 | 16 |
| < $10^4$ | 22.5 | 9 | 2 | 3 | - | - | - | - | - | - |
| | 30 | 10 | 2 | 3 | | | | | | |

[a] pH values are given for 25°C. Formulation of gel systems is given in Annexure.

gel systems used in disc gel electrophoresis are summarized in the Table 6.10.

*SDS-Polyacrilamide Gel Electrophoresis:* The oligomeric proteins containing two or more subunits are dissociated into its subunits and separated as separate bands when subjected to SDS-Polyacrylamide gel electrophoresis (SDS-PAGE). SDS polyacrylamide electrophoresis of proteins is carried out in presence of sodium dodecyl sulphate, an anionic detergent that readily binds and dissociates oligomeric proteins in presence of a reducing agent, 2-mercaptoethanol into their subunits. The number of SDS molecules bound to a polypeptide chain is approximately half the number of amino acid residues in that chain. The protein-SDS complex carries net negative charges, and hence moves towards the anode.

*Procedure:*

*Disc Gel Electrophoresis* (Davis, 1964)

1. Prepare **solution A** (lower gel buffer) by dissolving 56.75g Tris buffer in 200 ml water. Adjust the pH to 8.9 using cone. HCl. Add sufficient water to make up the volume to 250 ml.
2. Prepare **solution B** (acrylamide) by dissolving 93.75g acrylamide and 2.5g N, N'-methylene bis (acrylamide) in sufficient distilled water to yield a final volume of 250 ml.
3. Prepare a solution of tracing dye by dissolving 5.0 mg bromophenol blue in 10 ml distilled water (0.05% bromophenol blue solution).
4 Prepare upper reservoir buffer by dissolving 6g Tris buffer and 28.8g glycine in sufficient distilled water so as to make final volume of 1 litre. Adjust pH to 8.3 with a small quantity of Tris or glycine.
5. Prepare the lower reservoir buffer by dissolving 908g Tris buffer in 2 litres of distilled water. Add sufficient conc. HCl to get pH 8.9. Add sufficient water to yield a final volume of 4 litres.
6 Prepare **solution C** (ammonium persulphate) by dissolving 17.5mg ammonium persulphate in 10 ml distilled water.
7. Prepare **solution D** (upper gel buffer) by dissolving 8.9g Tris buffer and 40ml 1M $H_3PO_4$ in sufficient distilled water to get a final volume of 250 ml (pH 6.5-6.7).

8. Prepare **solution E** (acrylamide) by dissolving 31.25g acrylamide and 7.8g N, N'-methylene bis (acrylamide) in sufficient distilled water to yield a final volume of 250 ml. Store, all acrylamide solutions at 4°C in brown bottles.

9. Prepare **solution F** (riboflavin) by dissolving 2.5mg riboflavin in 100 ml distilled water. Store at 4°C in brown bottles.

10. Before preparing the gels, warm all the solutions to room temperature. It is also advisable to remove gas bubbles from acrylamide solutions.

11. Cover the ends of the glass gel tubes (12 × 0.6 cm ID) with rubber caps, paraffin or serum vial caps. Place these tubes in a convenient rack.

12. The separating gel of different strength are prepared by mixing the following solutions

| *Solution* | *5%* | *7.5%* | *10%* |
|---|---|---|---|
| A | 0.4ml | 0.4 ml | 0.4ml |
| B | 1.33ml | 2.0ml | 2.67ml |
| C | 1.0ml | 1.0ml | 1.0ml |
| TEMED | 0.016 ml | 0.016 ml | 0.016 ml |
| Water | 7.27 ml | 6 60 ml | 5.93 ml |

Mix the above reagents so as not to produce bubbles in the solution. Otherwise, de-gas the solution immediately. Once ammonium persulphate has been added to the reaction mixture, the gels start to polymerize. Therefore, mix the solution and transfer to the gel tubes as quickly as possible. If polymerisation occurs too rapidly, decrease the quantity of ammonium persulphate.

Transfer 2 ml of the reaction mixture prepared above to each of the tubes with the help of Pasteur pipette.

13. Gently pipette 0.5 - 1.0 ml of diluted solution A on the top of the gel. Take extreme care not to mix the buffer with the polymerizing gel. This may be accomplished by placing a Pasteur pipette with the tip drawn to a fine diameter just below the surface of the liquid. The purpose of this buffer addition is to flatten the surface of the gel. If allowed to polymerize in absence of buffer, the gel would have a concave surface

owing to the presence of a meniscus resulting in crescent shaped protein bands after electrophoresis.

14. Very shortly after the interface appears between the buffer and the gel surface, it disappears. When the gel has fully polymerized (15-20 minutes), the interface again becomes visible.
15. Prepare upper stacking gel by mixing the following solutions,

| *Solution* | *Quantity (ml)* |
|---|---|
| D | 1.0 |
| E | 1.0 |
| F | 1.0 |
| Water | 4.0 |

Mix the solutions in such a way that there are no bubbles in the solution. Otherwise, de-gas immediately and pour it over the top of the separating gel. Overlayer the gel solution with water as before and place the tubes under a fluorescent light or desk lamp for polymerization (20-50 minutes). Finally, cover the gel with a layer of electrode buffer (pH 8.3).

16. Remove the buffer A and rinse the top of each gel with a small quantity of the solution prepared above. Transfer sufficient quantity of solution onto the top of the gels to yield a depth of 1.0 - 1.3 cm. Repeat the above steps with water instead of buffer.
17. Place the gel tubes in close proximity to a strong fluorescent light source. Allow the gels to polymerize until the upper layer becomes opalescent. Carefully remove the water from the top of the gel.
18. Prepare a sample of freshly drawn haemolymph by mixing 1.0 ml haemolymph, 50% aqueous glycerol and 0.1 ml tracing dye.
19. Gently pipette 0.02 - 0.05 ml of this solution onto the top of the gel. Place electrophoresis apparatus and power supply in a cold room (4°C). Assemble the gel tubes into the upper buffer reservoir and remove coverings from the tube bottoms.
20. Using the method described earlier, fill the gel tube completely with the upper reservoir buffer, which was diluted by 1 part of buffer with 4 parts of water.

21. Fill the lower reservoir with buffer and lower the tubes into it. If bubbles are observed on the bottom interface of the gels, remove them.
22. Gently fill the upper reservoir with buffer and connect the electrodes to the electrophoretic apparatus. Be sure that the anode is at the bottom.
23. Energize the power supply and adjust it to an output of 1 to 2 mA/gel. The voltage should be held constant at no more than 75V.
24. Continue the electrophoresis until the tracing dye has moved 0.5 – 1.0 cm from the bottom of the tube.
25. Turn off the power and unplug the power supply from the wall outlet. Now, the electrodes may be safety removed from the apparatus.
26. Disassemble the apparatus and pour off the reservoir buffers.
27. Remove the gels from the gel tubes by gently injecting distilled water between the wall of the tubes and the gel using a syringe while rotating the tube.
28. Rinse the gels with water and place them in 13 × 100 mm test tubes. Fill each tube with a solution composed of the following ingredients - sodium lactate, 20 ul 60% syrup/ml; NAD, 0.7 mg/ml and 100 mM Tris-HCl buffer adjusted to pH 8.0. The ingredients of the solution may be prepared before hand and kept frozen. Even in this condition, the phenazine methosulfate and nitroblue tetrazolium containers should be wrapped with aluminium foils. Incubate one of the gel tubes to have a large amount of protein in a reaction mixture identical to that just described above but omit lactate.
29. Cover the tubes with paraffin and gently invert them several times to ensure that any water adhering to their surfaces is replaced with the enzyme reaction mixture.
30. Incubate the tubes at 37°C until 3-5 blue bands of insoluble formazan are clearly visible.
31. When the gels have stained to the desired intensity, remove them from the reaction mixture and wash them thoroughly with water.

32. Make a pencil sketch of the protein bands on graph paper or photograph them.

*SDS-Polyacrylamide Gel Electrophoresis* (Weber & Osborn, 1969)

1. For a typical run of 6 gels, 7.5 ml of gel buffer (Dissolve 7.8g mono-sodium dihydrogen phosphate, 38.6g $Na_2HPO_4.7H_2O$ and 2g sodium dodecyl sulphate in water and dilute to 1 litre; pH 7.0) is deareated and mixed with 6.75 ml of acrylamide solution (Dissolve 22.2g acrylamide and 0.6g bis-acrylamide in 100 ml water and filtered to remove the insoluble material). It is further deareated and mixed with 0.75 ml of freshly prepared ammonium persulphate (15 mg/ml) and 0.02 ml of TEMED.
2. Properly cleaned glass tubes (0.5 × 7 cm size) supported on rubber stoppers are filled with the mixed solution upto the mark (about 2 ml of the mixture), on the top of which is placed a layer of 0.01 ml water.
3. The gels are allowed to polymerize for 30 minutes in day light fluorescent lamp. Water layer is removed after polymerization and the tubes are fixed to the electrophoretic apparatus in such a way that the tubes are immersed about 1/4th in the tray buffer in the lower compartment.
4. The protein is incubated at 37°C for 2 hours in 10mM sodium sulphate buffer pH 7.0 containing 1% SDS and 1% β-mercaptoethanol.
5. After incubation, the protein solution is dialyzed for several hours at room temperature against 500 ml of the 10 mM phosphate buffer, pH 7.0 containing 1% SDS and 1% (b-mercaptoethanol.
6. For each gel tube, 3 μl of tracing dye (0.05% bromophenol blue), 1 drop of glycerol and 5 μl of dialysis buffer are mixed in a small tube followed by addition of 10-50 μl of protein solution.
7. After mixing, the solution is applied on gels. The gel buffer diluted with equal volume of distilled water is carefully layered over each sample. The two compartments of electrophoresis apparatus are filled with the same buffer.
8. The electrophoresis is performed at a constant current of 8 mA per gel tube with anode in lower compartment.

9. The process is stopped when the tracing dye approached the bottom of the gels.
10. The gels are then removed from the tubes and stained with Coomassie blue (Dissolve 1.25g Coomassie blue in a mixture of 454 ml of 50% methanol and 46 ml of acetic acid Insoluble material is removed by filtration.) for 2-5 hours.
11. The gels are then removed from staining solution, rinsed with distilled water and de-stained with a mixture of 75 ml of acetic acid, 50 ml of methanol and 875 ml of water for 12 hours.
12. Measure the length of the gels and position of protein bands. Calculate the mobility relative to the tracing dye using the following relationship -

$$\text{Mobility} = \frac{\text{Distance moved by the protein (mm)}}{\text{Distance moved by tracing dye (mm)}}$$

13. Plot the mobility of marker proteins against the molecular weight on a semilogarithmic scale or against $\log_{10}$ molecular weight and calculate the molecular weight of the unknown from the calibration curve.

## 6.4 NUCLEIC ACIDS

### 6.4.1 Separation of RNA and DNA

*Schmidt-Thannhauser-Schneider Method*

*Step 1:* Mix 1 ml of tissue homogenate with 2.5 ml of cold 10% tri-chloroacetic acid (TCA) and centrifuge. Re-suspend the precipitate in 2.5 ml of cold 10% TCA and centrifuge again. Pool the supernatants, which constitute the acid soluble phosphorus fraction.

*Step 2:* Suspend the tissue residue in 1 ml water, mix 4 ml of 95% ethanol and centrifuge. Re-suspend the residue in 5 ml of ethanol and centrifuge. Extract the residue thrice with three portions of alcohol-ether (3:1) at room temperature with brief stirring. The combined extract is phospholipids fraction.

*Step 3:* Treat the residue with 1N KOH for 16-20 hours at 37°C, approximately 10 ml per gram of fresh tissue, which results in solution of tissue. Neutralize the solution with 6N HCl or $HClO_4$ or by a strong acid cation exchanger in the hydrogen form.

The DNA and protein are precipitated by 1 volume of 5% TCA or PCA and the extracts are combined to give RNA fraction phosphorus derived from phosphoprotein. Inorganic phosphate may be precipitated from this fraction as calcium salt.

*Step 4:* Suspend the residue in 5ml of 5% TCA/PCA, heat for 15 minutes at 90°C, cool and centrifuge at 2000 rpm for 15-20 minutes. Re-suspend the residue in 5ml of 5% TCA and centrifuge. Combine the extracts to get the DNA fraction. The residue is "protein" fraction.

***Non-hydrolytic Isolation Method (Tyner, Heidelberger & LePage, 1953)***

The homogenate is acidified with 0.4N PCA. Centrifuge the precipitate and wash it twice with 0.4N PCA and twice with ethanol. The lipids are extracted by three washings with ethanol-ether (3:1) at 40-50°C. The residue is suspended in 10% NaCl solution and the

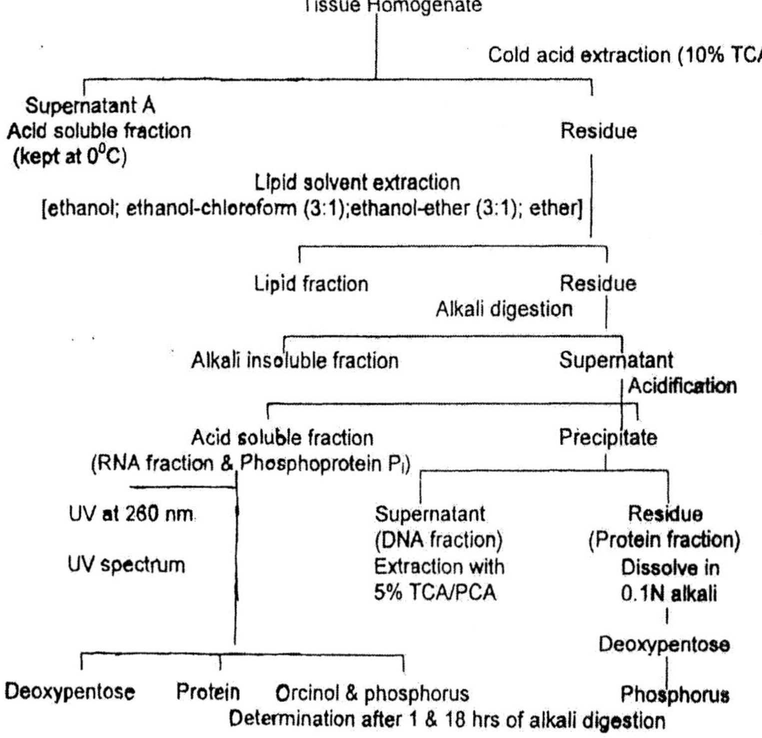

solution is neutralized with NaOH. After centrifugation, the supernatant is discarded. The residue is re-suspended in 5 volumes of 10% NaCl (neutral or slightly alkaline $NaHCO_3$) and the mixture is heated for 1 hour at 100°C to extract the nucleic acids. After centrifugation, the supernatant is filtered through glass wool and the process is repeated on suspending the residue with 3 volumes of NaCl and heating at 100°C for 30 minutes. The NaCl extract is treated with 3 volumes of cold ethanol and allowed to stand at 0°C overnight. The precipitated sodium nucleate are obtained by centrifugation and subjected to alkaline hydrolysis to separate RNA as mono-nucleate from un-degraded DNA. The acid precipitated DNA is dissolved with dilute alkali and subjected to double enzymatic degradation with DNase and diesterase to yield deoxyribonucleotides or degraded to bases with formic perchloric acid or to purine bases and pyrimidine nucleotides with HCl.

### *Smille and Krotkov Method*

Homogenize the tissue with ice-cold perchloric acid (25 ml; 0.5N) and centrifuge at 5000g for 10 m inutes. Wash the residue with 0.5N PCA, delipidize by repeated extraction with ethanol-ether-chloroform (2:1:2; v/v) mixture and hydrolyze with 20 ml of 0.3N KOH for 1 hour at 37°C. Acidify the hydrolysate with 9N PCA to pH 7.0. Add 1 volume of 95% ethanol, keep at 0°C for 15 minutes and then centrifuge. Adjust the pH of the supernatant to 8 with KOH, chill and centrifuge. The supernatant contains RNA. Hydrolyze the residue with 5 ml of 0.5N $HClO_4$ on a boiling water bath at 90°C for 10 minutes and centrifuge at 5000g for 15 minutes. The supernatant contains DNA.

## 6.4.2 Estimation of RNA

***Orcinol method (Merchant et al., 1969).***

*Reagents:*

1. *Orcinol reagent* - Dissolve 100 mg orcinol and 100 mg ferric chloride in conc. HCl (Sp. Gr. 1.19) and make up the volume to 100 ml with conc. HCl [or dissolve 13.5g ferric ammonium sulphate and 20g orcinol in 500 ml of glass distilled water. Keep in cold (at 4°C) and just before use, mix 25 ml of stock solution to 415 ml of conc. HCl and dilute to 500 ml with water].

2. *Standard RNA solution* - Dissolve 100 mg RNA in distilled water and make up the volume to 100 ml. A working standard containing 100 mg RNA per ml is prepared by diluting 10 ml of stock solution to 100 ml with *distilled water*.

*Procedure:*

*Method 1:* Measure the absorbance of RNA extract at 260 nm and calculate the concentration of RNA from a standard curve prepared with yeast RNA.

*Method 2*: Pipette out 3.2 ml of the extract in a test tube and add 2 ml of orcinol reagent. Cover the test tubes with glass stoppers and heat for 8 minutes on a water bath maintained at 100°C. Cool the test tubes under running tap water and measure the absorbance against reagent blank at 665 or 670 nm. If the colour intensity is strong, dilute the contents to 10 ml with n-butanol. Calculate the concentration of pentose present from a standard curve prepared with ribose (10 µg/ml) and express the results as ribose equivalents.

*Method 3*: Number the 16x150mm test tubes and place them into a test tube rack. In each tube, carefully, pipette out one of the following volumes of standard RNA solution (10 mg/100 ml) - 0, 0.1, 0.2, 0.3, 0.4, 0.5, 0.6, 0.7, 0.8, 0.9 and 1 ml. Add appropriate amount of distilled water to each tube to give a final volume of 1 ml. Take 1 ml of test solution in another test tube and add 3 ml of diluted reagent in each tube. Mix the solution thoroughly and heat them for 20 minutes in a boiling water bath. Cool to room temperature and read at 660 nm. Set the instrument with blank reagent and prepare standard curve with standard RNA solution. Calculate RNA concentration with the help of standard curve.

### 6.4.3 Estimation of DNA

***Burton Method (1968)***

*Reagents:*

1. *Diphenylamine reagent* - Dissolve 1.5g pure white crystalline diphenylamine in 100 ml glacial acetic acid followed by addition of 1.5 ml of conc. sulphuric acid (Sp. Gr. 1.84). Just before use, 0.1 ml of 1.6% aqueous acetaldehyde is mixed with 20 ml of the reagent. Or dissolve 1g diphenylamine (pure white, crystallized in 70% alcohol twice) in 98ml of redistilled glacial

acetic acid and then add 2ml of concentrated sulphuric acid (Sp. Gr. 1.84). Stable for 1-2 days in cold.

2. *Standard DNA solution,* - Dissolve 3 mg Calf Thymus DNA in 10 ml of 5mM NaOH (0.3 mg / ml). The working solution is prepared by mixing equal volumes of stock solution with 1N perchloric acid and heating it for 15 minutes at 70°C. The solution is stable for three weeks.

*Procedure:*

*Method 1*: Suspend the residue obtained earlier in suitable aliquots of 5% TCA at 0°C, centrifuge at 2000 rpm for 15 minutes and discard the supernatant. Repeat the process and wash the residue once with absolute alcohol and then with ethanol-ether mixture. The residue is suspended in 0.5N $HClO_4$ and incubated at 90°C for 7 minutes in a constant temperature water bath. Centrifuge and collect the supernatant. Combine the supernatants and make up the volume. Add equal volume of 1N KOH to a known aliquot to precipitate excess perchlorate as $KClO_4$. Centrifuge and collect the supernatants. Measure the absorbance of the DNA extract at 260 nm and calculate the concentration from a standard curve.

*Method 2*: Dilute the extract with 0.5N perchloric acid so that the sample for analysis contains 0.02 - 0.25 μ mole of DNA-phosphate / ml. Pipette 2 ml of the sample in test tubes and add 4 ml of diphenylamine reagent. Maintain appropriate standards and blank tubes containing the same amount of perchloric acid. Incubate the tubes for 15-17 hours at 25-30°C and then measure the absorbance at 600 nm.

*Method 3:* Take 1 ml of test solution and 2.5 ml of the reagent in a test tube and heat for 5 minutes in boiling water bath. Cool and read at 540 nm. Calculate the concentration from standard curve.

## 6.5 LIPIDS

### 6.5.1 Extraction of Lipids

*Bloor's Method* - The suitable quantity of tissues (say 1g) is homogenized in 10 ml of ethyl ether-ethanol mixture (3:1; v/v) till complete extraction is effected. The extracts are pooled together, centrifuged at 2,000 rpm for 10 minutes and the supernatant is taken in a separating funnel. To this, 2 ml of 0.05M KCl solution is added and shaken well. The water present in KCl solution helps in separation of layers. The lower layer is taken for quantitative determination of lipids.

*Bligh and Dyer Method* - The haemolymph/tissue is first homogenized in motar and pestle with distilled water and then the homogenized material is extracted with chloroform-methanol mixture (2:1; v/v; 20 ml solvent per g tissue or ml of haemolymph). For complete extraction, this is kept overnight at room temperature and then another 20 ml of chloroform-methanol mixture is added. The resulting solution is subjected to centrifugation. Three clear layers are obtained after centrifugation - lower layer of chloroform containing all the lipids, a coloured aqueous layer of methanol with all water soluble materials and the residue. The lower layer is carefully collected

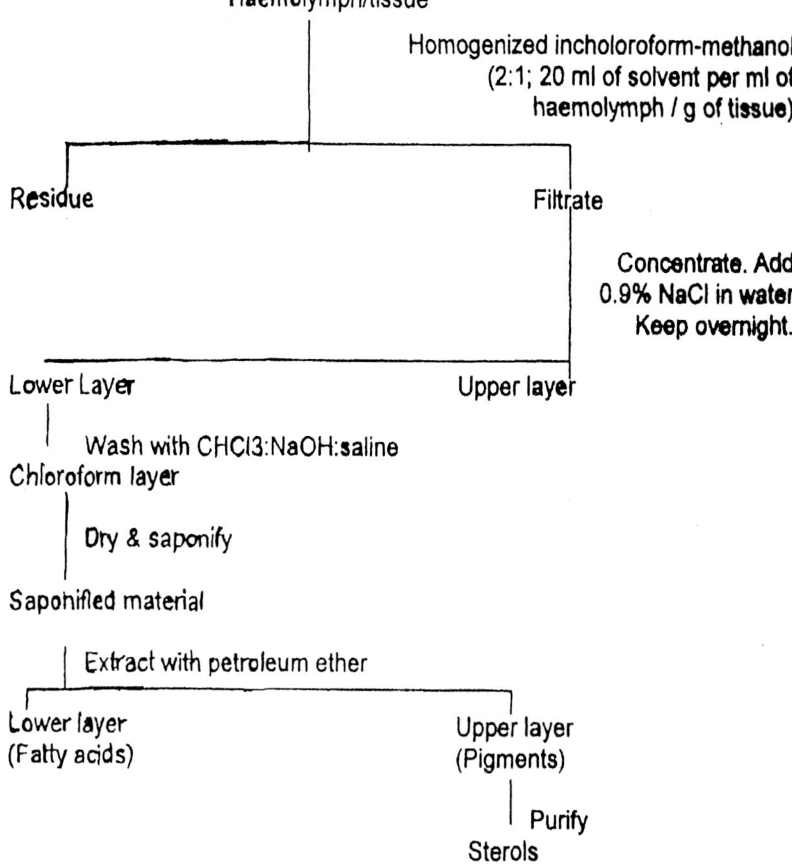

in a pre-weighed beaker or crucible and concentrated. It is advisable to keep the sample covered with a dark paper to protect from light otherwise there are chances that some of the lipids may get polymerized or decomposed. The weight of dried materials gives the weight of total lipids.

Add 0.9% NaCl in water (1:5 volume of original extract) and keep overnight. The lower layer is taken, which is washed with $CHC_3$:NaOH:saline (15:47:48). The chloroform layer is dried and saponified with 50 ml of ethanol-ether (3:1) and 0.5 ml of 10N KOH. The saponified material is extracted with petroleum ether. The lower layer contains fatty acids while the upper layer the pigments. The pigments are purified.

### 6.5.2 Quantitative Estimation of Lipids

#### *6.5.2.1 Esterified Fatty Acids*

The lipid extract (2ml) is taken in a test tube and evaporated to dryness. The residue is dissolved in 2 ml of isopropanol. Pipette 0.5 ml of this solution in separate test tube and add 1 ml alkaline hydroxylamine (0.2% in isopropanol). Mix well and allow to stand at room temperature for 30 minutes. Then add 0.5 ml of 3.8N HCl followed by 0.5 ml of ferric chloride (10% v/v in 0.1N HCl). Mix well and read at 575 nm after 30 minutes.

#### *6.5.2.2 Cholesterol*

*Method 1:* The lipid extract (2ml) is taken in a test tube and evaporated to dryness. The residue is dissolved in 2 ml of ethylacetate-absolute alcohol (1:1; v/v). Add 2.5 ml of ferric chloride (0.1% v/v in ethylacetate) and mix well. Then, add 2 ml of concentrated sulphuric acid, mix well and cool. Read at 550 nm after 30 minutes.

*Method 2* (Liebermann-Burchard test): The cholesterol in haemolymph is estimated as follows - Take 20 ml of ether-alcohol (3:1; v/v) in a 25 ml-glass stoppered flask and add slowly 0.5 ml of haemolymph Mix well so that the precipitate gets finely divided. Immerse the flask in boiling water for few seconds, cool and centrifuged at 2,000 rpm for 10 minutes. The residue is extracted again and the supernatants are pooled together. The volume is made up to 25 ml with ether-alcohol mixture. Transfer 5 ml of the filtrate to a beaker and evaporate to dryness over a steam bath. Add 1 ml of anhydrous chloroform, bring to boiling and then carefully pour off the liquid in the calorimeter tube. Repeat the process with 1 ml more of chloroform. Allow the combined extract to cool and make up a final volume of 5 ml with chloroform. A blank is prepared with 5 ml of chloroform to which 1 ml of acetic anhydride is added, mixed well and the calorimeter is adjusted to 0 reading. To the contents of unknown tubes, add exactly 1 ml of acetic anhydride, mix and read in calorimeter at 420 ($R_1$). A standard is also prepared with 5 ml of cholesterol solution (Dissolve 160 mg cholesterol in 100 ml of chloroform and dilute 1 ml of this solution to 25 ml with chloroform). Now, add 0.1 ml of concentrated sulphuric acid to both the standard and unknown. Mix well and place in dark for 15 minutes. Read against blank tubes set at zero reading ($R_2$). Calculate cholesterol concentration with the help of following formula,

mg cholesterol in 5 ml of ether-alcohol filtrate =

$$\frac{\text{mg cholesterol in 5 ml}}{\text{reading of standard}} \times (R_2 - R_1)$$

*Method 3:* The samples to be assayed (containing 0.1 to 0.5 mg cholesterol) are adjusted to 80% (v/v) ethanol : acetone (1:1) in order to deproteinize the reaction mixtures and extract cholesterol. For this, take the reaction mixture into a centrifuge tube and add the desired amount of ethanol : acetone in small aliquots. Centrifuge it and decant the supernatant in a clean test tube. Extract the precipitate with approximately 0.5 ml of warm ethanol : acetone for 5-10 minutes in water bath (40 to 50°C) to effect complete recovery of cholesterol. Centrifuge the sample again and add the supernatant to the first extract. Evaporate the combined extracts to dryness in a water bath under air jets. When the samples are completely dry, add 5 ml of chloroform in each tube followed by 2 ml of cold acetic anhydride-sulphuric acid reagent (Four parts of cold reagent-grade acetic anhydride and one part of cold concentrated sulphuric acid are mixed slowly with constant stirring in a ice-cooled flask just prior to use). Place the tubes in water bath (16-18°C) and allow the colour to develop for 15 minutes. Read the blue-green colour at 625 nm. Calculate cholesterol concentration with the help of standard curve prepared with standard cholesterol solution (Dissolve 100 mg cholesterol in chloroform and adjust to 100 ml at refrigerator temperature. Dilute this stock solution 1:10 with chloroform at room temperature for a working solution).

### 6.5.2.3 Phospholipids

Phospholipids are a group of lipid molecules containing phosphorus. They are similar to triglycerides in being fatty acid esters of glycerol except that in addition they possess phosphoric acid moieties. Lecithin, cephalin, cardiolipin and phosphatidyl inositol are some of the phospholipids. Phospholipids are frequently found in association with proteins. For estimation of phospholipids, they are first digested with perchloric acid to liberate the organic phosphorus as free inorganic phosphorus and then estimate the phosphorus content by Fiske-Subbarao method.

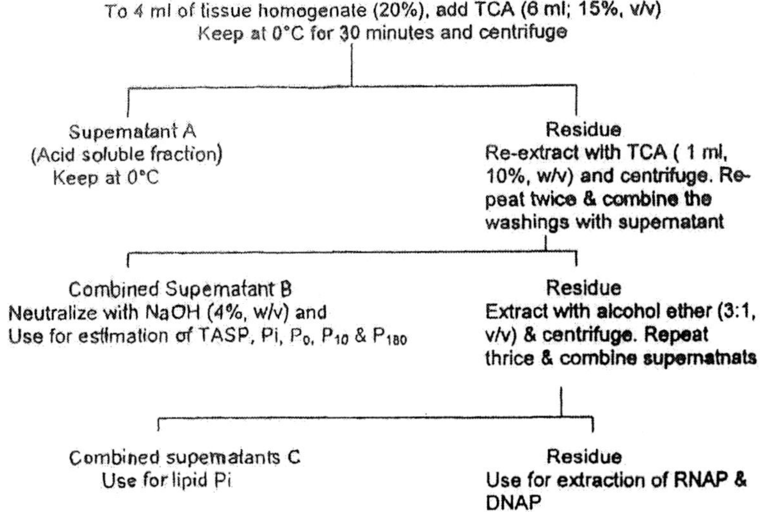


The suitable aliquots of combined supernatant C (5.0 - 7.5 ml) are evaporated to dryness and digested with perchloric acid (2.2 ml, 60%). The supernatant is utilized for estimation of lipid phosphorus by Fiske-Subbarao method.

*Fiske-Subbarao Method*

*Reagents:*

1. Double distilled water
2. 10M sulphuric acid: Add 450 ml of conc. sulphuric acid to 1300 ml of distilled water.

3. Standard phosphorus solution: Dissolve 0.351g of pure dry monopotassium phosphate in distilled water and transfer quantitatively to a one-litre volumetric flask. Add 10 ml of 10N sulphuric acid, dilute to mark and mix well. This solution contains 0.4 mg phosphorus in 5 ml.
4. Molybdate solution: Dissolve 25g ammonium molybdate in 200 ml of distilled water. Transfer to a flask containing 500 ml of 10N sulphuric acid and bring the final volume to 1 litre using distilled water. Mix well and store in a brown bottle.
5. Amino naphthosulphonic acid (ANSA) reagent: Take 195 ml of 15% sodium bisulphate solution in a glass-stioppered cylinder. Add 0.5g 1,2,4-amino naphthosulfonic acid. Add 5 ml of 20% sodium sulphite solution and shake the contents until the powder is dissolved. If the solution is not complete, add more sodium sulphite (1 ml at a time) with shaking. Transfer the solution in a brown bottle and store in cold.

*Procedure:*

1. Pipette out 2 ml of the extract into a test tube and make the volume to 4.2 ml with distilled water.
2. Take 0.2, 0.4, 0.6, 0.3 and 1.0 ml of standard phosphorus solution into a series of test tubes and make up the volume to 4.2 ml with distilled water. The tube with no standard solution is treated as blank.
3. Add 0.6 ml of molybdate solution in each tube, mix well and then add 0.2 ml of ANSA reagent.
4. After 10 minutes, read the absorbance at 660 nm against the blank.
5. Calculate the amount of phosphorus present in the sample using the standard curve and express it as percentage.
6. The phospholipids content is determined by multiplying the phosphorus content by 25.

### 6.5.2.4 Neutral Lipids

Neutral lipids comprise the most abundant group of lipids in nature. These are nothing but esters of fatty acids with glycerol or esters of glycerol, commonly known as acylglycerols or glycerides One, two or all the three hydroxyl groups of glycerol may be esterified with fatty acids to give mono-, di-, and triglycerides. They are also called fats or oils depending on whether they are solid or liquid at room temperature. The glycerides are of two types - simple glycerides and mixed glycerides. **Simple glycerides** are those in which the **glycerol hydroxyls** are esterified by identical fatty acids. The mixed glycerides contain two or three different fatty acids.

The neutral lipid content is obtained by substracting quantity of phospholipids from the quantity of esterified fatty acids.

## 6.5.3 Separation of Lipids

### *6.5.3.1 Ascending Paper Chromatography of Fatty Acids (Kennedy & Barker, 1951)*

In this method, the volatile fatty acids are applied to the paper in the form of non-volatile salts and the chromatogram is developed in solvents containing free ammonia. After the ammonium salts of the fatty acids are applied to the paper, it is developed with ammonium hydroxide-ethanol solvent mixture. The paper is then dried and sprayed with bromophenol blue-water-citric acid solution. The locations of fatty acids are identified as intense blue spots, while the background is yellow.

*Reagents:*

1. 1% oxalic acid
2. Acetic acid, n-butanol, n-caproic acid, 0.1M, each neutralized with $NH_4OH$
3. Solvent mixture: Mix 1 ml of conc. ammonium hydroxide with 100 ml of 95% ethyl alcohol.
4. Developing spray solution: Dissolve 50 mg of bromophenol blue in 100 ml of water containing 200 mg citric acid.
5. Hydrolysed and purified fatty acids

*Procedure:*

1) Cut Whatman No. 1 filter paper to measure 12" x 16" (31 × 41 cm).
2) Immerse the paper in a bath of 1% oxalic acid for 15 minutes; wash thoroughly in a constantly changing water bath and dr in a hood with the exhaust fan on.

3) Draw a line 2.5 cm from the bottom of the paper and make 3' × 5' about 3" apart on the starting line.
4) Label each point below the line with the name of the appropriate fatty acid.
5) Spot about 0.01 ml of each fatty acid and sample in its proper place.
6) Set the chromatography jar with 100 ml solvent mixture at the bottom.
7) Seal the glass cover on the jar and develop the chromatogram until the solvent front reaches 1-2 inches from the upper edge of the paper. Take the chromatogram out of the jar.
8) Dry the paper in a 100°C oven for 10 minutes and spray with the developing spray solution.
9) Locate the spot by blue colour.
10) Use n-caproic acid for the control spot and calculate relative $R_f$ values of the fatty acids.

### 6.5.3.2 Reverse Phase Paper Chromatography of Fatty Acids

In this method, an organic solvent, a 'stationary phase', is adsorbed to an inactive supporting material while the other aqueous solvent is a mobile phase, which is immiscible with the first solvent. During chromatography, the fatty acids are partitioned between two immiscible phases. Paraffin, purified hydrocarbon, silicon oil and kerosene can be used as stationary phase.

*Reagents:*

1) 10% liquid paraffin in petroleum ether
2) Standard fatty acids: 2 - 5 mg each of saturated fatty acids (lauric acid, myristic acid, palmitic acid, and stearic acid). 1 - 3 mg each of unsaturated fatty acids (linolenic acid, oleic acid and linoleic acid).
3) Fatty acid samples dissolved in known volume of petroleum ether.
4) Solvent 1: acetic acid-acetonitrile (1:1, v/v)
5) Solvent 2: acetone-water (80:20, v/v)
6) 10% Phosphomolybdic acid in ethanol.

*Procedure:*

1) Dip a Whatman No. 1 or 3 filter paper (20 x 20 cm) in 10% liquid paraffin in petroleum ether. Dry the paper at room temperature.
2) Apply approximately 20-25 mg of fatty acid samples with the help of micropipette approximately 3 cm from the bottom edge of the paper.
3) On the same paper, apply the standard fatty acids as above.
4) Develop the paper in chromatographic chamber saturated with solvent 1 and solvent 2.
5) Remove and hold the paper at room temperature till all the solvent is evaporated.
6) Spray the paper with 10% solution of phosphomolybdic acid and heat in hot air oven at 120°C for 10-15 minutes or until the blue spot appears.
7) Identify the fatty acid composition of the sample by comparing with the $R_f$ values of standard fatty acids.

### 6.5.3.3 Gas Liquid Chromatography of Fatty Acids

Fatty acids are made volatile by converting them into methyl esters through a process called trans-esterification (from glycerol esters to methyl esters). The esters are identified and quantified by gas liquid chromatography.

*Reagents:*

1) 10% barium chloride or fluoride in methanol
2) Saturated sodium chloride solution
3) Anhydrous sodium sulphate
4) Hexane
5) Nitrogen gas
6) Gas Liquid Chromatography

Column: Pretested 10% silar 10°C on gas chromatograph Q 100 to 200 mesh (6' x ¼" - internal diameter 4 mm)

Detector: Flame ionization detector

Injector lamp: 280°C

Carrier gas: Nitrogen at 50 ml/minute

Column/oven temperature: 165°C

*Procedure:*

1) Take 150 to 300 mg of oil in a culture tube and add 3 ml of 10% barium chloride or fluoride
2) Add boiling chips, cover the tube with aluminium foil and heat at 83°C for 6 minutes.
3) Transfer the contents and washings of the tube to a 30 ml separating funnel. Washing is done with hexane. Shake and allow to separate.
4) Add 4 ml of saturated sodium chloride solution, shake and collect hexane layer over anhydrous sodium sulphate.
5) Rinse the funnel with hexane, collect hexane layer and combine the hexane extracts.
6) Filter the hexane extract through Whatman No. 4 filter paper.
7) Reduce the volume of the filtrate to 2-3 ml by drying with a slow stream of nitrogen.
8) Inject an aliquot to pre-conditioned ,GLC.
9) Inject the standard methyl esters separately and calculate the retention time for individual esters.
10) The computerized GLC model gives concentration of fatty acids directly.

### 6.5.3.4 Thin Layer Chromatography of Lipids

Lipids in biological material are present as a complex mixture and are first fractionated into a number of groups by solvent extraction. Resolution of the compounds within each group is then carried out by thin layer Chromatography. The neutral lipids are separated with non-polar solvents and charged lipids with polar solvents. A number of supporting media are used for lipid fractionation but silica gel is widely used.

*Reagents : Material*

1) Thin layer plates of silica gel
2) Separating chambers
3) Solvent (petroleum ether, BP 60-70°C:diethylether:glacial acetic acid: 80:20:1)

4) Hydrocarbons
5) Cholesterol esters
6) Triacetylglycerols
7) Free fatty acids
8) Sterols
9) Naturally occurring oils
10) 2', 7'-dichlorofluorescin (2g/l in 95% v/v ethanol)
11) Sulphuric acid (50%, v/v)
12) 0.5% Rhodamino G in ethanol
13) Oven at 110°C
14) Ultraviolet lamp

*Procedure:*

Clean the glass plates with ethanol, then pour an aqueous slurry of the silica gel onto their surface 250 μm thick. Activate the plates by heating at 110°C for 1 hour and allow to cool. Spot about 20-50 μl of an approximately 1% w/v solution of each lipid in the solvent at two places and develop the chromatogram in the solvent. Locate the lipids by spraying with the dichlorofluorescin solution and view the plates in ultraviolet light (270 nm). The lipids look as green spots against a dark background. Alternately, spray the plates with 0.5% Rhodamine G solution and observe the spots of orange fluorescence under UV light. Alternately, the spots can be viewed by spraying with 50% v/v sulphuric acid followed by heating at 110°C for 10 minutes. The lipids are seen as black spots.

Phospholipids are also separated by the same procedure but solvent is chloroform-methanol-acetic acid-water (25:15:4:2, v/v). Detection of phospholipids is made by exposure to iodine vapours. The lipid spots absorb iodine and are seen as brown spots on yellow background.

### 6.5.3.5 Column Chromatography of Lipids

Set up a column with 5g silicic acid slurried in benzene. Load the lipid sample (about 3-4 mg) and elute them by the following schedule: (a) 20 ml benzene (b) 20 ml of 5% chloroform in benzene (c) 20 ml of 10% chloroform in benzene (d) 20 ml of 25% chloroform in benzene (e) 20 ml of 50% chloroform in benzene, (f) 20 ml of chloroform and (g) 20 ml of methanol. Collect the elluent in 2 ml fractions. The order of elution will be hydrocarbons, fatty acids, esters, triglycerides, sterols, diglycerides, free fatty acids and phospholipids.

Run under identical conditions, some compounds in individual columns. Using specific tests, locate where they are eluted. Under the same conditions, if an unknown sample is chromatographed, the compound will be eluted at the same position.

## 6.6 ENZYMES

### 6.6.1 Extraction of Enzymes

*Extraction in acetone* - Weigh the tissue, cut into 1-2 cm pieces, transfer to a blender and add chilled acetone (-20°C) enough to cover the tissues (5 ml buffer for each g of tissue). Blend at high speed for 3-5 minutes, filter the extract through Buckner funnel using Whatman No. 1 filter paper and wash the powder with chilled acetone at least for thrice. Wash again with cold diethyl ether. Dry the powder on the funnel with suction and spread the powder on Whatman No. 1 filter paper and air dry for about 1 hour. Store the powder in containers with tight cap in a freezer.

Just before use, weigh 0.1g of the powder, grind in 5 ml of phosphate buffer (0.1M) at pH 6.6 at 4°C for 10-15 minutes in a mortar with pestle. Centrifuge for 30 minutes at 2000 r.p.m. at 4°C temperature. Decant the supernatant and use the clear extract for enzyme assay. Determine the protein content of the extract by Lowry's method to express the results as specific activity per unit of protein.

*Extraction in buffer* - Weigh the tissue, cut into 1-2 cm pieces, transfer to a blender and add chilled (2-4°C) phosphate buffer (0.1M) at pH 6.5 (5 ml buffer for each g of tissue). Grind at low speed for 2-3 minutes, squeeze the extract through cheese cloth to remove pulp and centrifuge the extract for 30 minutes at 4000 r.p.m. at 4°C temperature. Decant the supernatant and use the clear extract for assay of nucleases. Determine the protein content of the extract by Lowry's method to express the results as specific activity per unit of protein.

*Extraction in water* - Weigh the tissue, cut into 1-2 cm pieces, transfer to a blender/mortar-pestle and add distilled water enough to cover the tissues (5 ml buffer for each g of tissue). Blend for 7-10 minutes, filter the extract through cheese cloth and centrifuge at 2000 r.p.m. for 30 minutes at 4°C temperature. Decant the supernatant and dialyze it against several volumes of distilled water at 2-4°C for 24 hours. The enzyme may also be extracted in acetic acid-acetone buffer at pH 5.2. Determine the protein content of the extract by Lowry's method to express the results as specific activity per unit of protein.

### 6.6.2 Assay of Enzymes

#### *Acid & Alkaline Phosphatase*

These are enzymes, which liberate inorganic phosphate from organic phosphate esters. There are two major classes of phosphatases -'**alkaline phosphatase**' and '**acid phosphatase**'.

*Assay:* Acid and alkaline phosphatase activities are determined with appropriate blank and buffered β-glycerophosphate.

*Incubated sample* - Measure 9 ml of alkaline / acid phosphate substrate (Alkaline hydroxylamine - 0.2 ml of hydroxylamine is made up to 100 ml with isopropanol; 0.2%) into a glass-stoppered cylinder and place it in an incubator or water bath at 37°C until the fluid reaches the incubator temperature. Add 1 ml of homogenate, mix and incubate for exactly 60 minutes. Remove, cool in ice water for several minutes and add 2 ml of 30% TCA. Mix well, let stand it for some time and filter through a low-ash filter paper.

*Control* Measure 9 ml of substrate into a cylinder and add 2 ml of 30% TCA. With mixing, add 1 ml of homogenate, shake and filter.

Now transfer 8 ml of each filtered solution in two test tubes. Take 8 ml of standard phosphate solution containing 0.04 mg phosphorus in third tubes and 8 ml of 30% TCA in fourth tube. Add to each tube 1 ml of molybdate solution (Dissolve 25g ammonium molybdate in about 200 ml of distilled water. In a one litre flask, place 300 ml of 10 N sulphuric acid. Add molybdate solution and dilute to one litre with distilled water) and mix well. Add 0.4 ml of amino-naphthosulfonic acid reagent (Place 195 ml of 15% sodium bisulfite solution. Add 0.5g 1,2,4-aminonaphthol sulfonic acid. Add 5 mi of 20% sodium sulfite. Stopper and shake until the powder is dissolved. If solution is not complete, add 1 ml sodium sulfite at a time with shaking. Store in brown bottles), dilute immediately to 10 ml with distilled water and mix well. Allow to stand for 5 minutes for colour development. Measure OD at 660 nm.

*Calculation*

mg inorganic phosphorus/ml =

$$\frac{\text{Optical density of Unknown}}{\text{Optical density of standard}} \times 0.04 \times \frac{3}{2} \times 100$$

**Transaminase**

*Assay:* Mix 0.1 ml of homogenate and 0.5 ml of GOT/GPT substrate (**GOT substrate** — Take 0.292g α-ketoglutamic acid and 26.6g aspartic acid and add 1N NaOH solution slowly with mixing. Adjust pH 7.4. Add sufficient buffer {13.97g $K_2HPO_4$ and 2.69g $KH_2PO_4$, made up to one litre} to make one litre; **GPT substrate** - Take 0.292g α-ketoglutamic acid and 17.8g DL-alanine and add 1N NaOH solution slowly with mixing. Adjust pH 7.4. Add sufficient buffer to make one litre) and incubate at 37°C for 60 minutes. Add 0.5 ml of 2,4-dinitrophenylhydrazine solution (0.198g 2,4-dinitrophenylhydrazine in enough 1N HCl to make one litre) and let stand for 15 minutes at room temperature. Add 5 ml of 0.4N NaOH, mix well and let stand for 20 minutes. Read in a photometer at 500-540 nm setting the instrument to zero with distilled water. Standard curve is prepared as follows.

| *Standard solution (ml)* | *Distilled water (ml)* | *Unit equivalent* |
|---|---|---|
| 0.0 | 2.2 | 0 |
| 0.1 | 2.1 | 20 |
| 0.2 | 2.0 | 55 |
| 0.3 | 1.9 | 95 |
| 0.4 | 1.8 | 148 |
| 0.5 | 1.7 | 216 |

Then, add 1 ml of 2,4-dinitrophenylhydrazine solution and let stand for 15 minutes at room temperature. Add 10 ml of 0.4N NaOH, mix well and let stand for 20 minutes. Read in a photometer at 500-540 nm setting the instrument to zero with distilled water.

## Aldolase

*Assay:* Take 100 µmoles of Tris-HCl buffer (pH 8.6), 12.6 µmoles of FDP, 140 µmoles of hydrazine and 0.02 ml of enzyme preparation in a test tube and make a final volume of 2.5 ml with distilled water. Incubate at 37°C for 10 minutes. The reaction is stopped by adding 2 ml of 20% TCA and the mixture is centrifuged. To 1 ml of TCA supernatant, add 1 ml of 0.75M NaOH. After 10 minutes, add 1 ml of 0.1% 2,4-dinitrophenylhydrazine prepared in 2N HCl. Incubate the solution at 37°C for 30 minutes. Then add 7ml of 0.75M NaOH and measure the colour intensity at 540 nm. One unit of aldolase is that amount, which catalyses the cleavage of one µmoles of FDP or FPD into 2 µ moles of trioses in 10 minutes at 37°C.

## Proteolytic enzyme

These enzymes hydrolyze the peptide bonds of proteins liberating amino acids. Some proteolytic enzymes are exo-proteases liberating amino acids starting from the end and some are endo-proteases attacking peptide bonds in the middle.

*Assay:* Incubate 1 ml of casein solution with 0.2 ml of enzyme at 37°C for 60 minutes. Then, add 1 ml of 12% TCA and cool the mixture rapidly in ice. Centrifuge and collect the supernatant.

Make the solution of the precipitated protein and pipette out 0.1 to 0.2 ml in different tubes. Make the final volume to 4 ml with distilled water. Add 6 ml of biuret reagent (Dissolve 3g copper sulphate and 9g sodium potassium tartarate in 500 ml of 0.2N NaOH and then add 5g potassium iodide and make up the volume to 1 litre with NaOH). Keep the tubes at 37°C for 10 minutes. Measure the optical density at 520 nm.

### *Urease*

The enzyme splits urea molecule liberating ammonia and carbon dioxide.

*Assay:* Pipette out 1 ml of the substrate solution (3% urea solution). Add 1 ml of 0.2M phosphate buffer pH 7.0. Add 1 ml of enzyme extract and incubate at 55°C for 15 minutes, at the end of which quickly place the tube in ice. Now, add 1 ml of 0.66 N sulphuric acid to stop the reaction and 1 ml of sodium tungstate solution to precipitate out the proteins. Filter or centrifuge to remove the precipitate. An aliquot of the supernatant is assayed for ammonium sulphate by Nessler's method. Measure the protein content of the enzyme extract and calculate the specific activity of the enzyme in terms of moles of ammonia liberated / minute / mg protein.

### *Amylase*

*Assay.* Pipette out different sets of solutions into different test tubes following the schedule given below.

| | *Quantity (ml)* | | | | |
|---|---|---|---|---|---|
| | *Tube 1* | *Tube 2* | *Tube 3* | *Tube 4* | *Tube 5* |
| Phosphate buffer (pH 6.7, 0.1 N | 2.5 | 2.5 | 2.5 | 2.5 | 2.5 |
| Starch soution | 2.5 | 0.0 | 0.0 | 2.5 | 2.5 |
| 1% NaCI solution | 1.0 | 1.0 | 1.0 | 1.0 | 1.0 |
| Mix well and keep test tubes for 10 minutes at 37°C | | | | | |
| Distilled water | 1.0 | 1.0 | 0.5 | 0.5 | 0.5 |
| Diluted enzyme solution | 0.0 | 0.0 | 0.5 | 0.5 | 0.5 |

Prepare tubes in duplicate.

Immediately after addition of enzyme solution, add 0.5 ml of 2N NaOH to tube 5 to stop the reaction. This is called 'zero' time control. The rest tubes are incubated at 37°C for 15 minutes. Then, add 0.5 ml of 2N NaOH.

Now, add 0.5 ml of Dinitrosalicylic acid reagent (Dissolve 1g 3,5-dinitrosalieylate, 30g sodium potassium tartarate and 1.6g sodium hydroxide in water and dilute to 100 ml), mix and heat the tubes in a boiling water bath for 5 minutes. Cool the tubes to room temperature and measure optical density at 520 nm using tube 1 as blank. The tube 2 is also a control since this does not contain enzyme / substrate. The tube 3 contains the enzyme but not the substrate (Starch solution - Mix 1g soluble starch with 200 ml of 0.1 N phosphate buffer pH 6.7). Its reading has to be substracted from that of tube 4 & 5. Comparing tube 4 and 5, it is seen that they contain complete mixtures except that the reaction in tube 5 has been stopped immediately on addition of enzyme. In other words, it is zero time control.

Amount of maltose formed in 15 minutes =

(OD tube 4 - OD tube 5) ÷ (OD tube 5 - OD tube 3)

From the standard curve of maltose, calculate the corresponding amount of maltose formed per ml of enzyme.

### *Lipase*

Lipases are the enzymes, which hydrolyze the triglycerides liberating fatty acids.

$$RCOOR' + H_2O \longleftrightarrow RCOOH + R'OH$$

*Assay* A mixture of emulsion, enzyme extract and buffer of known pH is taken in a beaker and the electrodes of a pH meter are kept dipped in it. At frequent intervals, or as the pH drops by about 0.2 units, 0.1N sodium hydroxide is added to bring the pH back to the original level. The amount of NaOH added is the measure of the fatty acid produced.

### ***Esterase***

The enzyme hydrolyzes acetyl choline into acetic acid, which is estimated as carbon dioxide manometrically by adding bicarbonate.

*Assay:* Mix bicarbonate buffer pH 7.4 containing 0.9% NaCl, 1.26% sodium carbonate and 1.76% $MgCl_2$ in the ratio of 100:3:2. The mixture is gassed for 15 minutes with 95% nitrogen and 5% carbon dioxide mixture. Now, take 1 ml of the above mixture, 1.7 ml of enzyme extract and 0.3 ml acetyl choline iodide solution (10 mM in fresh bicarbonate buffer). The substrate is taken in side arm. The thermobarometer contains 3 ml buffer, while the blank has 2.7 ml buffer and 0.3 ml substrate. After gassing all the tubes with 5% $CO_2$ in $N_2$ for 10 minutes, the flasks are closed. After equilibrium of levels, the substrate is tipped and amount of $CO_2$ liberated is followed against time.

### *Arginase*

Take 100 µmoles of arginine, 40 µmoles of glycerine-NaOH buffer pH 9.5 and 0.05 ml of diluted enzyme solution in a test tube. Make up the volume to 1 ml with distilled water. Incubate at 37°C for 10 minutes. Add 2 ml of 10% TCA to stop the reaction and estimate urea.

### *ATPase*

Take 3 µmoles of disodium salt of ATP, 5 µmoles of Tris-HCl buffer (pH 7.4), 0.5 µmoles of EDTA, 0.1 ml of enzyme preparation and when incubated 140 µmoles of NaCl, 14 µmoles of KCl and 1 µmole of Ouabain in a final volume of 1 ml. Incubate at 37°C for 30 minutes. Add 1 ml of 10% TCA. Determine content of orthophosphate by the method of Fiske & Subba Rao.

### *5'-nucleotidase*

$$\text{5-AMP} + H_2O \longrightarrow \text{Adenosine} + \text{Pi}$$

Take 100 µmoles of sodium citrate buffer pH 6.5, 1 µmoles of neutralized substrate (5'-AMP) and various aliquots of enzyme extract in the test tubes. Make up the volume 1 ml with distilled water. Incubate at 37°C for 30 minutse and the released orthophosphate is determined by Fiske & Subbarao method. A unit of enzyme activity corresponds to the liberation of 1 micromole of inorganic phosphorus (Pi) per hour.

### *Cellulase*

Pipette 4 ml of carboxy-methyl cellulose solution (Dissolve 0.5g carboxmethyl cellulose in 100 ml of sodium acetate-acetic acid buffer at pH 5.2 at 50-60°C kept in a blender. Blend for 3-5 minutes at low speed, stir the contents and again blend at high speed for 3-5 minutes. Filter through Whatman No. 1 filter paper and add 2 ml of 1% aqueous merthiolate solution. Store at 4°C.) in a test tube follows by 1 ml of buffer and 2 ml of enzyme extract (water extract). Incubate the tube in water bath, withdraw the aliquots of 1 ml from each tube at pre-fixed intervals and determine the amount of reducing sugars released by Nelson Somogyi's method. Express the enzyme as the amount of glucose released/ml of enzyme extract/unit time.

### *Cellobiase*

Pipette 1.5 ml of the buffer (sodium acetate-acetic acid, pH 5.8), 2.5 ml of 5mM cellchiose (Dissolve in buffer) and 1 ml of enzyme extract (water extract) into a test tube. incubate at 30°C for 2 hours. Terminate the reaction by placing the tube in a boiling water bath for 10 minutes. Measure the amount of glucose released as per Nelson's method.

*Aryl-β-glucosidase*

To 1 ml of 1 mM *p*-nitrophenyl-β-D-glucoside in sodium acetate buffer (pH 5.0; 0.05 M). add 2 ml of enzyme extract (water extract) and 1 ml of the buffer. Incubate the reaction mixture at 37°C for 1 hour. Add 20 ml of 1M sodium carbonate and raise the volume to 30 ml with distilled water. Measure the amount of *p*-nitrophenol released by measuring the absorbance of the solution at 425 nm.

***Deoxyribonuclease***

Pipette aliquots of 2 ml of the enzyme extract (buffer extract) into test tubes. Add 0.5 ml of 0.05% calf thymus DNA solution in 0.05M phosphate buffer pH 6.5 followed by 0.5 ml of 0.2% magnesium chloride solution. After incubating the reaction mixture at 37°C for 4 hours in a water bath, add 0.5 ml of chilled perchloric acid to terminate the reaction and precipitate the unhydrolyzed nucleic acid. Maintain a zero time blank. Cool the contents of the tubes in tap water and centrifuge it at 2000 r.p.m. for 15 minutes. Measure the absorbance of the supernatant at 260 nm in a spectrephotometer. Consider an increase of 1.0 $A_{260}$ /hour as one unit of the enzyme.

***Ribonuclease***

Pipette aliquote of 0.2 ml of yhe enzime extract (buffer extract) into test tubes followed by 1 ml of 0.15% yeast RNA solution in 0.05M phosphate buffer pH 6.5. Incubate tubes containing reaction mixture in a water bath at 37°C for 2 hours; add 0.5 ml of chilled 25% perchloric acid containing 0.75% uranyl acetate to terminate reaction and precipitate the unhydrolyzed ribonucleic acid. Similarly zero time blank. Cool the contents of the tubes in a running tap water and at 2000 r.p.m. for 15 minutes. Measure the absorbance of the supernatant at 260 nm bonus against zero time blank. Consider an increase of 1.0 $A_{260}$ /hour as one enzyme

$$\text{Unit / ml} = \frac{A_{260} \times (\text{ml assay solution} + \text{ml preciplita ry}) \times \text{disolution factor}}{\text{ml enzyme extract solution} \times \text{min utes}}$$

*Succinate Dehydrogenase*

The enzyme catalyzes the dehydrogenation of succinate to fumarate.

*Assay:* The assay mixture consists of - 5 ml of 0.4M potassium phosphate buffer pH 7.2, 0.2 ml of 0.15M sodium succinate pH 7.0, 0.2 ml of 0.2M sodium azide and 0.2 ml of DCPIP (6 mg/ml). The optical density of the mixture is measured at 600 nm. Now, add 0.4 ml of the enzyme extract (containing 0.5 to 1.0 mg protein) to the assay mixture, mix well and measure the optical density at 600 nm. This is the zero value. Incubate the tube at the desired temperature and measure the optical density after 5 minute intervals till there is no change.

The time course is drawn on a graph sheet. Generally, the graph will be exponential. The rate is calculated from the linear portion of the curve. The molar extinction coefficient of pure DCPIP is of the order of $19.1 \times 10^3$. Thus, from the change in optical density the amount of DCPIP reduced in mole units can be calculated.

***Glucose Oxidase***

The enzyme catalyzes the conversion of D-glucose to D-glucose-r-lactone, which spontaneously hydrolyzes to gluconic acid non-enzymatically forming hydrogen peroxide. The enzyme has very high specificity for the substrate, D-glucose.

*Assay:* Take two test tubes or cuvettes and mark them as control and test. To the control tube, add 2.6 ml of dye-buffer solution (add 0.1 ml of 1% aqueous solution of o-dianisidine to 12 ml of 0.1 M phosphate buffer), 0.3 ml of 18% glucose solution and 0.1 ml of peroxidase enzyme. To the other tube, add 2.5 ml of dye-buffer, 0.3 ml of glucose and 0.1 ml of peroxidase. Adjust the optical density of test solution to zero against the control solution at 460 nm. Now, add 0.1 ml of enzyme extract to the test solution, mix well and measure the increase in optical density at 460 nm at 30-second intervals for 3-4 minutes. Plot a graph and calucate the rate.

7

# ECOLOGICAL STUDIES

## 7.1 ECOSYSTEMS

### 7.1.1 Grassland Ecosystem

Grasslands are one of the most important ecosystems in the world, which provide a natural source of food for grazing animals. The legumes in grasslands help in maintaining nitrogen content of the soils and the grass cover is a protection against soil erosion. They are also the systems, which can be easily converted to an agro-system. The abiotic environment of the grasslands constitutes the soil and overlying air. The grassland soils have the same features as are common to forest soils except that they have a high rate of humification but slow rate of mineralization (Odum 1971). The grassland communities comprise of producers, herbivores, carnivores and decomposers. The main producer component is the grasses, which are tall, mid or short in height.

### 7.1.2 Forest Ecosystem

The forest is an ecosystem comprising of mainly the trees and varying assemblage of herbs and shrubs. They occur in both tropical and temperate regions of the world and are usually classified based on climate, external appearance (physiognomy), floristics or the dominant vegetation. Forests accounts for about 23% of the total land area and are one of the most productive ecosystems in the world.

### 7.1.3 Desert Ecosystem

Deserts account for one fifth of the land surface of the world. They occur in the regions having less than 25 cm rainfall. They are either hot deserts or cold deserts. The extreme environment of a desert leads to development of highly adapted organisms.

### 7.1.4 Aquatic Ecosystem

Aquatic ecosystem is primarily comprise of oceans, seas, lakes, ponds, rivers, springs and transition zones between oceans and the rivers. The organisms found in an aquatic ecosystem can be grouped on the basis of their life forms and their position in food chain. Major organisms are planktons, bentghos, periphytons, neustons and nektons. However, the kind, distribution and diversity of the aquatic organisms are affected by many factors, such as, temperature, movement of currents and concentration of oxygen, carbondioxide, nutrients and salts.

## 7.2 METHODS FOR GRASSLAND FOREST STUDIES

### 7.2.1 Estimation of Frequency, Density Biomass and Coverage

These are the non-destructive parameters for describing the essential characteristics of the vegetation. There are three sampling units to assess these characteristics.

*Quadrats* - A quadrat is a sampling unit having four arms or a small enclosure for the community. It may be circular, rectangular or square in shape. These units are used to enclose an area and the plants falling in it are studied. The heterogeneous nature of the vegetation demands that the size and number of quadrat must be adequate to provide a correct sampling.

*Transects* - This method is employed when the vegetation has sudden spatial variations as on a hill top or along any gradient. Line transect is a long calibrated string put across the vegetation and plants touching the lines are recorded. In belt transect, a strip of same width is selected and divided into several segments and the vegetation is studied in these segments.

*Point Frame* - In this method, a frame is used which contains 10 holes. The frame is fixed with legs and 60 cm pins hang through these holes. The frame is put to various places in the ecosystem and the pins are lowered. The plants which are touched by these pins are recorded.

#### *7.2.1.1 Estimation of Frequency*

*Frequency denotes the homogeneity of distribution of various species in the ecosystem. It is expressed as the percentage of units in which species occurred out of all the units studied.*

$$\text{Frequency} = \frac{\text{No. of units in which the species occurred}}{\text{No. of units studied}} = 100$$

*Quadrat Method:* Lay out the required number of quadrats (circular; 6-12 m radius for forest ecosystem) in the ecosystem and count the presence and absence of the species. For example,

| *Name of the species* | *Occurrence in quadrats* | | | | | | | | | | *Frequency* |
|---|---|---|---|---|---|---|---|---|---|---|---|
| | *1* | *2* | *3* | *4* | *5* | *6* | *7* | *8* | *9* | *10* | *%* |
| A | ✗ | ✓ | ✓ | ✗ | ✓ | ✗ | ✗ | ✓ | ✓ | ✓ | 60% |
| B | | | | | | | | | | | |
| C | | | | | | | | | | | |
| D | | | | | | | | | | | |
| E | | | | | | | | | | | |
| F | | | | | | | | | | | |

The frequency is calculated using the formula given above. From this frequency data, various species can be distributed to different frequency classes and a histogram can be drawn to understand the distribution pattern of the vegetation.

### *Line Transect Method*

Before sampling, it should be ensured that the area is at least ecologically uniform. A 100 m surveyor's tape, conspicuously marked at 10 m intervals is laid on the ground in different directions. Then, walk along the chain and count ground cover plant and tree seedlings whose projections touch one edge of the chain. The number of plants is recorded in a special form carried along to the field. The number is recorded separately for each segment of 10 m. The data is recorded in the following table.

| *Species* | *Length of the tape (cm)* | | | | | | | | | *Frequency* |
|---|---|---|---|---|---|---|---|---|---|---|
| | 10 | 20 | 30 | 40 | 50 | 60 | 70 | 80 | 90 | 100 |
| A | 1 | 2 | 3 | - | 2 | - | 4 | 2 | 2 | 180 |
| B | 3 | - | 5 | 2 | 2 | 2 | 5 | 4 | - | 380 |
| C | | | | | | | | | | |
| D | | | | | | | | | | |
| E | | | | | | | | | | |

The frequency is calculated as -

$$\text{Frequency} = \frac{\text{No. of segments in which species is present}}{\text{Total number of segments studied}} \times 100$$

*Point Frame Method :* Put the point frame at a number of places all over the field and note the hits recorded by a particular species. Use the following formula to calculate the frequency.

$$\text{Frequency} = \frac{\text{Total No. of hits of a species occurred}}{\text{Total No. of hits used}} \times 100$$

Total number of hits = Number of holes in the frame – Number of times the frame was used

### 7.2.1.2 Estimation of Density

*Density is defined as the number of individuals of a species in an unit area.* While the frequency is concerned only with the presence or absence of a species, the density takes into account the number of individuals. A species may have high frequency but the density may be less.

*Quadrat Method:* Lay the required number of quadrats (10 × 10 m or larger) or belt transects as per the frequency estimation procedure and count the number of individuals of each species in all the quadrats or in the segments of the belt transect. The density from the quadrat data can be calculated as follow:

$$\text{Density (D)} = \frac{\text{Total number of individual s of the species}}{\text{Total number of quadrats taken}}$$

Density is represented as individuals or plants per square meter. Take a quadrat of 10 × 10 m or larger. Measure the circumference of all the trees present in this quadrat by using a tape and record the data in the following manner.

| *Plant* | *Circumference of the individuals in cm* | | | | | | | | | |
|---|---|---|---|---|---|---|---|---|---|---|
| *species* | *1* | *2* | *3* | *4* | *5* | *6* | *7* | *8* | *9* | *10* |
| A | | | | | | | | | | |
| B | | | | | | | | | | |
| C | | | | | | | | | | |
| D | | | | | | | | | | |

The radius of the bole of the individual tree is calculated as

r = Circumference ÷ 2 A

where r = radius of the tree, and A = 3.14

Basal area = A $r^2$

% basal area = [Total basal area ÷ Area of study] × 100

Relative dominance of species (R.D.) = [Total basal area of the species in all the quadrats ÷ Total basal area of all the species in all the quadrats] × 100

### 7.2.2 Vegetation Mapping

*The vegetation mapping is the recording of the composition and abundance of a particular type of vegetation on a map, which shows the vegetation pattern and distribution in the area.*

The vegetation maps can be prepared on various scales depending on the aims of the study. Usually, the maps of 1:50,000 are prepared where details are necessary. The first step is to get an aerial photograph or topographic survey sheet of that area. If an aerial survey photograph is not available, a map is prepared using any ground survey methods. Different vegetations are shown by different symbols.

### 7.2.3 Estimation of Biomass and Primary Production

The following relationship exists in a terrestrial ecosystem between the production and other components (Chapman, 1976),

$P_n$ = B + L + G

Where,

$B_1$ = Biomass of a plant community at time $t_1$

$B_2$ = Biomass of a plant community at time $t_2$

$\Delta t$ = $t_2 - t_1$

$\Delta B$ = Change in biomass during the period $t_2 - t_1$

L = Plant losses by death and shedding during the period $t_2 - t_1$

G = Plant losses by grazing etc during the period $t_2 - t_1$

$P_n$ = Net Production by the community during the period $t_2 - t_1$.

There are several methods for measurement of primary production in an ecosystem.

### *7.2.3.1 Harvest Method*

It is the simplest method in which the density of vegetation (number of plants per square meter) and average weight of one plant is estimated and primary production is determined by multiplying the density with the weight of one plant. It is represented by g dry weight per $m^2$. This method can be used in the stands where the density is low, the number of the species present is low and all plants are of more or less the same age.

In case only above ground vegetation is to be considered, the vegetation may be cut by using hand shears, secatures or some other mechanical device from all the quadrats selected for the study. The plant material from each quadrat is collected in separate polythene bags with a tag of quadrat number. The dead material fallen on the ground is also collected in separate bags for each quadrat. The plants are brought to the laboratory, washed in a jet of water and kept on a slope to drain off the excess water. The fresh weight is taken on a balance. If the biomass is to be estimated for each species, the species are separated and the weights are taken. A weighed portion of the plant material is dried at 105°C temperature overnight in a hot air oven and dry weight is recorded next day. The dead material is also treated in the same way. This represents the biomass. For calculating annual primary productivity, the biomass is taken at monthly intervals and net annual primary productivity is calculated by summing up all positive values. For example,

| *Month* | *Biomass ($g/m^2$)* | *Increase or decrease in biomass* | *Net Primary Productivity* |
|---|---|---|---|
| July | 274.9 | + 74.9 | Total of all positive values = 236.8 $g/m^2$ |
| August | 344.9 | + 70.0 | |
| September | 416.1 | + 71.2 | |
| October | 436.7 | + 20.6 | |
| November | 379.3 | -59.4 | |
| December | 325.6 | -53.7 | |

| | | |
|---|---|---|
| January | 278.2 | -47.4 |
| February | 243.4 | -34.8 |
| March | 236.4 | -7.0 |
| April | 205.8 | -30.6 |
| May | 199.8 | -6.0 |
| June | 200.0 | - |

In case of trees, 5 or 10 sample plots (quadrats) are marked in a selected forest on random basis. The height, diameter at breast height, depth and spread of the tree is noted. These observations are made for each species. Now, 2 or 3 trees of each species are selected and felled. The trees are cut into various portions, like roots, boles, leaves and fruits. Weigh and carry a suitable sample to the laboratory for dry weight estimation. Record the data once before the growing season and then at the end of the growing season. The difference between two gives annual increment. The data is represented in kg dry weight per $m^2$.

Determination of an average tree

| *Tree* | *Height (m)* | *Diameter (m)* | *Canopy spread (m)* |
|---|---|---|---|
| Tree A | 10 | 1.2 | 20 |
| Tree B | 10 | 1.8 | 18 |
| Tree C | 7 | 1.2 | 12 |

Average tree height = 9.0 m

Diameter = 1.4 m

Canopy spread = 17 m

Determination of biomass and primary production

Quadrat 1

| *Sl. No.* | *Species* | *Oven Dry Weight* | | | | | |
|---|---|---|---|---|---|---|---|
| | | *Roots* | *Bole* | *Branches* | *Leaves* | *Fruits* | *Total* |
| | | | | | | | |

Average of all the quadrats is biomass in kg dry weight per plant. This is multiplied by density to get values in kg dry weight per $m^2$.

#### *7.2.3.2 Indirect Method*

It is usually not possible to harvest the sample area in a forest and hence an indirect estimation technique is employed in which relationship between diameter at breast height and biomass is established and regression model are developed. For example,

$$\text{Log e } Y = a + b \log e\, X$$

Where,

Y = Biomass

X = Some readily measured parameter of the tree

a & b are constant, which are to be derived

### 7.2.4 Vegetation Analysis

#### *7.2.4.1 Association*

**It refers to the characteristic assemblage of species and also implies to the measurement of the similarity of occurrence of two species.** The association may be positive when two species likely to occur together or negative when the species tend to occur separately. Changes in degree of association may be caused by several factors. The association may be evaluated by use of 2 x 2 contingency table and $\chi^2$ (chi square) test. In this test, data is collected on the number of quadrats out of the total quadrats studied in which one or both the species are found. The data is presented in 2 x 2 contingency table as below.

| | | *Species A* | | |
|---|---|---|---|---|
| | | *Present* | *Absent* | |
| Species B | Present | a | b | a + b |
| | Absent | c | d | c + d |
| | | a + c | b + d | *n* |

In the above case, there are two species A and B whose association is to be worked out and *n* is the total number of quadrats studied. a, b, c and d denote the frequencies as:

a = Total number of quadrats out of n in which both the species are present b = Total number of quadrats out of n in which only species B is present.

c = Total number of quadrats out of n in which only species A is present,

d = Total number of quadrats out of n in which only both the species are absent.

Suppose 100 quadrats are studied in a community and the results are presented as follows.

| | | Species A | | |
|---|---|---|---|---|
| | | *Present* | *Absent* | |
| Species B | Present | 32 | 16 | 48 |
| | Absent | 10 | 42 | 52 |
| | | 42 | 58 | *100* |

The calculated $\chi^2$ is obtained by the following formula.

$$\chi^2 = \frac{(ad - bc)^2 n}{(a+b)(c+d)(a+c)(b+d)}$$

$$\chi^2 = \frac{[(32 \times 42) - (16 \times 10)]^2 \times 100}{(32+16)(10+42)(32+10)(16+42)}$$

$$\chi^2 = [(1344 - 160)^2 \times 100] \div (48 \times 52 \times 42 \times 58)$$

$$\chi^2 = 1401185600 \div 6080256$$

$$\chi^2 = 23.05$$

See the table value of critical $\chi^2$ from the $\chi^2$ distribution table at 1 degree of freedom (degree of freedom for 2 x 2 contingency table is always 1 at 5%). If the calculated value is higher than the critical value, then the species has positive association. If the critical value is lower than the critical $\chi^2$, it indicates negative association between the species.

### *7.2.4.2 Correlation*

Besides mere presence or absence of the species, the relationship between the quantities of two species is also important in the community structure. When the data are in the form of quantitative terms such as cover, frequency, abundance or density, a correlation coefficient is the simplest measure of relationship between species, which is calculated by the following formula.

$$\text{Correlation coefficient } (x_r) = \frac{(\sum XY - X\sum y)}{\sqrt{\phi\sum x^2 - x\sum x \sum y^2 - y\sum y}}$$

where, x and y are the values of either species and x and y are their means. The calculated value of r varies between +1 and -1, which signify as,

+1 = Perfect positive correlation

0 = No correlation

-1 = Perfect negative correlation

### *7.2.4.3 Diversity*

Total number of species present in any ecosystem indicates its richness in species, while the number of individuals of a species in relation to the total number of individual of all the species denotes the dominance. These parameters are important aspects of an ecosystem structure and can be determined by mathematical expressions like diversity indices. The most common index used for quantification of the ecosystem structure is Shannon and Wiener's diversity index, which is calculated by the following formula.

Diversity Index (H) = $\sum$ pi. $\log_2$ pi where,

pi = ni / N

(ni is the number of individuals of a species and N is total number of all the individuals of all the species) $\log_2$ can be calculated from the values of $\log_{10}$ or $\log_{ln}$ multiplying by the following factors,

$x \log_2 = x \log_{ln} \times 1.44296$

$x \log_2 = x \log_{10} \times 3.3219281$

The diversity index is always positive and is denoted by *bits per individual*. For example,

| Species | *Situation A* | *Situation B* | *Situation C* | *Situation D* | *Situation E* | *Situation F* |
|---|---|---|---|---|---|---|
| A | 10 | 20 | 30 | 40 | 60 | 80 |
| B | 10 | 20 | 20 | 10 | 20 | 0 |
| C | 10 | 10 | 5 | 5 | 5 | 0 |
| D | 10 | 10 | 5 | 5 | 5 | 0 |
| E | 10 | 5 | 5 | 5 | 0 | 0 |
| F | 10 | 5 | 5 | 5 | | 0 |
| G | 10 | 5 | 5 | 5 | 0 | 0 |
| H | 10 | 5 | 5 | 5 | 0 | 0 |
| 8 | 80 | 80 | 80 | 80 | 80 | 80 |
| Diversity Index | 3 | 2.75 | 2.531 | 2.375 | 1.424 | 0 |

The table shows six hypothetical situations where the species are distributed with different dominances. The maximum possible diversity can be obtained by the formula,

Maximum possible diversity for N number of species = N $\log_2$

In the present example, the maximum possible diversity with 8 species is 8 $\log_2$ = 3 bits.

**8**

# BIOSTATISTICS

In present-day competitive and dynamic world, there is an ever-increasing importance for decision-making processes, which require sophisticated methods of data analysis and hence there is wide applicability of the quantitative techniques. Quantitative techniques refer to the group of statistical and operation research techniques. Statistics is defined as "a **body of scientific methods used to acquire systematic knowledge in every sphere of life".** These scientific methods are referred as statistical methods, which are used to collect, organize or classify, present, analyze and interpret the information effectively for the purpose of taking wise decisions. The statistical methods broadly fall into three categories - descriptive statistics (data collection and presentation), inductive statistics (statistical inference and estimation) and statistical decision theory (analysis of business decision). The **descriptive statistics** helps in summarizing the results of an investigation or experiment in charts and graphs, in well arranged tables or in terms of mean, range and standard deviation, i.e., re-arranging, grouping and summarizing sets of data to obtain better information of facts and thereby better description of situation that can be made. It include the methods of collection and presentation of data, measure of central tendency and dispersion, trends, index numbers etc. **Inductive statistics** is concerned with the development of some criteria, which can be used to derive information about the nature of the members of entire group (population) from the nature of small portion (sample) of the given group. The specific values of the population are called parameters and that of sample 'statistics'. Inductive statistics include the methods like probability and probability distribution, testing of hypothesis, correlation, regression, factor analysis, time series analysis etc. The **statistical decision theory** deals with analyzing complex business problems with alternative course of action and possible consequences. Statistics have got wide applicability in biological sciences and hence it is also referred as Biostatistics. Biostatistics can be defined as **the body of methods used in design, analysis and interpretation of biological experimentation.**

## 8.1 ARRANGING DATA: TABLES AND GRAPHS

**Data are collections of any number of related observations. A collection of data is called a data set and a single observation is a data point.** When it means statistical data, it refers to numerical description of quantitative aspects of things. These descriptions may be in the form of counts or measurements. But not all numerical data is statistical data. The statistical data possesses the following characteristics -

- Data must be aggregate of facts,
- They should be affected to a marked extent by multiplicity of causes,
- They must be enumerated or estimated according to reasonable standard of accuracy,
- They must be collected in a systematic manner for a pre-determined purpose,
- They must be placed in relation to each other,
- They must be numerically expressed.

Thus, statistical data is a set of facts expressed in quantitative form. The data is collected from a sample or the population that the sample represents. **A population is a collection of all the elements or items on which information is required and about which we are trying to draw conclusion. A sample is a collection of some of the elements of the population or any group of measurements selected from the population.** A representative sample contains the relevant characteristics of the population in the same proportion, as they are included in that population. Statistical data can be broadly grouped in two categories - primary data and secondary data. **Primary data** are those, which are collected afresh and for the first time under the control and supervision of the investigators or enumerators and thus happen to be original in character. It may either be collected through the observation method or through the questionnaire method (personal interview, mail questionnaire or telephone). The **secondary data,** on the other hand, are those that have been collected and analyzed by some other agency. The sources of secondary data may be published (national and international publications) or unpublished sources (internal records of different agencies). Statistical data may also be classified as micro and macro. **Micro data** relate to a particular unit or region whereas **macro data** relate to the entire industry, region or economy.

### 8.1.1 Classification of Data

After data has been systematically collected, it is called **raw data.** Arranging data in descending or ascending order is referred as data array, which is the simplest way to present data. Data array has several advantages over raw data,

- We can quickly find the lowest and highest values in the data.
- We can easily divide the data into sections.
- We can see whether any value comes more than once in the array.
- We can observe the distance between succeeding values in the data.

However, a data array is not useful many a times for interpretation and decision-making. It needs to be further compressed and analyzed for better presentation. The next step in presentation of data is classification. **Classification of data is the process of arranging the data according to the points of similarities and dissimilarities.** The man objectives of classification of data are -

- to condense the mass of data in such a way that its salient features can easily be seen,
- to facilitate comparisons between attributes of variables,
- to prepare data which can be tabulated, and
- to highlight the significant features of the data at a glance.

The characteristics of an ideal classification are -

- It should be unambigious
- It should be stable
- It should be flexible

There are four types of classification methods - geographical (according to area or region), chronological (according to occurrence of an event in time), qualitative (according to attributes) and quantitative (according to magnitude). In quantitative classification of data, the data represents the value of a variable that is measurable. Quantitative data may further be classified unto two types -discrete or continuous. The discrete data is the quantitative data that is limited to certain numerical value of the variable. For example, number of employees in an organization or number of machines in a factory is discrete data. Continuous data takes all values of the variable. For example, weight, distance, volume etc are continuous data.

**Tabulation is an arrangement of data in columns and rows.** It involves the systematic presentation of data to elucidate the problem under investigation. It is a process between the collection of data on the one hand, and its final analysis on the other. In fact, tabulation is meant to properly arrange the answers relating to the questions posed in any investigation, and is very helpful in analysis of the collected data as also in drawing inference from them. Tabulation is the final stage in collection and compilation of data, and is sort of stepping stone to the analysis and interpretation of figures.

Quantitative classification forms the basis for frequency distribution. **When the data is arranged into groups or classes or categories accordingly to conveniently established divisions or the range of the observation, the process is called frequency distribution and tabular form of data is known as frequency table.** In a frequency distribution, raw or array data is represented by distinct groups, which are known as **classes.** The number of observations that fall into each of the classes is known as **frequency.** When data is described by a continuous variable, it is called continuous data and when it is described by discrete variable, it is called discrete data. A frequency distribution shows the number of observations from the data set that fall into each of the classes and frequency of each value is expressed as a fraction or aa percentage of the total number of observations. **A relative frequency distribution presents frequencies in terms of fractions or percentages.**

### 8.1.2 Construction of A Discrete Frequency Distribution

The process of constructing a discrete frequency table is very simple. In case of discrete data, place all the possible values of the variable in ascending order in one column, and then prepare another column of Tally' mark to count the number of times a particular value of the variable is repeated. To facilitate counting, block of five Tally' marks are prepared and some space is left in between blocks. The frequency column refers to the number of Tally' marks, a particular class will contain. For example, a sample study of 50 families is done to find the number of children per family and the data of the study is as follows.

3, 2, 2, 1, 3, 4, 2, 1, 3, 4, 5, 0, 2, 1, 2, 3, 3, 2, 1, 1, 2, 3, 0, 3, 2, 1,4, 3, 5, 5, 4, 3, 6,5,4,3,1,0,6,5,4,3,1,2,0,1,2,3,4,5

To condense the data into a discrete frequency distribution, we will take the help of Tally' marks as shown below.

| *No. of Children* | *No. of Families* | *Frequency* |
|---|---|---|
| 0 | IIII | 4 |
| 1 | IIIII IIII | 9 |
| 2 | IIIII IIIII | 10 |
| 3 | IIIII IIIII | 12 |
| 4 | IIIII II | 7 |
| 5 | IIIII II | 6 |
| 6 | II | 2 |
| Total | | 50 |

### 8.1.3 Construction of A Continuous Frequency Distribution

In construction of the frequency distribution for a continuous data, the following steps are involved.

#### *8.1.3.1 Decide upon the type and number of classes for dividing the data*

*Class Limits* - Class limit denotes the lowest and highest value that can be included in the class. The two boundaries of a class are known as the lower limit and the upper limit of the class.

*Class intervals* - It represents the width (span or size) of a class. The width may be determined by subtracting the lower limit of one class from the lower limit of the following class (alternately successive upper limits may be used).

*Class Frequency* - The number of observations falling within a particular class is called its class frequency or frequency. Total frequency (sum of all the frequencies) indicates the total number of observations considered in a given frequency distribution.

*Class Mid-point* - Mid-point of a class is defined as the sum of two successive lower limits divided by two. Therefore, it is the value lying halfway between the lower and upper class limits.

*Type of Class Interval* - There are different ways in which limits of class intervals can be shown such as: mutually exclusive and all-inclusive, and open-ended. When all the data fit into one class or another i.e. no data point falls into more than one category, such classes are called *all-inclusive* and *mutually exclusive.* In exclusive method, the upper limit of one class is the lower limit of the next class but upper limit is excluded for determining the class frequency in that class. In inclusive method, the upper limit of one class is included in that class itself. In open-ended classes, either the upper or the lower end of a quantitative classification is limitless. *Discrete* classes are separate entities that do not progress from one class to the next without a break while *continuous* classes do progress from one class to the next without break.

To classify the data, the first step is to decide how many different classes to use and the range each class should cover. The following guidelines are followed in choosing the classes and the class intervals -

1. The number of classes should not be too small or too large but shall depend upon the number of data points and the range of the data collected. Normally, number of classes should be between 5 and 15 classes. However, the more the data points or the wider the range of the data, there would be more number of classes.
2. The range should be divided by equal classes, i.e., the width of the class intervals from beginning of one class to the beginning of the next class needs to be the same for every class.
3. As far as possible, the widths of the intervals should be numerically simple like 5, 10, 25 etc. Values like 3, 7, 19 etc. shall be avoided.
4. The starting point of a class should be 0, 5, 10, or multiples thereof.

5. The class interval should be determined after taking into consideration the minimum and maximum values and the number of classes to be formed.

Width of the class intervals =

$$\frac{\text{Next unit value after largest value in data - smallest value in data}}{\text{Total number of class intervals}}$$

### *8.1.3.2 Sort the data points into classes and count the number of points in each class*

Each data point is sorted out into different classes so that it fits into at least one class and no data point fits into more than one class. It should be ensured that the smallest data point corresponds with the lower boundary of starting class and the largest data point with the upper boundary of the last class. The class frequency thus obtained can be expressed in different ways - cumulative frequency or relative frequency. Cumulative frequency cumulates the frequencies, starting at either the lowest or highest value. The cumulative frequency of a given class represents the total of all the previous class frequencies including the class against which it is written. Relative frequency shows the percentage of frequency for each class and it is obtained by dividing the frequency of each class by total number of observations (total frequency).

### *8.1.3.3 Illustrate the data in a chart*

The frequency distribution can be presented by different methods of charting, which enable a quick interpretation of the data. There are four common methods of charting frequency distribution.

- Bar diagram
- Histogram
- Frequency Polygon
- Ogive or Cumulative Frequency Curve

To construct a diagram or graph, the class intervals are taken on the X-axis while the frequency on the Y-axis.

*Bar Diagram:* In a simple bar diagrams, one bar represents only one figure and as such there will be as many bars as the number of figures. Such diagrams represent only one particular type of data.

For example, the number of students in a university year after year can be represented by such bars. Each bar would represent their number in a particular year.

Multiple bar diagrams are prepared on the basis of simple bar diagrams. These diagrams represent more than one type of data at a time.

The sub-divided bar diagrams are those in which each bar is divided into certain parts and each part represents a particular phenomenon.

*Histogram:* A histogram is a set of vertical bars whose areas are proportional to the frequencies represented. While constructing histogram, the variable is always taken on the X axis and the frequencies depending on it on the Y axis. Each class is then represented by a distance on the scale that is proportional to its class interval. The distance for each rectangle on the X axis shall remain the same in case the class interval is uniform throughout. The area of the histogram represents the total frequency as distributed throughout the classes.

*Frequency Polygon:* A frequency polygon is a graph of frequency distribution. It has more than four sides. It is particularly effective in comparing two or more frequency distribution. There are two methods of constructing frequency polygon.

We may draw a histogram of the given data and then join by straight lines the midpoints of the upper horizontal side of each rectangle with the adjacent ones. The figures so formed is called frequency polygon. Another method of constructing frequency polygon is to take the midpoints of the various class intervals and then plot the frequency corresponding to each point and then to join all these points by straight lines. Frequency polygon has a special advantage over the histogram. The frequency polygons of several distributions may be plotted on the same axis, thereby making certain comparisons possible, whereas histograms cannot be usually employed in the same way.

*Smoothed frequency curve:* A smoothed frequency curve can be drawn through the various points of the polygon. The curve is drawn freehand in such a manner that the area included under the curve is approximately the same as that of the polygon. The objective

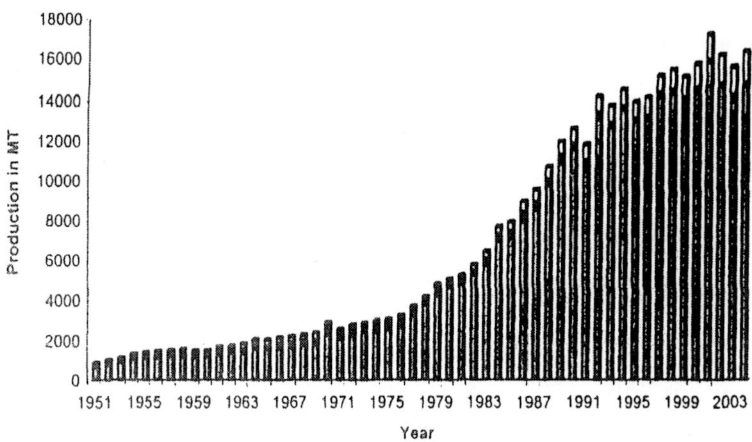


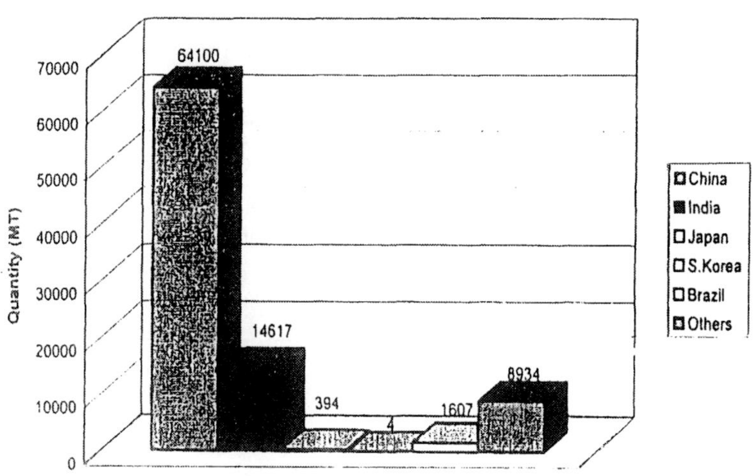


Fig 8.1 Methods of charting of Data

of drawing a smoothed frequency curve is to eliminate, as far as possible, accidental variations that might be present in the data. The following points should be kept in mind while smoothing a frequency curve.

1. Only frequency distribution based on samples should be smoothed.
2. Only continuous series should be smoothed.
3. The total area under the curve should be equal to the area under the original.

*Cumulative frequency curves or Ogives:* The curve obtained by plotting cumulative frequencies is called the cumulative frequency curve or an ogive. There are two methods of constructing ogive - Less than ogive and More than ogive.

In the less than ogive method, we start with the upper limits of the classes and go on adding the frequencies. When these frequencies are plotted, we get a rising curve. In the more than method, we start with the lower limits of the classes and from the frequencies we subtract the frequency of each class. When these frequencies are plotted, we get a declining curve.

For example, the ages of 50 employees of a company is given below.

22, 21, 37, 33, 28, 42, 56, 33, 32, 59, 40, 47, 29, 65, 45, 48, 55, 43, 42, 40, 37, 39, 56, 54, 38, 49, 60, 37, 28, 27, 32, 33, 47, 36, 35, 42, 43, 55, 53, 48, 29, 30, 32, 37, 43, 54, 55, 47, 38, 62

In order to form the frequency distribution of this data, we take the difference between 60 and 21 and divide it by class interval 10 to form 5 classes as follows.

| *Classes (Age, years)* | *Tally Marks* | *Frequency* | *Cumulative frequency* | *Relative frequency* |
|---|---|---|---|---|
| 20-30 | IIII II | 7 | 7 | 0.14 |
| 30-40 | IIII IIII IIII I | 16 | 23 | 0.32 |
| 40-50 | IIII IIII IIII | 15 | 38 | 0.30 |
| 50-60 | IIII IIII | 9 | 47 | 0.18 |
| 60-70 | III | 3 | 50 | 0.06 |
| | | 50 | | 1.00 |

**Variety-wise share of Raw Silk Production during the year 2004-2005 (Pie Chart)**

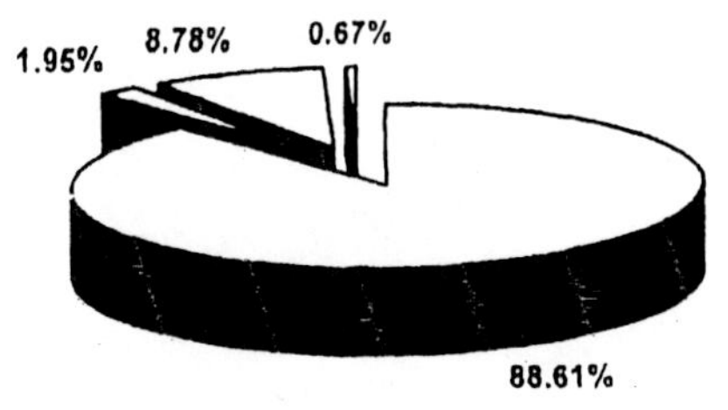


☐ Mulberry ☐ Tasar ☐ Eri ☐ Muga

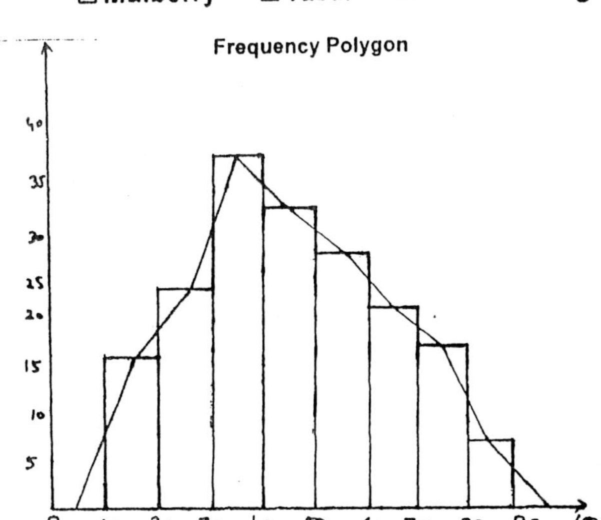


## 8.2 DATA ANALYSIS AND INTERPRETATION

### 8.2.1 Percentages

Percentage refers to a special kind of ratio, which are used in making comparison between two or more series of data. Percentages are used to describe relationships. Percentages are also used to compare the relative terms, the distribution of two or more series of data. It is calculated as frequency of a particular cell or value multiplied by 100 and divided by total value.

### 8.2.2 Growth Rates

When an entity increases in magnitude over a period of time, it is referred to as have grown over that period. The growth is measured as a ratio of the increase in magnitude of the entity to its initial magnitude. Growth rate is the growth in one unit of time, expressed in percentage, i.e.,

$$P_n = P_o\ (1+r)^n$$

$$\text{or, } r = [\{\sqrt[n]{(P_n/P_o)}\}-1] \times 100$$

where,

$P_o$ = Magnitude of entity in the beginning of the unit time

$P_n$ = Magnitude of entity at the end of the unit time

$r$ = Growth rate

$n$ = Period of time.

### 8.2.3 Measures of Central Tendency

One of the most important objectives of statistical analysis is to get one single value that describes the characteristics of the entire mass of the unwidely data. Such a value is called a central value.

#### *8.2.3.1 Measures of Location or Averages*

The measure of Central Tendency i.e. condensing the mass of data in one single value enables us to get an idea of the entire data. It also enables us to compare two or more sets of data to facilitate comparison. A good measure of central tendency should possess the following properties.

- It should be easy to understand.
- It should be simple to compute.
- It should be based on all the observations.
- It should be uniquely defined.
- It should be capable of further algebraic treatment.
- It should not be unduly affected by extreme values.

There are seven types of averages, which are commonly used in business, industry and research as well.

1. Arithmetic average or mean
2. Weighted Arithmetic Mean
3. Median
4. Quantiles
5. Mode
6. Geometric Mean
7. Harmonic Mean

**Arithmetic Mean or Average**

The arithmetic mean (mean or average) is the most commonly used and readily understood measure of central tendency. It is defined as the sum of the numerical values of all the items (the scores or measures or values of variables recorded in raw data) divided by the total number of values added or items or observations. It is expressed mathematically as,

$$X = \sum xi/n \quad i = 1 \;.......n \text{ (n: total number of observations)}$$

where, $\Sigma x_i$ indicates the sum of the values of all the observations and n is the total number of observations. If, for ungrouped data, $x_1$, $x_2$, $x_3$, ....$x_n$ are the n values of a variate $X_i$, then the arithmetic mean is calculated as follows.

$$\text{Mean} = \frac{X_1 + X_2 + X_3 + \,..........X_n}{n}$$

$$\text{or } \frac{\sum x_i}{n}$$

When the observations are classified into a frequency distribution and if the value $X_1$ occurs $f_1$ times, the value $X_2$ occurs $f_2$ times ......... the value $x_n$ occurs $f_n$ times, then

$$\text{A. M.} = \frac{f_1x_1 + f_2x_2 + \,....... + f_nx_n}{N}$$

$$= \frac{\sum f_iX_i}{n}$$

where,

$N = f_1 + f_2 + \ldots\ldots\ldots\ldots f_n$

= Total frequency

For example, calculate the average height of plants of *T. arjuna* from the following frequency distribution.

| *Height of plants* | *No. of Plants* $(x_i)$ $(f_i)$ | $f_ix_i$ |
|---|---|---|
| 5 | 1 | 5 |
| 4 | 2 | 8 |
| 3 | 3 | 9 |
| 2 | 2 | 4 |
| 1 | 1 | 1 |
| | N = 9 | $\sum f_ix_i = 27$ |

$X = \sum f_ix_i / N = 27/9 = 3$

For the grouped data, the midpoint of the class interval would be treated as the representative average value of that class and hence, the arithmetic mean is defined as follows.

$$\text{Mean} = \sum fX / N$$

where X is the midpoint of various classes, f the frequency for corresponding class and N is the total frequency.

Calculate the average length of branch of *T. tomentosa* plant from the following grouped data.

| *Interval (inches)* | *Mid value (xi)* | $f_i$ | $f_ix_i$ |
|---|---|---|---|
| 21-25 | 23 | 2 | 46 |
| 16-20 | 18 | 3 | 54 |
| 11-15 | 13 | 5 | 65 |
| 6-10 | 8 | 3 | 24 |
| 1-5 | 3 | 2 | 6 |
| | | N = 151 | $\sum f_ix_i = 195$ |

$X = \sum f_ix_i / N$ $\quad = 195/15 \quad = 13$

To simplify calculations, the following formula for arithmetic mean may be more convenient to use.

$$X = A + (\sum fd \ / \ N) \ i$$

where, A is an arbitrary point, d = (X-A)/i and i is the size of the equal class intervals. This formula makes the computations very simple and takes less time.

| *Interval (inches)* | *Mid value* $(X_i)$ | $f_i$ | $d = (X\text{-}13)/5$ | *fd* |
|---|---|---|---|---|
| 21-25 | 23 | 2 | 2 | 4 |
| 16-20 | 18 | 3 | 1 | 3 |
| 11-15 | 13 | 5 | 0 | 0 |
| 6-10 | 8 | 3 | -1 | -3 |
| 1-5 | 3 | 2 | -2 | -4 |
| | N | = 15 | $\sum$ fd | 0 |

$$X = A + (\sum fd / N) i$$

$$= 13 + (0/15) \times 5$$

$$= 13 + 0 \times 5$$

$$= 13 \text{ or } 13''$$

*Merits and Demerits of the Arithmetic mean -*

- It is simplest average to understand and easiest to compute,
- It is affected by the value of every item in the series,
- It is defined by the rigid mathematical formula with the result that everyone who computes the average gets the same answer,
- Being determined by a rigid formula, it lends itself to subsequent algebric treatment better than median or mode,
- It is a calculated value, and not based on position in the series,
- The mean is typical in the sense that it is the centre of gravity balancing the values on either side of it,
- In a distribution with open-end classes, the value of mean cannot be computed without making assumptions regarding the size of the class interval of the open-end classes,

- The arithmetic mean is not always a good measure of central tendency.

## Weighted Arithmetic Mean

The arithmetic mean gives equal importance or weight to each observation. In some cases, all observations do not have the equal importance and hence the weighted arithmetic mean is computed as follows.

$$X_w = \sum WX / \sum W$$

where, Xw represents the weighted arithmetic mean, W are the weights assigned to the variable X. The weighted mean enables us to calculate an average that takes into account the importance of each value to the overall total.

For example, a company wants to know the average cost of labour per hour for two products produced by three grades of labour - unskilled, semiskilled and skilled.

| *Grade of labour* | *Hourly wage (X)* | *Labour hour per unit of output* | |
|---|---|---|---|
| | | *Product A* | *Product B* |
| Unskilled | 50 | 1 | 4 |
| Semiskilled | 75 | 2 | 3 |
| Skilled | 100 | 5 | 3 |

To produce one unit of Product A, 1 hour of unskilled labour, 2 hours of semiskilled and 5 hours of skilled labour are required (total 8 hours). Hence, the weight attached to each grade of labour would be -

Unskilled labour = 1/8

Semiskilled labour = 2/8

Skilled labour = 5/8

Weighted mean cost (Xw) = $\sum WX / \sum W$

= [(50 × 1/8) ÷ (75 × 2/8) + (100 × 5/8)]/(1/8 + 2/8 + 5/8)

= (700/8) / (8/8) = 700/8 = 87.50

The average cost of labour for production of Product A would be Rs 87.50 per hour. Similarly, for product B, the average cost will be Rs 72.50 per hour.

The computation of weighted mean is same as used for averaging ratios and determining the mean of grouped data. Weighted mean is specifically useful in problems related to the construction of index numbers.

## Median

The median is that value which divides the distribution into two equal parts i.e. which lies exactly half-way between the lowest and highest value when the data is arranged in an ascending or descending order. **It is the value of the middle observation when the series is arranged in order of size or magnitude.** The median is a single value from the data set that measures the central item in the data. The median is not affected by the value of the observation but by the number of observations.

To find out the median from ungrouped data, first the data is arrayed in ascending or descending order. If the data set contains an odd number of items, the middle item of the array is the median. If there is an even number of items, the median is the average of the two middle items. Thus, median is,

Median = the $(n + 1)/2^{th}$ item in a data array

where, $n$ is the number of items in the array. For example, if the income of seven persons in rupees is 1100, 1200, 1350, 1500, 1550, 1600 and 1800, then the median income will be Rs 1500/- since it is the middle value of data array, which exactly divides it into two equal parts. However, if there is one more person with income of Rs 1850/-, then the median income would be (1500 + 1550)/2 = 1525.

For grouped data, the median is determined by using the following formula.

Median = L + [(N/2 - pcf) / f] × f

where, L is the lower limit of the median class, pcf is the preceding cumulative frequency to the median class, f is the frequency of the median class and i is the size of the median class.

For example, the age distribution of 1000 workers in an industrial establishment is as under.

| *Age (Years)* | *No. of workers* |
|---|---|
| Below 25 | 120 |
| 25 - 30 | 125 |
| 30 - 35 | 180 |
| 35 - 40 | 160 |
| 40 - 45 | 150 |
| 45 - 50 | 140 |
| 50 - 55 | 100 |
| 55 and above | 25 |

The location of the median value is facilitated by the use of a cumulative frequency distribution as shown below.

| *Age (Years)* | *No. of workers (f)* | *Cumulative frequency (cf)* |
|---|---|---|
| Below 25 | 120 | 120 |
| 25-30 | 125 | 245 |
| 30-35 | 180 | 425 |
| 35-40 | 160" | 585 |
| 40 -45 | 150 | 735 |
| 45- 50 | 140 | 875 |
| 50-55 | 100 | 975 |
| 55 and above | 25 | 1000 |

In this example, the median value is equal to N/2th observation, that is, 500th observation, which lies in the class 35-40. Now,

Median class = 35 – 40
Lower limit of median class (L) = 35
Size of the median class (i) = 5
Frequency of median class (f) = 160
Preceding cumulative frequency (pcf) = 425

$$\text{Median} = L + [(N/2 - pcf)/f] \times f = 35 + [(500 - 425)/160] \times 5$$
$$= 35 + [(75) / 160] \times 5 = 35 + (375 / 160)$$
$$= 35 + 2.34 = 37.34 \text{ years}$$

Hence, the median age is approximately 37 years, which suggests that half of the workers are below the age of 37 years and other half are above the age of 37 years.

### *Merits and Demerits of Median*

- It is not influenced by the magnitude of extreme deviations from it,
- It is the most appropriate average in dealing with qualitative data,
- The value of median can be determined graphically whereas the value of mean cannot be graphically ascertained,
- It is especially useful in case of open-end classes since only the position and not the values of items must be known,
- The value of the median is affected more by sampling fluctuations than value of arithmetic mean,
- It is erratic if the number of items is small,
- It is not capable of algebric treatment.

### Mode-

The mode is the typical or commonly observed value in a set of data. **Mode is defined as the value, which occurs most often or with the greatest frequency.** For ungrouped data, the mode is determined by arranging in data into ascending or descending order and finding out the value that is repeated most often. In a series of numbers 3, 4, 5, 5, 6, 77, 8, 8, 8, 8, the mode is 8 because it occurs maximum number of times (4 times).

For group data, first the mode class is determined, which is the class with the maximum frequency, and then the following formula is used,

$$\text{Mode} = L + [d_1 / (d_1 + d_2)] \times i$$

where, L is the lower limit of the mode class, $d_1$ is the difference between the frequency of the mode class and the frequency of the preceding class, $d_2$ is the difference between the frequency of the mode class and the frequency of the succeeding class and i is the size of the mode class. For example,

| *Daily sales (Rs thousands)* | *No. of Firms* |
|---|---|
| 20 - 30 | 15 |
| 30 - 40 | 23 |
| 40 - 50 | 27 |
| 50 - 60 | 20 |
| 60 - 70 | 35 |
| 70 - 80 | 25 |
| 80 - 90 | 5 |

Since the maximum frequency 35 is in the class 60 - 70, it is the mode class. Now,

Lower limit of mode class (L) = 60

$d_1$ = (35-20) =15

$d_2$ = (35-25) =10

Mode = $L + [d_1 / (d_1 + d_2)] \times i$

$= 60 + [15 / (15 + 10)] \times 10$

$= 60 + (15/25) \times 10$

$= 60 + 6$

$= 66$

Hence the mode daily sale is Rs 66 thousands.

### *Merits and Demerits of Mode*

- It is not affected by extremely large or small items,
- Its value can be determined in open-end distribution without ascertaining the class limits,
- It can be used to describe qualitative phenomenon,
- By definition, mode is the most typical or representative value of the distribution,
- The value of the mode cannot be always determined,
- It is not capable of algebric manipulations,
- The value of mode is not based on each and every item of the series.

## Geometric Mean

**Geometric mean is defined as the Nth root of the product of the given value observations,** i.e., geometric mean of a series of numbers $X_1$, $X_2$,..... $X_n$ is given by the formula -

Geometric mean = $N \int X_1, X_2, X_3, .... X_n$

To simplify the computation, the above formula is transformed into logarithms as under.

$$\text{Log GM} = \frac{1}{N} (\log X_1 + \log X_2 + ..... \log X_n) N$$

$$GM = \text{Antilog} (\sum \log X / N)$$

For grouped data,

$$GM = \text{Antilog} (\sum \log X / N)$$

Geometric mean is usually useful in construction of index numbers. It is the average most suitable when large weights have to be given to small values of observations and small weights to the large values of observations. This is also useful in measuring the growth of population, averaging ratios and percentages, in determining the rates of increase and decrease etc.

For example, a machine was purchased for Rs. 50,0007 in 1994. Depreciation on the diminishing balance was charged @ 40% in the first year, 25% in the second year and 15% in the next three years. What is the average depreciation charged during the whole period?

| *Year* | *Diminishing value (for a sum of Rs 100 /-)* | *Log X* |
|---|---|---|
| 1994 | 100 -40 = 60 | 1.77815 |
| 1995 | 100-25 =7 5 | 1.87506 |
| 1996 | 100-15 = 85 | 1.92941 |
| 1997 | 100 -15 = 85 | 1.92941 |
| 1998 | 100 -15 = 85 | 1.92941 |
| $\sum \log X$ | | 9.44144 |

GM = Antilog ($\sum$ log X / N) = Antilog (9.44144 / 5)

= Antilog 1.8883 = 77.32

The diminishing value being Rs 77.32, the depreciation will be 100 - 77.32 = 22.68%.

*Merits and Demerits of Geometric mean*

- It is based on each and every item of the series,
- It is rigidly defied,
- It is capable of algebric manipulation,
- It gives less weight to large items and more to small ones than does the arithmetic average,
- It is difficult to understand,
- It cannot be computed when there are both negative and positive values in a series or one or more of the values are zero,
- It is difficult to compute and interpret.

**Harmonic Mean**

The harmonic mean is a measure of central tendency for data expressed as rates such as kilometers per hour, tones per day, kilometers per litre etc. The **harmonic mean is defined as the reciprocal of the arithmetic mean of the individual observations.** If $X_1$, $X_2$,....$X_n$ are N observations, then harmonic mean can be represented by the following formula -

$$\text{H. M.} = \frac{N}{1\ X_1 + 1/X_2 + \ldots\ldots\ldots\ldots 1/X_n}$$

$$= \frac{N}{\sum(1/Xn)}$$

For the grouped data, the formula is -

$$\text{H. M.} = \frac{N}{\sum(f/X)}$$

The harmonic mean is useful for computing the average rate of increase of profits, or average speed at which a journey has been performed, or the average price at which an article has been sold.

*Merits and Demerits of Harmonic Mean*

- Its value is based on every item on every item of the series,
- It tends itself to algebric manipulation,
- It is not easily understood,
- It gives largest weight to smallest items,
- It can be computed when there are both positive and negative items in a series or when one or more items are zero.

### *8.2.3.2 Measures of Dispersion or Spread*

**The property, which denotes the extent to which the values are dispersed about the central value is termed as dispersion.** A measure of variation describes the spread or scattering of the individual values around the central value. Measuring variation is important for the following purposes.

- Measuring variability determines the reliability of an average by pointing out as to how far an average is representative of the entire data.
- Measuring variability determines the nature and cause of variation in order to control the variation itself.
- Measures of variation enable comparisons of two or more distributions with regard to their variability.

A good measure of variation should possess, as far as possible, the same properties as those of a good measure of central tendency. The following are some of the well known measures of dispersions, which provide a numerical index of the variability of the given data.

- Range
- Quartile deviation or semi-quartile deviation
- Average deviation
- Standard deviation

**Range** -

The simplest possible measure of dispersion is the range, which is **defined as the difference between the highest and the lowest values of the variables in a data set**

$R = H - L$

*Ungrouped data* - The following distribution shows the number of labourers kept engaged per day during tasar silkworm cultivation.

80, 90, 60, 63, 68, 61, 67, 65, 100, 75, 89, 84, 86

Arranging the series in ascending order, we get,

60, 61, 63, 65, 67, 68, 75, 80, 84, 86, 89, 90, 100

Hence, the Range is, R = H – L = 100 – 60 = 40

Range is the absolute measure of dispersion.

*Grouped data* - In case of the grouped data, the range is the difference between the upper boundary of the highest class and the lower boundary of the lowest class. No consideration is given to the frequencies.

**Average Deviation -**

**Average deviation is defined as the mean of deviations from the mean or mode.** All the deviations are treated as positive regardless of sign. It tells an average distance of any observation in the data set from the mean of the distribution. Mathematically, average deviation can be represented by the following formula.

Average deviation = I (X – X) / N or I (X – median) / N

For grouped data, the formula to be used is given as.

Average Deviation = $\sum$f (X – X) / N

For example, consider the following grouped data,

| *Sales (Rs thousand)* | *No. of Days* |
|---|---|
| 40 -50 | 10 |
| 50-60 | 15 |
| 60 -70 | 25 |
| 70-80 | 30 |
| 80 -90 | 12 |
| 90 -100 | 8 |

The Average deviation is calculated as follows.

| *Sales (Rs thousand)* | *Mid Point (X)* | *No. of days (f)* | *fX* | *IX-XI* | *f IX-XI* |
|---|---|---|---|---|---|
| 40 -50 | 45 | 10 | 450 | 24.3 | 243 |
| 50-60 | 55 | 15 | 825 | 14.3 | 214.5 |
| 60 -70 | 65 | 25 | 1625 | 4.3 | 107.5 |
| 70-80 | 75 | 30 | 2250 | 5.7 | 171 |
| 80 -90 | 85 | 12 | 1020 | 15.7 | 188.4 |
| 90 -100 | 95 | 8 | 760 | 25.7 | 205.6 |
| | 100 | 6930 | | 1130 | |

Mean = $\sum f / N$ = 6930/100 = 69.3

Average Deviation = $\sum f \; (X - X) / N$

= 1130 /100

= 11.30

The **relative measure of average deviation is called the coefficient of average deviation** and is obtained by dividing average deviation by the particular average used in computing the average deviation, i.e.,

Coefficient of Average Deviation = Average deviation / mean or median

In the above example,

Average deviation = 11.30

Average or mean = 69.3

Coefficient of Average deviation = 11.30 / 69.3

= 0.16

**Standard Deviation**

It is the most widely used and important measure of variation. It is used to find out the variation in the sample or population. **Standard deviation, also known as root mean square deviation,**

**is the square root of the mean of the squares of the deviations** and calculated by using the following formula,

$$\sigma = \sqrt{\sum (X_n - X)^2 / N} \text{ population}$$

$$s = \sqrt{\sum (X_n - X)^2 / n - 1} \text{ sample}$$

where,

$\sigma$ = Standard deviation

$X_n$ = nth item or observation

$X_i$ = Average or mean

N = Number of items or observations

**The square of the standard deviation is called variance.** Therefore,

Variance = $\sigma^2$

The standard deviation and variance become larger as the variability or spread within the data becomes greater. The greater the standard deviation, the greater is the variability. For grouped data, the formula for standard deviation is as under.

$$\sigma = \sqrt{\sum (X_n - X)^2 / N}$$

$$= \sqrt{\sum (fx^2 / N)} - (\sum (fx / N)^2)$$

$$= \sqrt{\sum (fx^2 / N) - x^2}$$

$$= \sqrt{\sum (fd^2 / N)} - (\sum (fd / N)^2 \times 1$$

where,

$d = (x - A) / i$

If data represent a sample of size n from a population N, then it can be proved that the sum of the square deviations are divided by (n–1) instead of N. However, for large samples, there is very little difference in the use of (n–1) or N.

**Relative Dispersion (Coefficient of Variation)**

The standard deviation is an absolute measure of dispersion that expresses variation in the same units as the original data. But it can not help in comparison of two distributions. The coefficient of variation is a relative measure of dispersion that enables in comparison of two distributions and expresses standard deviation as percentage of the mean. For a population, the coefficient of variation is calculated by the following formula -

$$\text{Coefficient of Variation} = \frac{\sigma}{\bar{x}} \times 100$$

The Coefficient of Variation indicates the degree of precision with which the treatments are compared and is good of the reliability of experiements. The higher the CV value, the lower is the reliability of the experiement (< 10%).

*Example:*

Consider the shell ratio of 50 cocoons of *Bombyx mori* and calculate mean, standard deviation, standard error and coefficient of variation.

17.67, 18.6, 16.5, 18.1, 16.4, 21.5, 17.8, 17.1, 18.4, 18.1, 16.9, 17.6, 18.1, 19.6, 17.4, 18.5, 15.4, 17.9, 16.6, 18.2, 18.3, 17.7, 13.3, 16.7, 17.6, 15.4, 11.8, 18.0, 17.6, 18.2, 18.8, 17.6, 19.2, 18.7, 19.8, 16.9, 16.7, 17.0, 18.8, 17.8, 18.6, 20.7, 17.6, 17.6, 17.8, 18.4, 18.1, 19.8, 15.1, 18.1

| *Class* | *fi* | *xi* | *fixi* | *fixi*$^2$ |
|---|---|---|---|---|
| 11.1-13.0 | 1 | 12.05 | 12.05 | 145.2025 |
| 13.1-15.0 | 1 | 14.05 | 14.05 | 197.4025 |
| 15.1-17.0 | 11 | 16.05 | 176.55 | 31169.9025 |
| 17.1-19.0 | 31 | 18.05 | 559.55 | 313096.2025 |
| 19.1-21.0 | 5 | 20.05 | 100.25 | 10050.0625 |
| 21.1 -23.0 | 1 | 22.05 | 22.05 | 486.2025 |
| $\sum$ | 50 | 102.3 | 884.5 | 355144.975 |

Mean $= \sum f_i x_i / \sum f_i = (884.5 / 50) = 17.69$

S. D. $= \sqrt{1/(N/N-1) \times (N\sum f_i x_i - \sum (f_i x_i)^2}$

$= \sqrt{(1/(50\times 49) \times ((50 \times 355144.975 - (8845.5\ 884.5))}$

$= \sqrt{(1/2450) \times (17757240.75 - 782340.25)}$

$= \sqrt{0.0004081 \times 16974908.5}$

$= \sqrt{6927.4602}$

$= 83.33$

Now, S.E. $= \sigma \div \sqrt{N}$

$= 83.23 \div \sqrt{50}$

$= 83.23 \div 7.07 = 11.77$

### 8.2.4 Measures of Variation

For making statistical inferences, estimation is made for the characteristics of populations from the information contained in samples. There are two types of estimates - point estimate and interval estimates. The point estimate is a single number that is used to estimate an unknown population parameter while interval estimate is a range of values within which a population parameter is likely to lie. The interval estimates of the population are determined by Student 't' Distribution if the sample size is relatively small and the population standard deviation is not known.

### 8.2.5 Sampling and Sampling Distribution

#### *8.2.5.1 Sampling Techniques*

A sample is a part of a group or aggregate selected with a view to obtaining information about the whole group, also known as population. The total number of units in the population and sample are known as population size and sample size respectively. Sampling technique is used in traditional research and management problems.

Sampling has got a number of advantages as compared to complete enumeration -

- It is less expensive owing to study of selected units from the population,
- It takes less time due to smaller units to be studied,
- It is possible to achieve greater accuracy and the sampling errors can be studied, controlled and probability statements can be made about their magnitude, and
- It is must when the enumeration is destructive.

The sampling methods are broadly classified under two categories - probability sampling and non-probability sampling.

***Probability sampling*** - A probability sample is chosen in such a way that each unit of the group has a known chance of being selected, i.e., some probability is attached to each unit of the population to be included in the sample and hence the sample becomes representative of the population.

*Simple Random Sampling:* In simple random sampling, a known and equal probability is attached to each unit of the population for being included in the sample. The simple random sample can be chosen both with and without replacement. In simple random sampling with replacement, an observation is elected from the population in such a manner that every unit of the population has an equal chance of 1/N to be included in the sample if the population consists of N units and a sample of size n is to be selected. After first unit is selected, its value is recorded and put back in the population. The second unit is drawn exactly in the same manner as the first unit and the procedure is continued till nth unit of the sample is selected. In the simple random sampling without replacement, when the first unit is chosen every unit has a chance of 1/N to be included in the sample but afterwards first unit is no longer replaced in the population and the second unit has a chance of 1/(N–1) to be included in the sample. The procedure is continued till nth unit selected with probability of 1/(N–n+1).

The random number table is used for selecting the sample in this method. Depending upon the size of the population, a number from 0 to Nth is assigned to each unit of the population. Now any number is selected from anywhere of the random number table and and continued to choose numbers according to the given sequence, row-wise or column-wise. The procedure is continued till the last unit of the sample (nth unit) is selected.

*Systematic Sampling:* In this method, the numbers are chosen in a systematic manner from the population but it is not necessarily that each unit has an equal chance of being included in the sample but has a known chance. Suppose the population consists of ordered N units (numbered from 1 to N) and a sample of size n is selected from the population in such a way that N/n = k (rounded to the nearest integer). Here, k is called sampling interval. This method, though involves relatively less time and work, is less representative.

*Stratified sampling:* When considerable heterogeneity is present in the population with regards to the subject matter under study, stratified sampling method is used to draw the sample. In this method, the population is divided into segments or strata on the basis of measurable characteristics of its members and a certain number of sampling units are selected from each strata thus ensuring representation from all the relevant segments. Suppose the population of N units is sub-divided into k sub-populations or strata the ith sub-population having $N_i$ units (i = 1, 2, .... k). These sub-populations are non-overlapping so that they comprise the whole population such that $N_1 + N_2 + \ldots N_k = N$ A simple random sample (with or without replacement) is selected independently in each stratum, the sample size in the ith stratum being $n_i$ (i = 1, 2, .... k). Thus, the total sample size is $n = n_1 = n_2 = \ldots\ldots\ldots = n_k$

The stratification is done in such a manner that the strata are homogenous within themselves with respect to the characteristics to be studied but heterogenous between them. In the stratified sampling, the sample size of each stratum is determined taking into consideration the followings -

- The total number of units in the stratum i.e. stratum size,
- The variability within the stratum, and
- The cost of taking observation per sampling unit each stratum.

Practically, two types of allocation are made - proportional allocation and optimum allocation. In proportionate allocation, the number of sampling units $n_i$ allocated to the ith stratum is proportionate to the number of units in the population. Symbolically,

$N_i = n(N_i \div N) : (i = 1, 2, ......k)$

In optimum allocation, the number of sampling units $n_i$ allocated to the ith stratum is not proportionate to the number of units in the population. The sample size in a particular stratum is larger if (i) the stratum size is larger (ii) the stratum has larger variability and (iii) the sampling cost in the stratum is lower.

*Cluster sampling:* In this method, various units comprising the population are grouped in clusters and the sample is selected randomly in such a way that each cluster has a known chance of being included in the sample. A cluster sample with clusters based on geographical subdividions is known as **area sample** and the procedure is known as **area sampling.** A cluster sample is useful in two situations - (i) when there is incomplete information on the composition of the population and (ii) when it is desirable to save time and costs by limiting the study to specific geographical areas.

*Multistage sampling:* This method is used when large geographical area such as a state or country has to be covered. In this technique, the whole geographical area is divided into different units referred as first stage units. The second stage units may have second stage units, second stage units have third stage units and so on. The appropriate sample size is drawn after randomly selecting units at different stages, which represents each stage of the population. The multistage sampling procedure has the advantage that the frame of second stage units is necessary only for the selected first stage units. Likewise, the frame of the third stage units is necessary only for the selected second stage units. The procedure is quite flexible and it permits the use of different selection procedures in different stages.

***Non-Probability Sampling*** - It is the sampling procedure, which does not provide any basis for estimating the probability that each unit in population possesses to be included in the sample. In such a case, the sampling error is not measurable. In general, there are three non-probability sampling methods.

*Judgement sampling:* It is also referred as purposive sampling. Under this procedure, a person knowledgeable about the population under study deliberately or purposively draws a sample from the population, which he thinks is a representative of the population. In this method, all the members of population are not given chance to be selected in the sample.

*Covenience sampling:* In this method, the sample units are chosen on the basis of convenience of the researcher or investigator. There is no way of determining the representativeness of the sample resulting into biased estimates. The standard error cannot be estimated since the difference between sample estimate and population parameter is unknown both in terms of magnitude and direction. This method is generally used in exploratory designs as a basis for generating hypotheses or useful in testing questionnaire etc.

*Quota sampling:* This is most commonly used method in marketing research studies. In this method, the sample is selected on the basis of certain basic parameters such as age, sex, income and occupation that describe the nature of a population so as to make it representative of the population. The investigator has to choose a sample that conforms to these parameters. The investigator is assigned quotas of the numbers of units satisfying the required characteristics on which data should be collected. Before collecting data on these units, the investigator is supposed to verify that the units qualify these characteristics. The selection of units is made non-randomly.

### 8.2.5.2 Sampling Distribution

For a given population, a probability distribution of all the possible values a statistic may take on for a given sample size is called sampling distribution of a statistic while a probability distribution of all the possible means of samples of a given size n from a population is referred as sampling distribution of mean. The distribution of sample means have its own mean and standard deviation. A distribution of sample means that is less spread out (that has a small standard error) is a better estimate of the population mean than a distribution of sample means that is widely dispersed and has a larger standard error. The sampling distribution of a mean of a normally distributed population demonstrates the following important properties -

- The sampling distribution has a mean equal to the populationm mean,
- The sampling distribution has a standard deviation (standard error) equal to the population standard deviation divided by the square root of the sample size, and

- The sampling distribution is normally distributed.

As the standard error decreases, the value of any sample mean will probably be closer to the value of population mean.

### 8.2.6 Test of Significance

A set of data is collected or an experiment is carried out not simply for summarizing the results and estimating suitable parameters but in order to test an idea. This idea or hypothesis may arise by purely theoretical reasoning or it may be based on the earlier similar type of experiments conducted.

For testing a hypothesis, the data is collected from a random sample and then statistical significance is determined. The maximum probability or levels of significance that have been more or less established as acceptable rejection points are 0.05 and 0.01. The degree of freedom is defined as the difference between total number of items and total number of constraints and represented as d.f. (n-k). In almost all the cases, k is equal to 1 except while estimating the correlation.

#### *8.2.6.1 Student's* t *Distribution*

The Student's *t* distribution is a probability distribution distinguished by its individual degree of freedom, similar in form to normal distribution and used when the sample size is small and the population standard deviation is not known. In general, the *t* distribution is flatter than normal distribution and there is a different *t* distribution for every possible sample size. When sample size increases, its flatness is lost and it tends to become quite similar to normal distribution. On comparing normal and *t* distribution, it is found that t distribution is lower at the mean and higher at the tails than a normal distribution.

The *t* table is more compact and shows areas and *t* values for only a few percentages (10, 5, 2 and 1 percent). It measures the chances that the population parameter being estimated will not lie within the confidence limits. For using table, the degree of freedom should be known.

For example, if the *t* value at 13 degree of freedom and at 90% confidence level is 1.771, it shows that the area under the curve between stnadrad error of mean plus 1.771 and stnadrad error of mean minus 1.771 will be 90 percent and the area outside these limits will be 10 percent.

**Student't' Test**

The Student 't' test is used to test the significance of the difference between two sample means. This test is applied when,

- The populations' standard deviation is not known,
- The sample size is small, and
- Only sample data is available on both groups.

*To test the significance of a population mean*

When the sample size is small and the standard deviation of the sample is known, the significance is found out using the following formula -

$$t = \frac{X - \mu}{\sigma} \times \sqrt{n}$$

a where,

$X$ = sample mean

$\mu$ = population mean

$\sigma$ = standard deviation of sample

$n$ = sample size

The standard deviation can be calculated by the formula -

$$\sigma = \sqrt{\sum (X_n - X)^2 /(n - 1)}$$

The computed and tabular values of t are compared and if the computed t value is higher than the tabular value, null hypothesis is rejected and it can be inferred that there exists significant difference.

*To test the significance of difference between two sample means*

The significance between two sample means is calculated by employing the following formula -

$$t = \frac{X_1 - X_2}{s} \times \sqrt{n_1 n_2 /(n_1 + n_2)}$$

where,

$X_1$ = Mean of the first sample

$X_2$ = mean of the second sample

s = Combined estimate of standard deviation

$n_1$ = Size of the first sample

$n_2$ = Size of the second sample

The combined estimate of standard deviation is determined by the following formula -

$$s = \frac{\sqrt{[S_1 - X_1)^2 + (S_2 - X_2)^2}}{n_1 + n_2 - 2}$$

The test is based on $(n_1 + n_2 - 2)$ degree of freedom. Here, $S_1$ and $S_2$ are standard deviation of two samples.

### 8.2.6.2 Chi Square Test

Chi square test is a test concerning the goodness of fit of the observed data to any hypothesized distribution, which decides whether, in any sample, the observed frequencies are in apparent with the theoretical or expected frequencies based on some specified law of distribution.

$$\chi^2 = \sum (O - E)^2 \div E; \text{ d.f. } N = 1$$

where,

O = Observed frequency

E = Expected frequency

N = Total number of observations

If Chi (calculated) is greater than Chi (tab) at specified level of significance for N-1 degree of freedom, then there is significant difference between expected and observed values; if not there is no significant difference between observed and expected values.

For example, the genetic model assumes that yellow color in mulberry cocoons is inherited as a simple dominant trait and that white color is inherited as a recessive trait. A cross between pairs of heterozygous yellow silkworm produces on $F_2$ generation of 440 yellow and 120 white cocoons. Test the hypothesis that the ratio falls in 3:1 pattern.

| | *Color of cocoons* | | |
|---|---|---|---|
| | *Yellow* | *White* | *Total* |
| Observed | 440 | 120 | 560 |
| Expected | 420 | 140 | 560 |

Thus,

$$\chi^2 = \{(440\text{-}420)^2 / 420\} + \{(120\text{-}140)^2 / 140\}$$

$$= (400/420) + (400/140)$$

$$= 0.952 + 2.857$$

$$= 3.809$$

Thus

$\chi^2$ (calculated) = 3.809

$\chi^2$ (tab) = 3.841 at 5% significance level

Since the Chi (calculated) is lower than Chi (tab), there is no significant difference between observed and expected values and hence the ratio of 3:1 is satisfied.

### 8.2.6.3 Analysis of Variance

Analysis of Variance (ANOVA) is used to test for significance of difference between more than two sample means. It can be used in every type of experimental design. This test is applied when each sample is drawn in a random manner from the normal population, the sample statistics tend to reflect the characteristics of the population and the populations from which the samples are taken have identical means and variances. There are two methods of using analysis of variance - one-way classification and two-way classification.

*One-way Classification*

In one-way classification, data are classified according to one criterion. The following steps are involved in computing the analysis of variance -

- Compute the variance between samples

a. Calculate the mean of each sample

b. Calculate the grand average mean

c. Take the difference between means of various samples and grand mean

d. Square these deviations and obtain total, which will give sum of squares between samples

e. Divide the total obtained above by the degree of freedom.

- Calculate the variance within the samples

a. Calculate mean of each sample

b. Take the deviation of various items in a respective sample

c. Square these deviations and obtain the total, which gives the sum of squares within samples

d. Divide the total obtained above by the degree of freedom.

- Calculate the F ratio as follows -

$$F = \frac{\text{Variance between the samples}}{\text{Variance within samples}}$$

- Compare the calculated F value for the degree of freedom at certain level of significance. If the calculated value exceeds the tabular value, then the difference in sample means is significant, that is, it is not due to sampling fluctuations or samples do not come from the same population.

Sum of all items of various samples (T) =

$$\sum X_1 + \sum X_2 \ldots\ldots + \sum X_n$$

Correction factor = $T^2$/Total sum of items in all samples (N)

Total sum of squares (SST) =

$$\sum X_1^2 + \sum X_2^2 \ldots\ldots + \sum X_n^2 - T^2 / N$$

Sum of squares between samples (SSC) =

$$[\sum (X_1)^2 / N_1 + (\sum X_2)^2 / N_2 \ldots\ldots (\sum X_n)^2 / N^n] - T^2 / N$$

Sum of squares within samples (SSE) = SST – SSC

ANOVA Table

| *Source of variation* | *S.S.* | *d.f.* | *Mean Square* | *F* |
|---|---|---|---|---|
| Between samples | SSC | (k-1) | MSC=SSC/(k-1) | MSC/MSE |
| Within samples | SSE | (n-k) | MSE=SSE/(n-k) | |

For example, the cocoon weights obtained in four races are as under,

| *Race* | *Cocoon weight (g)* | *No. of observations* |
|---|---|---|
| 1 | 1.954, 2.094, 1.953, 1.936, 2.100 | 5 |
| 2 | 1.653, 1.764, 1.805, 1.665, 1.753, 1.900 | 6 |
| 3 | 1.79, 1.986, 1.876, 1.805 | 4 |
| 4 | 2.173, 2.106, 2.113, 1.987, 1.993 | 5 |

Test the hypothesis that the races are significantly different.

| *Race* | *Cocoon weight (g)* | *No. of observations* | *Total (TC)* | *Mean (Xi)* |
|---|---|---|---|---|
| 1 | 1.954, 2.094, 1.953, 1.936, 2.100 | 5 | 10.037 | 2.0074 |
| 2 | 1.653, 1.764, 1.805, 1.665, 1.753, 1.900 | 6 | 10.54 | 1 .7567 |
| 3 | 1.79, 1.986, 1.876, 1.805 | 4 | 7.457 | 1.8643 |
| 4 | 2.173, 2.106, 2.113, 1.987, 1.993 | 5 | 10.372 | 2.0744 |

Grand Total (G) = 38.406 Correection factor (CF) = 1475.0208 / 20 = 73.751

Total Sum of Square (SST) =

$$\sum X_1^2 + \sum X_2^2 \ ...... + \sum X_n^2 - T^2 / N$$

= 74.200 – 73.751

= 0.449

Sum of Squares between Samples (SSC) =

$$[\sum(X_1)^2/N_1+(\sum X_2)^2/N_2 .... (\sum X_n)^2/N"]-T^2/N$$

= 74.0809 – 73.751

= 0.330

Sum of Sqaures within samples (SSE) = SST – SSC

= 0.449 – 0.330

= 0.119

ANOVA Table

| *Source of variation* | *S.S.* | *d.f.* | *Mean Square* | *F* |
|---|---|---|---|---|
| Between races | 0.330 | 3 | 0.11 | 14.8649 |
| Within cocoon weights | 0.119 | 16 | 0.0074 | |

F (tab; 3,16) = 3.24 at 5% and 5.29 at 1%

Since the calculated F value is greater than the table F value both at 5% and 1% level of significance, there is significant difference between the cocoon weights of the races.

*Two-way Classification* -

In this method, the effect of two factors is studied in the same experiment. For each factor, there will be a number of classes or levels or replications. The following steps are involved in the analysis of variance of a two factor experiement involving three factors, five treatments and four replications in a randomized complete block (RCB) design -

- Denote the number of replications by *r*, the number of factors as a, and number of treatments as *b*. Construct the outline of the analysis of variance as:

| *Source of Variation* | *Degree of freedom* | *Sum of square* | *Mean square* | *Computed F* | *tablular F* |
|---|---|---|---|---|---|
| Replication | r – 1 = 3 | | | | |
| Treatment | ab – 1 = 14 | | | | |
| Factor A | a – 1 = 2 | | | | |
| Treatment B | b – 1 = 4 | | | | |
| A × B | (a–1) (b–1) = 8 | | | | |
| Error | (r–1) (ab–1) = 42 | | | | |
| Total | rab-1 = 59 | | | | |

- Group the data by treatments and replications and calculate the treatment total (T), replication total (R) and grand total (G) Compute the correction factor and various sum of squares as follows:

a. Calculate the correction factor as,

C. F. = $G^2/r\ a\ b$

b. Calculate the total sum (SS) by substracting the correction factor from the sum of square of all the data points,

Total SS = $\sum X^2 -$ C. F.

c. Calculate the replication sum of square by substracting correction factor from the sum of square of replication totals divided by total number of replications and multiplied by number of factors,

Replication SS = $(\sum R^2\ /\ ab)$ - C. F.

d. Calculate the treatment sum of square by substracting correction factor from the sum of square of treatment totals divided by number of replications,

Treatment SS = $(\sum T^2\ /\ r)$ - C. F.

e. Calculate error sum of square as,

Error SS = Total SS - Replication SS - Treatment SS

- Construct the factor A × factor B two-way table of totals, with factor A totals and factor B totals computed.

| *Source of variation* | *Degree of freedom* | *Sum of squares* | *Mean square* | *Computed F value* | *Tabular F* |
|---|---|---|---|---|---|
| Replication | | | | | |
| Treatment | | | | | |
| Total | | | | | |
| Total | | | | | |

- Compute the three factorial components of the treatment sum of squares as:

  Factor A SS = $(\sum A^2/rb)$-C.F.

  Factor B SS = $(\sum B^2/ra)$-C.F.

  Factor A × B SS = Treatment SS – Factor A SS – Factor B SS

- Compute the mean square for each source of variation by dividing the sum of square by its degree of freedom:

  Factor A MSS = Factor A SS ÷ (a – 1)

  Factor B MSS = Factor B SS ÷ (b – 1)

  Factor A x B MSS = (Factor A × B SS) ÷ (a – 1) (b – 1)

  Error MSS = Error SS ÷ (r – 1)(ab – 1)

- Compute the F value for each of the three factorial components as:

  F value (factor A) = Factor A MSS ÷ Error MSS

  F value (factor B) = Factor B MSS ÷ Error MSS

  F value (factor A × B) = Factor A × B MSS ÷ Error MSS

- Compare each of the computed F values with the tabular F value with $f_1$ = d.f. of the numerator MSS and $f_2$ = d.f. of the denominator MSS at a prescribed level of significance. If the calculated value exceeds the tabular value, then the difference in sample means is significant, that is, it is not due to sampling fluctuations or samples do not come from the same population.

- Compute the coefficient of variation as:
  *cv* - V Error MSS - Grand Mean × 100

For example, the yields of three rice varieties teated with five levels of nitrogeb in a RGB design are as under -

| *Nitrogen level (kg/ha)* | *Grain yield (tons/ha)* Replication 1 | Replication 2 | Replication 3 | Replication 4 | *Treatment total (T)* |
|---|---|---|---|---|---|
| Variety$_1$ | | | | | |
| $N_0$ | 3.852 | 2.606 | 3.144 | 2.894 | 12.496 |
| $N_1$ | 4.788 | 4.936 | 4.562 | 4.608 | 18.894 |
| $N_2$ | 4.576 | 4.454 | 4.884 | 3.924 | 17.838 |
| $N_3$ | 6.034 | 5.276 | 5.906 | 5.652 | 22.868 |
| $N_4$ | 5.874 | 5.916 | 5.984 | 5.518 | 23.292 |
| Variety$_2$ | | | | | |
| $N_0$ | 2.846 | 3.794 | 4.108 | 3.444 | 14.192 |
| $N_1$ | 4.956 | 5.128 | 4.150 | 4.990 | 19.224 |
| $N_2$ | 5.928 | 5.698 | 5.810 | 4.308 | 21.744 |
| $N_3$ | 5.664 | 5.362 | 6.458 | 5.474 | 22.958 |
| $N_4$ | 5.458 | 5.546 | 5.786 | 5.932 | 22.722 |
| Variety$_3$ | | | | | |
| $N_0$ | 4.192 | 3.754 | 3.738 | 3.428 | 15.112 |
| $N_1$ | 5.250 | 4.582 | 4.896 | 4.286 | 19.014 |
| $N_2$ | 5.822 | 4.848 | 5.678 | 4.932 | 21.280 |
| $N_3$ | 5.888 | 5.524 | 6.042 | 4.756 | 22.210 |
| $N_4$ | 5.864 | 6.264 | 6.056 | 5.362 | 23.546 |
| Replication Total (R) | | 76.992 | 73.688 | 77.202 | 69.508 |
| Grand Total (G) | | | | | 297.390 |

Correction Factor $= G^2/rab$

$= (297.390)^2 \div (4)(3)(5)$

$= 1474.014$

Total Sum of Squares $\sum X^2 - \text{C.F.}$

$= [(3.852)^2 + .... (5.562)^2] - 1474.014 = 53.530$

Replication Sum of Squares ($\sum R^2$ /ab)– C.F.

= [{(76.992)² + .....+ (69.508)² ÷ 3)(5)] - 1474.014

= 2.599

Treatment Sum of Squares $\sum T^2$ – C. F.

= [{(12.496)² + ... + (23.546)²} ÷ 4] – 1474.014

= 44.578

Error Sum of Squares = Total SS – Replication SS – Treatment SS

= 53.530 – 2.599 – 44.578 = 6.353

| *Source of variation* | *Degree of freedom* | *Sum of squares* | *Mean square* | *Computed F value* | *Tabular F 5%* | *1%* |
|---|---|---|---|---|---|---|
| Replication | 3 | 2.599 | 0.866 | 5.74** | 2.83 | 4.29 |
| Treatment | 14 | 44.578 | 3.184 | 21.09** | 1.94 | 2.54 |
| Total | 42 | 6.353 | 0.151 | | | |
| Total | 59 | 53.530 | | | | |

** Significant at 1% level

| *Nitrogen* | *Yield Total (AB)* *Variety₁* | *Variety₂* | *Variety₃* | *Nitrogen Total (B)* |
|---|---|---|---|---|
| $N_0$ | 12.496 | 14.192 | 15.112 | 41.800 |
| $N_1$ | 18.894 | 19.224 | 19.014 | 57.132 |
| $N_2$ | 17.838 | 21.744 | 21.280 | 60.862 |
| $N_3$ | 22.868 | 22.958 | 22.210 | 68.036 |
| $N_4$ | 23.292 | 22.722 | 23.546 | 69.560 |
| Variety Total (A) | 95.388 | 100.840 | 101.162 | 297.390 |

Factor A ∫SS = ($\sum A^2$ / rb) – C.F.

= [(95.388)² + (100.840)² + (101.162)²] ÷ (4×5) – 1474.014

= 1.052

Factor B SS = $(\sum B^2 / ra)$ - C.F.

= $[(41.800)^2 + ... + (69.560)^2] \div (4 \times 3) - 1474.014$

= 41.234

Factor A × B SS = Treatment SS – Factor A SS – Factor B SS

= 44.578 – 1.052 – 41.234

= 2.292

Factor A MSS = Factor A SS/(a-1)

= 1.052/2

= 0.526

FactorB MSS = Factor B SS/(b-1)

= 41.234/4

= 10.308

Factor A x B MSS = Factor A × B SS / (a – 1)(b – 1)

= 2.292 / (2 × 4)

= 0.286

Error MSS = Error SS / (r – 1 )(ab – 1)

= 6.353 / (3 × 14)

= 0.151

F value (Factor A) = Factor A MSS + Error MSS

= 0.526 ÷ 0.151

= 3.480

F value (Factor B) = Factor B MSS + Error MSS

= 10.308 ÷ 0.151

= 68.260

F value (Factor A × B) = Factor A × B MSS + Error MSS

= 0.286 ÷ 0.151

= 1.89

### 8.2.7 Simple Regression and Correlation

In biological experiments, two or more variables are apparently associated or interdependent. In characterizing the association between variables, some statistical tools are required that can handle several variables at a time. If there are only two variables involved, the analysis of variance can evaluate only one character at a time, even though change in one variable may affect the other or the variables

Analysis of variance of data from a 3 × 5 Factorial Experiment in RCB design

| *Source of variation* | *Degree of freedom* | *Sum of squares* | *Mean square* | *Computed F value* | *Tabular F 5%* | *1%* |
|---|---|---|---|---|---|---|
| Repication | 3 | 2.599 | 0.866 | 5.74** | 2.83 | 4.29 |
| Treatment | 14 | 44.578 | 3.184 | 21.09** | 1.94 | 2.54 |
| Variety A | (2) | 1.052 | 0.526 | 3.48* | 3.22 | 5.15 |
| Variety B | (4) | 41.234 | 10.308 | 68.26** | 2.59 | 3.80 |
| A × B | (8) | 2.292 | 0.286 | 1.89$^{ns}$ | 2.17 | 2.96 |
| Error | 42 | 6.353 | 0.151 | | | |
| Total | 59 | 53.530 | | | | |

* Significant at 5% level: ** Significant at 1% level; ns Not siqnificant

$$CV = \sqrt{\text{Error MSS} + \text{Grand Mean}} \times 100$$

$$= \sqrt{0.151 \div 4.956 \times 100}$$

$$= 7.8\%$$

may simultaneously influence both the characters. Regression and correlation analysis examine the association between two or more variables, which are expressed quantitatively. Regression analysis describes the effect of one or more variables on a single variable while correlation provides a degree of association between the variables.

### *8.2.7.1 Correlation Analysis*

When it can be demonstrated that the variation in one variable is somehow associated with the variation in another, then two variables are said to be correlated. Correction may be positiove or negative. It is positive when X and Y both increase and negative when X increases and Y decreases.

Correlation analysis is a statistical tool, which is used to describe the degree to which an variable is linearly related to another variable. It is used to measure the degree of association between two or more variables. Two variables are said to be correlated if movements in another accompany the movements in one. **The correlation coefficient is a measure of correlation and summarizes the direction and degree of correlation.** In other words, correlation coefficient is a measure of the degree of linear association between two variables X and Y. Correlation coefficient denoted by 'r' may be positive or negative. The sign of coefficient indicates the direction of the relationship between two variables. It is the square root of the sample coefficient of determination -

$$r = \sqrt{R^2}$$

$$R^2 = \frac{a(\sum y) + b(\sum xy) - n(y)^2}{\sum y^2 - n(y)^2}$$

$$R^2 = \frac{\sum (y - y)^2}{\sum (y - y)^2}$$

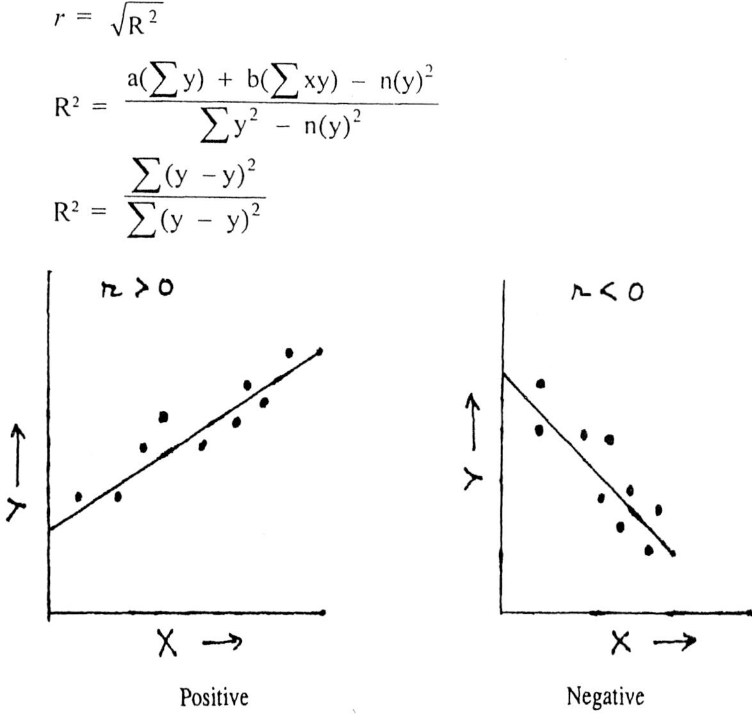


**Fig. 8.2: Positive and Negative Correlation**

Karl Pearson's coefficient of correlation was used to find out the relationship between various independent variables and dependent variables besides studying the relationship between growth in different varieties of silks produced in the world, country, state and project areas. Pearson's formula for correlation coefficient is given as

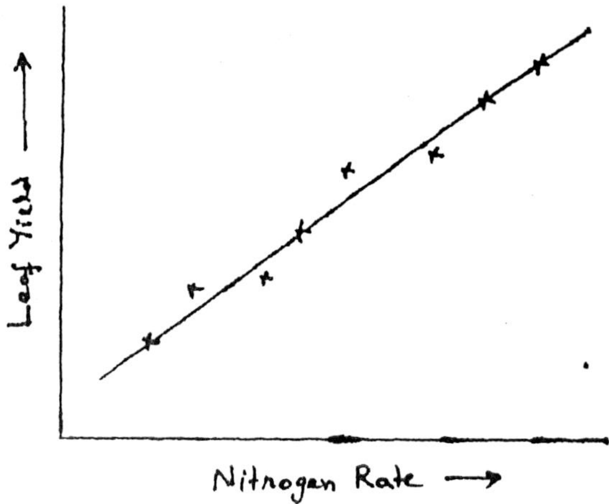


**Fig. 8.3: Regression Curve of leaf yield on nitrogen rate**

$$r = 1/n[\sum (X - \bar{X})(Y - \bar{Y})] \div \sigma_x \sigma_y$$

where,

r = correlation coefficient between X and Y

$\sigma_x$ = the standard deviation of variable X,

$\sigma_y$ = the standard deviation of variable Y,

n = the number of values of variables X and Y,

Alternately,

$$r = \frac{\sum xy}{\sqrt{\sum x^2}\sqrt{\sum y^2}}$$

where,

$x = X - \bar{X}$ = deviation of a particular X value from the mean

$y = Y - \bar{Y}$ = deviation of a particular Y value from the mean

The value of r lies within the range of -1 and +1, with extreme value indicating perfect linear association and mid value of zero indicating no linear association between two variables. The procedure for the estimation and test of significance of a simple linear correlation coeffiocient between two variables X and Y are

• Compute the mean X⁻ and Y⁻, the corrected sum of squares $\sum x^2$ and $\sum y^2$, and the corrected sum of cross products $\sum xy$ of the two variables as

$$X^- = \sum X \div n$$

$$Y^- = \sum Y \div n$$

$$\sum x^2 = \sum_{i=1}^{n} (X_i - X)^2$$

$$\sum Y^2 = \sum_{i=1}^{n} (Y_i - Y)^2$$

$$\sum YX = \sum_{i=1}^{n} (X_i - X)^2 (Y_i - Y^-)^2$$

• Compute the simple linear correlation coefficient as

$$r = \frac{\sum xy}{\sqrt{\sum x^2}\sqrt{\sum y^2}}$$

• Test the significance of the simple correlation coefficient by comparing the computed r value to the tabular r value with degree of freedom (n–2). The simple linear correlation coefficient is declared significant at the a level of significance of the absolute value of the computed r value is greater than the corresponding tabular r value at the a level of significance.

For example,

Calculate the r value and test of significance for the following data,

| | *1* | *2* | *3* | *4* | *5* | *6* | *7* | *8* | *9* | *10* |
|---|---|---|---|---|---|---|---|---|---|---|
| Cocoon weight (g) X | 1.75 | 1.88 | 1.93 | 2.04 | 2.02 | 1.77 | 1.58 | 1.85 | 1.97 | 1.88 |
| ERR% Y | 75.6 | 84.9 | 89.0 | 89.0 | 85.4 | 86.7 | 84.2 | 85.8 | 91.8 | 91.7 |

Step 1: Compute the mean, corrected sums of squares and corrected sum of cross products as follows,

| *Sample No.* | *Cocoon wt. (X)* | *ERR (Y)* | *(X-X)* | *(Y–Y)* | *(X–Y)* | *(Y–Y)²* | *xy* |
|---|---|---|---|---|---|---|---|
| 1 | 1.75 | 75.6 | 0.117 | -10.81 | 0.0137 | 116.856 | -1.2648 |
| 2 | 1.88 | 84.9 | 0.013 | -1.51 | 0.0002 | 2.280 | - 1 .4970 |
| 3 | 1.93 | 89.0 | 0.063 | 2.59 | 0.0040 | 6.708 | 0.1632 |
| 4 | 2.04 | 89.0 | 0.173 | 2.59 | 0.0299 | 6.708 | 0.4481 |
| 5 | 2.02 | 85.4 | 0.153 | -1.01 | 0.0234 | 1.020 | -0.8570 |
| 6 | 1.77 | 86.7 | -0.097 | 0.29 | 0.0094 | 0.084 | -0.2767 |
| 7 | 1.58 | 84.2 | -0.287 | -2.21 | 0.0824 | 4.884 | 0.6343 |
| 8 | 1.85 | 85.8 | -0.017 | -0.61 | 0.0003 | 0.372 | 0.0104 |
| 9 | 1.97 | 91.8 | 0.103 | 5.39 | 0.0106 | 29.052 | 0.5552 |
| 10 | 1.88 | 91.7 | 0.013 | 5.29 | 0.0002 | 27.984 | 0.0688 |
| Total | 18.67 | 864.1 | 0.234 | 0.000 | 0.1740 | 195.949 | -2.0155 |
| Mean | 1.867 | 86.41 | | | | | |

Step 2: Compute the simple linear correlation coefficient r,

$$r = \frac{\sum xy}{\sqrt{\sum x^2}\sqrt{\sum y^2}}$$

= –2.0155 ÷ (V0.174 × V195.949)

= –2.0155 ÷ 3377 (0.4171331 × 13.998178)

= –2.0155 ÷ 5.8391

= – 0.3452

Step 3: Compare the absolute value of the computed r value to the tabular values with (n–2) = 8 degree of freedom, which are 0.632 at 5% level of significance and 0.765 at 1% level of significance. Because the tabular r values are higher than the computed r values, we can conclude that there is no significant correlation between cocoon weights and ERR %.

### 8.2.7.2 Regression Analysis

Correlation analysis provides a measure of intensity of association between two random variables, neither of which is under the control of the researcher. However, regression analysis predicts the quantum of change in Y variable as a result of unit change in variable X. **Regression analysis is a statistical tool, which is used to find out the values of the dependent variables from those of independent variables.** It is used to determine the degree of association or correlation between two variables. It studies the nature of relationship, not the degree of relationship between variables.

### *8.2.7.2.1 Simple Linear Regression Analysis*

The simple linear regression analysis deals with the estimation and tests of significance concerning two variables X and Y in the equation, usually called as regression equation, which is represented as -

$y = a + bX$ where,

y = Estimated value of dependent variable

X = Value of independent variable

a = Intercept of regression line (y - bX)

b = Slope on regression line

$$= [\sum XY - (\sum X \sum Y / n)] \div \{(\sum X^2)/n\}$$

Regression analysis is simple as well as multiple. Multiple regression analysis is used to determine the contribution of selected independent variables on dependent variables while Step-down regression analysis is employed to obtain information about the order of importance of independent variables in predicting the dependent variables.

The step-by-step procedure for the estimation and test of significance of a simple linear regression between two variables X and Y is -

- Compute the mean $X^-$ and $Y^-$, the corrected sum of squares $\sum x^2$ and $\sum y^2$, and the corrected sum of cross products $\sum xy$ of the two variables as

$$X^- = \sum X \div n$$

$$Y^- = \sum Y \div n$$

$$\sum x^2 = \sum^n_{i=1} (X_i - X^-)^2$$

$$\sum Y^2 = \sum^n_{i=1} (Y_i - Y^-)^2$$

$$\sum YX = \sum^n_{i=1} (X_i - X^-)^2(Y_i - Y^-)^2$$

- Compute the estimates of the regression parameters $\alpha$ and $\beta$

$a = y - bx$

$b = \sum xy \div \sum x^2$

- Plot the observed points and draw a graphical representation of the estimated regression equation:

    Plot the n observed points

    Using the estimated linear regression, compute the y values and plot on (X,Y) plane

Test the siginificance of $\beta$:

- Compute the residual mean square as:

$$S^2_{y.x} = [\sum y^2 - \{(\sum xy)^2 / \sum x^2\}] + (n - 2)$$

- Compute the $t_\beta$ value as:
- Compare the computed tb value to the tabular t values with (n – 2) degrees of freedom
- Construct the (100 – $\alpha$) confidence interval for p, as:
- $C.I. = b \pm t_0 (\sqrt{S^2_{y.x}} / \sum x^2)$
- Test the hypothesis that $\alpha = \alpha_0$
- Compute the $t_\alpha$ value as:

$$t_\alpha = a - \alpha_0 + [\sqrt{S^2_{y.x}} / \{1/n \div (X^{-2} / \sum x^2)\}$$

- Compare the computed $t_\alpha$ value to the tabular t value with (n – 2) degree of freedom and at a prescribed level of significance.

For example,

Compute the single linear regression equation between leaf yield and nitrogen rate in mulberry.

| *Nitrogen rate (kgs / ha)* | *Leaf yield (kgs / ha)* |
|---|---|
| 0 | 4230 |
| 50 | 5443 |
| 100 | 6663 |
| 150 | 7150 |

Step 1: Compute the means, the corrected sums of squares and the corrected sum of the cross products as follows.

| | *Nitrogen rate(X)* | *Leaf yield(Y)* | *(X - X) =x* | *(Y - Y) = y* | *(X–X)²* | *(Y–Y)²* | *(x)(y)* |
|---|---|---|---|---|---|---|---|
| | 0 | 4230 | -75 | 1641.50 | 5625 | 2694522.25 | 123112.5 |
| | 50 | 5443 | -25 | - 428.50 | 625 | 183612.25 | 10712.5 |
| | 100 | 6663 | 25 | 791.50 | 625 | 626472.25 | 19787.5 |
| | 150 | 7150 | 75 | 1278.50 | 5625 | 1634562.25 | 95887.5 |
| Total | 300 | 23486 | 0 | 0.0 | 1250 | 5139169 | 249500 |
| Mean | 75 | 5871.50 | | | | | |

Step 2: Compute the estimates of regression parameters as,

$a = y - bX$

$b = \sum xy + \sum x^2$

$b = 249500 / 12500 = 19.96$

$a = 587.75 - (19.96 \times 75) = 4374.5$

Thus the estimated linear equation is

$y = a + bX$

$= 4374.5 + 19.96\ X$ for $0 \leq X \leq 150$

Step 3: Compute the estimated Y values and plot a graph as,

Y0 = 4374.5 + 19.96(0) = 4374.5

Y50 = 4374.5 + 19.96 (50) = 5372.5

Y100 = 4374.5 + 19.96(100) = 6370.5

Y150 = 4374.5 + 19.96(150) = 7368.5

Step 4: Test the significance of β, as:

$$S^2_{y.x} = [\sum y^2 - \{(\sum xy)2 / \sum x^2\}] + (n - 2)$$

$$= [5139169 - \{(249500)^2 / 125005\}] \div (4 - 2) = 789574.5$$

$$t_b = b \div (\sqrt{S^2_{y.x} / \sum x^2})$$

$$= 19.96/(\sqrt{79574.5/12500}) = 7.91$$

The tabular value at the 5% and 1% levels of significance, with 2 degree of freedom are 4.303 and 9.925 respectively. Because the computed $t_b$ value is greater than the tabular t value at the 5% level of significance, the linear response of leaf yield to changes in the nitrogen rate application within the range of 0 to 150 kg / ha is significant at 5% level of significance.

Step 5: Construct the (100 - α)% confidence interval for p as:

$$C.I. = b \pm t_a (\sqrt{S^2_{y.x} / \sum x^2})$$

$$= 19.96 \pm 4.303 (\sqrt{79574.5/12500})$$

$$= 19.96 \pm 10.81$$

$$= (9.15, 30.77)$$

Thus, the increase in kleaf yield for every kg/ha increase in the rate of nitrogen application within the range of 0 to 150 kg N/ha is expected to fall between 9.10 kg/ha and 30.82 kg/ha of the time.

### *8.2.7.2.2 Multiple Regression*

Regression analysis involving more than one independent variable is called multiple regression analysis. When all independent variables are assumed to affect the dependent variable in linear fashion and independently of one another, the procedure is called multiple linear regression analysis A multiple linear regression is said to be operating if the relationship of the dependent variable Y to the k independent variables $X_1$, $X_2$, ..... $X_k$ can be expressed as

$$Y = \alpha + \beta_1 X_1 + \beta_2 X_2 + ..... \beta_k X_k$$

The data required for application of multiple linear regression analysis involving k independent variables are the (n)(k + 1) observations as detailed below-

| *Observations* | *Observation Values* | | | | | |
|---|---|---|---|---|---|---|
| | Y | $X_1$ | $X_2$ | $X_3$ | ..... | $X_k$ |
| 1 | $Y_1$ | $X_{11}$ | $X_{21}$ | $X_{31}$ | ..... | $X_{K1}$ |
| 2 | $Y_2$ | $X_{12}$ | $X_{22}$ | $X_{32}$ | ...... | $X_{k2}$ |
| 3 | $Y_3$ | $X_{13}$ | $X_{23}$ | $X_{33}$ | ...... | $X_{k3}$ |
| . | . | . | . | . | ..... | . |
| n | $Y_n$ | $X_{1n}$ | $X_{2n}$ | $X_{3n}$ | .... | $X_{kn}$ |

The (k + 1) variables Y, $X_1$ $X_2$ .... $X_k$ must be measured simultaneously for each of the n units of observation. In addition, there must be enough observations to make n greater than (k + 1). The multiple regression procedure involves the estimation and test of significance of the (k + 1) parameters for a case where k = 2, using the following equation -

$$Y = \alpha + \beta_1 X_1 + \beta_2 X_2$$

Step 1: Compute the mean and the corrected sum of squares of each of the (k + 1) variables Y, $X_1$, $X_2$ .... $X_k$ and the corrected sum of cross products for all possible pair-combinations of the (k + 1) variables following the procedure described earlier.

| *Variable* | *Mean* | *Corrected sum of squares and cross products* | | | | |
|---|---|---|---|---|---|---|
| | | $X_1$ | $X_2$ | .... | $X_k$ | Y |
| $X_1$ | $x_1$ | $\sum x_1^2$ | $\sum x_1 x_2$ | ..... | $\sum x_2 y$ | $\sum x_1 y$ |
| $X_2$ | $X_2$ | | $\sum x_2^2$ | ..... | $\sum x_1^2$ | $\sum x_2 y$ |
| . | . | | ..... | | | |
| $X_k$ | $X_k$ | | | | $\sum x_k^2$ | $\sum x_k y$ |
| Y | Y | | | | | $\sum y^2$ |

Step 2: Solve for $b_1, b_2, \ldots b_k$ from the following k simultaneous equations, which are generally referred to as the normal equations -

$$b_1 \sum x_1^2 + b_2 \sum x_1x_2 + \ldots + b_k \sum x_1x_k = \sum x_1y$$

$$b_1 \sum x_1x_2 + b_2 \sum x_2^2 + \ldots + b_k \sum x_2x_k = \sum x_2y$$

. . . .

. . . .

$$b_2 \sum x_1x_k + b_2 \sum x_2x_k + \ldots + b_k \sum x_k^2 = \sum x_ky$$

If k = 2, the normal equations are -

$$b_1 \sum x_1^2 + b_2 \sum x_1x_2 = \sum x_1y$$

$$b_1 \sum x_1x_2 + b_2 \sum x_2^2 = \sum x_2y$$

and the solutions for $b_1$ and $b_2$ are -

$$b_1 = \frac{(\sum x_2^2)(\sum x_1y) - (\sum x_1x_2)(\sum x_1y)}{(\sum x_1^2)(\sum x_2^2) - (\sum x_1x_2)^2}$$

$$b_2 = \frac{(\sum x_1^2)(\sum x_2y) - (\sum x_1x_2)(\sum x_1y)}{(\sum x_1^2)(\sum x_2^2) - (\sum x_1x_2)^2}$$

Step 3: Compute the estimate of the intercept $\alpha$ as

$\alpha = Y^- b_1X_1 - b_2X_2^- \ldots - b_kX_k$

where, $Y, X_1, X_1, \ldots X_k^-$ are the means of the (k + 1) variables computed in Step 1. Thus, the estimated multiple linear regression is -

$Y = \alpha + b_1X_1 + b_2X_2 + \ldots b_kx_k$

Step 4: Compute

The sum of squares due to regression, as

$SSR = \sum^k_{i=1} (bi)(\sum x_iy)$

The residual sum of squares, as

$$SSE = \sum y^2 - SSR$$

The coefficient of determination as

$$R_2 = SSR - \sum y^2$$

The coefficient of determination measures the contribution of the linear function of k independent variables to the variation in Y. It is usually expressed as percentage. Its square root is referred to as the multiple regression coefficient.

Step 5: Test the significance of $R_2$ -

- Comput F value as SSR/K

$$F = \frac{SSR / K}{SSE /(n - k - 1)}$$

- Compare the computed F value to the tabular F values with $f_1 = k$ and $f_2 = (n - k - 1)$ degrees of freedom. The coefficient of determination is said to be significant if the computed F value is greater than the corresponding tabular F value at the prescribed level of significance.

### 8.2.8 Time Series Analysis

Forecasting or predicting is an essential tool in any decision making process. Time Series Analysis is one of the most powerful quantitative methods (moving averages, exponential smoothing, regression methods, causal models etc.) especially for short term forecasting purposes or determining patterns in data collected over time. From the historical data, this method attempts to obtain the underlying pattern so that a suitable model of the process can be developed.

#### *8.2.8.1 Decomposition Method*

Economic or business oriiented time series is made up of four components -trend, seasonally, cycle and randomness. It is also assumed that the relationship between these four components is multiplicative, which is shown by the following equation -

$X_t = T_t S_t C_t R_t$ where,

$X_t$ is the observed value of the time series

$T_t$ denotes trend

$S_t$ denotes seasonality

$C_t$ denotes cycle

$R_t$ denotes randomness

*Trend Analysis*

A set of values observed over a period of time is referred as a time series and a set of observations taken at regular intervals over a period of time is known as time series data. **Trend analysis is a statistical tool to reveal pattern of variations plotting the data on a time scale.** It indicates the mode of growth or decline of data over a period of time and projects the past patterns into the future.

If the time series represents a seasonal pattern of L periods, then by taking a moving average of L periods, we could get the mean value for the year. Such a value will obviously be free of seasonal effects, since high months will be offset by low ones. If $M_t$ denotes the moving average of equation, it will be free of seasonality and will contain little randomness. Thus, we can say,

$M_t = T_t C_t$

The trend and cycle components can be decomposed by assuming different kinds of trends such as -

- Linear trend, which implies a constant rate of change,
- Parabolic trend, which implies a varying rate of change,
- Exponential or logarithmic trend, which implies a constant percentage rate of change,
- An S curve, which implies slow initial growth, with increasing rate of growth followed by a declining growth rate and eventually saturation.

The linear trend is estimated using the linear regression analysis (Least Square Method). If a linear trend gives,

$T_t = a + bt$

The cycle component $C_t$ can now be isoloated from the trend $T_t$ by use of the above equation as follows,

$$\frac{M_t}{T_t} = \frac{M_t}{a + bt}$$

The seasonality is isolated in the following manner,

$$\frac{X_t}{T_t} = \frac{T_t\ S_t\ C_t\ R_t}{T_t\ C_t}$$

Finally randomness is eliminated by averaging the different values of the above equation. The averaging is done on the same months or seasons of different years. The result is a set of seasonal values free from randomness, called seasonal indices.

The steps involved in this method of forecasting are -

Step 1: Deseasonalizing the time series by constructing a moving average M and taking the ratio Xt/ Mt which represents the seasonality and randomness.

Step 2: Fitting a trend line of the type Tt = a + bt to the deseasonalized time series.

Step 3: Identifying the cyclical variation around the trend line.

Step 4: Use the above information for forecasting for the next year.

For example, the quarterly price of the mulberry raw silk over a period of last five years is as follows,

| *Year* | *Price of Mulberry raw Silk (Rs per kg)* | | | |
|---|---|---|---|---|
| | *Quarter I* | *Quarter II* | *Quarter III* | *Quarter IV* |
| 1996 | 1106 | 1196 | 1278 | 1324 |
| 1997 | 1200 | 1307 | 1313 | 1358 |
| 1998 | 1344 | 1264 | 1304 | 1337 |
| 1999 | 1304 | 1254 | 1300 | 1332 |
| 2000 | 1292 | 1282 | 1336 | 1386 |

Computation of moving averages and ratios of the original variable to the moving average is,

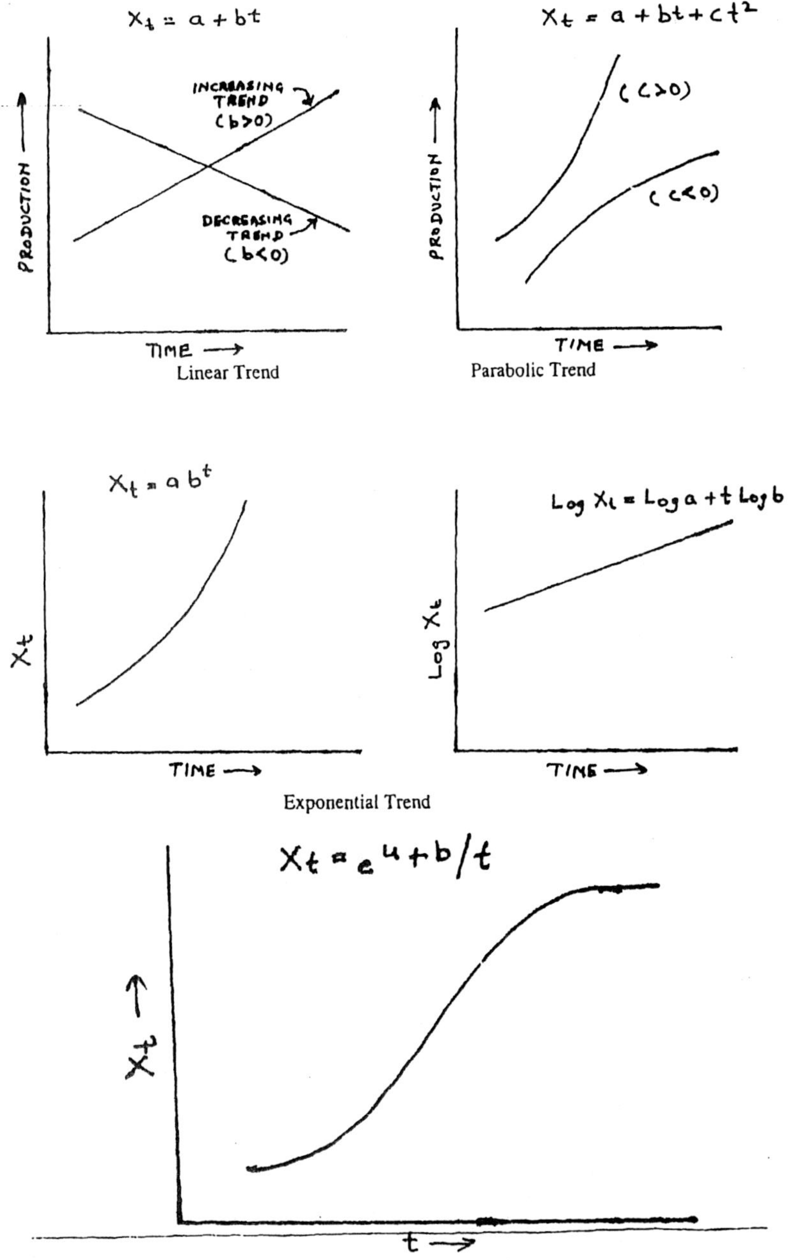


Typical S Curve

**Fig. 8.4: Kind of Trends in Decomposition Method**

| Year | Quarter | Actual price | 4 quarter moving total | Centred moving total | Centred moving average ($M_t$) | $X_t/M_t$ (3 / 6) |
|---|---|---|---|---|---|---|
| 1 | 2 | 3 | 4 | 5 | 6 | 7 |
| 1996 | I | 1106 | | | | |
| | II | 1196 | | | | |
| | III | 1278 | | 4951 | 1238 | 1.032 |
| | IV | 1324 | 4904 | 5054 | 1289 | 1.027 |
| 1997 | I | 1200 | 4998 | 5127 | 1282 | 0.936 |
| | II | 1307 | 5109 | 5161 | 1290 | 1.013 |
| | III | 1313 | 5144 | 5250 | 1313 | 1.000 |
| | IV | 1358 | 5178 | 5301 | 1325 | 1.025 |
| 1998 | I | 1344 | 5322 | 5275 | 1319 | 1.019 |
| | II | 1264 | 5279 | 5259 | 1315 | 0.961 |
| | III | 1304 | 5270 | 5229 | 1307 | 0.998 |
| | IV | 1337 | 5249 | 5204 | 1301 | 1.028 |
| 1999 | I | 1304 | 5209 | 5197 | 1299 | 1.004 |
| | II | 1254 | 5199 | 5193 | 1298 | 0.966 |
| | III | 1300 | 5195 | 5184 | 1296 | 1.003 |
| | IV | 1332 | 5190 | 5192 | 1298 | 1.028 |
| 2000 | I | 1292 | 5178 | 5224 | 1306 | 0.989 |
| | II | 1282 | 5206 | 5269 | 1317 | 0.973 |
| | III | 1336 | 5242 | | | |
| | IV | 1386 | 5296 | | | |

The seasonal indices for the quarterly price data is now computed by taking averages of the $X_t/M_t$ ratios of the respective quarters for different years as,

| *Year* | *Price of Mulberry raw Silk (Rs per kg)* | | | |
|---|---|---|---|---|
| | *Quarter I* | *Quarter II* | *Quarter III* | *Quarter IV* |
| 1996 | | | 1.032 | 1.027 |
| 1997 | 0.936 | 1.013 | 1.000 | 1.025 |
| 1998 | 1.019 | 0.961 | 0.998 | 1.028 |
| 1999 | 1.004 | 0.966 | 1.003 | 1.028 |
| 2000 | 0.989 | 0.973 | | |
| Mean | 0.987 | 0.978 | 1.008 | 1.027 |
| Seasonal index | 0.987 | 0.978 | 1.008 | 1.027 |

The seasonal indices are computed from the quarter means by adjusting these values of means so that the average over the year is unity. Thus, the sum of means in above table is ... and since there are four quarters, each mean is adjusted by multiplying it with the constant figure of 4 / 3.988 to obtain the indicated seasonal indices. These seasonal indices are then used to obtain the deseasonalized price by dividing the actual price by the corresponding index as,

| *Year* | *Quarter* | *Actual price* | *Seasonal index* | *Deseasonalized price (4 x 5)* |
|---|---|---|---|---|
| 1 | 2 | 3 | 4 | 5 |
| 1996 | I | 1106 | 0.987 | 1091.60 |
| | II | 1196 | 0.978 | 1169.70 |
| | III | 1278 | 1.008 | 1288.20 |
| | IV | 1324 | 1.027 | 1359.75 |
| 1997 | | | | |
| | I | 1200 | 0.987 | 1184.40 |
| | II | 1307 | 0.978 | 1278.25 |
| | III | 1313 | 1.008 | 1323.50 |
| | IV | 1358 | 1.027 | 1394.70 |
| 1998 | | | | |
| | I | 1344 | 0.987 | 1326.50 |
| | II | 1264 | 0.978 | 1236.20 |
| | III | 1304 | 1.008 | 1314.40 |
| | IV | 1337 | 1.027 | 1373.10 |
| 1999 | | | | |
| | I | 1304 | 0.987 | 1287.05 |
| | II | 1254 | 0.978 | 1226.40 |
| | III | 1300 | 1.008 | 1310.40 |
| | IV | 1332 | 1.027 | 1368.00 |
| 2000 | | | | |
| | I | 1292 | 0.987 | 1275.20 |
| | II | 1282 | 0.978 | 1253.80 |
| | III | 1336 | 1.008 | 1346.70 |
| | IV | 1386 | 1.027 | 1423.40 |

Now, to fit a straight line of the form Y = a + bX to the deseaonalized price, we proceed as,

| *Year* | *Quarter* | *Deseasonalized price (Y)* | *X* | $X^2$ | *XY* |
|---|---|---|---|---|---|
| 1996 | | | | | |
| | I | 1091.60 | -19 | 361 | - 20740.40 |
| | II | 1169.70 | -17 | 289 | -19984.90 |
| | III | 1288.20 | -15 | 225 | -19323.00 |
| | IV | 1359.75 | -13 | 169 | - 17676.75 |
| 1997 | | | | | |
| | I | 1184.40 | -11 | 121 | -13028.40 |
| | II | 1278.25 | -9 | 81 | -11504.25 |
| | III | 1323.50 | -7 | 49 | - 9264.50 |
| | IV | 1394.70 | -5 | 25 | - 6973.50 |
| 1998 | | | | | |
| | I | 1326.50 | -3 | 9 | - 3979.50 |
| | II | 1236.20 | -1 | 1 | - 1236.20 |
| | III | 1314.40 | 1 | 1 | 1314.40 |
| | IV | 1373.10 | 3 | 9 | 4119.30 |
| 1999 | | | | | |
| | I | 1287.05 | 5 | 25 | 6435.25 |
| | II | 1226.40 | 7 | 49 | 8584.80 |
| | III | 1310.40 | 9 | 81 | 11793.60 |
| | IV | 1368.00 | 11 | 121 | 15048.00 |
| 2000 | | | | | |
| | I | 1275.20 | 13 | 169 | 16577.60 |
| | II | 1253.80 | 15 | 225 | 18807.00 |
| | III | 1346.70 | 17 | 289 | 22893.90 |
| | IV | 1423.40 | 19 | 361 | 27044.60 |
| | 25831.25 | 0 | 2660 | 8907.05 | |

$a = \sum Y/n = 25831.25/20 = 1291.56$

$b = \sum XY/\sum x^2 = 16577.6 \ / \ 2660 = 6.232$

Hence the trend line is,

Y = 1291.56 + 6.232 X

Now, the cyclic variations are identified by measuring deseasonalized variation around the trend line, as the ratio of actual deseasonalized price to the value predicted by the trend line as,

| Year | Quarter | Deseasonalized price (Y) | Trend (a + bX) | Y / a + bX |
|---|---|---|---|---|
| 1996 | I | 1091.60 | 1173.15 | 0.930 |
| | II | 1169.70 | 1063.75 | 1.100 |
| | III | 1288.20 | 1198.10 | 1.075 |
| | IV | 1359.75 | 1210.50 | 1.123 |
| 1997 | I | 1184.40 | 1223.00 | 0.968 |
| | II | 1278.25 | 1235.50 | 1.035 |
| | III | 1323.50 | 1247.90 | 1.061 |
| | IV | 1394.70 | 1260.40 | 1.107 |
| 1998 | I | 1326.50 | 1272.90 | 1.042 |
| | II | 1236.20 | 1285.30 | 0.962 |
| | III | 1314.40 | 1297.80 | 1.013 |
| | IV | 1373.10 | 1310.25 | 1.048 |
| 1999 | I | 1287.05 | 1322.70 | 0.973 |
| | II | 1226.40 | 1335.20 | 0.918 |
| | III | 1310.40 | 1347.65 | 0.972 |
| | IV | 1368.00 | 1360.10 | 1.006 |
| 2000 | I | 1275.20 | 1372.60 | 0.929 |
| | II | 1253.80 | 1385.00 | 0.905 |
| | III | 1346.70 | 1397.50 | 0.964 |
| | IV | 1423.40 | 1410.00 | 1.009 |

The forecasting with the decomposed components of the time series can be done by using the trend line.

$$Y = 1291.56 + 6.232\ X$$

The estimates of the raw silk price for III quarter of the year 2002 can be obtained by using the trend line,

$$Y\ (35) = 1291.56 + 6.232\ (35) = 1509.70$$

### 8.2.8.2 Auto-correlation Method

Auto-correlation describes the association or mutual dependence between values of the same but at different time periods. The auto-correlation coefficients provide important information about the structure of a data set. If data consists of monthly figures, a twelve month time lag will show how values of the same month but of different years correlate with each other. If the auto-correlation coefficient is positive, it implies that there is a seasonal pattern of twelve months duration. On the other hand, a near zero auto-correlation indicates the absence of a seasonal pattern. Simlarly, if there is a trend in the data, values next to each other will relate in the sense if one increases the other too will increase.In case of completely random data, all auto-correlations will tend to zero.

The auto-correlation coefficient is calculated using the following formula for time lag k,

$$Y_k = \frac{\sum^{n-k}{}_{t=1}(X_1 - X)(X_{t+k} - X)}{\sum^{n}{}_{t=1}(X_t - X)^2}$$

where,

$r_k$ = auto-correlation coefficient for time lag k

k = the length of time !ag

n = number of observations

$X_t$ = value of variable at time t

X = mean of all data

# BIBLIOGRAPHY


Aizawa K., Kobayashi M. and Abe F. (1980). Electrophoresis on filter paper of blood proteins of the 'Duer pupae' in the silkworm *Bombyx mori.* **J. Sericult. Sci. Japan, 29(3):** 197-202.

AOAC (1965). Official Methods of Analysis of the Association of Official Agricultural Chemists, 10th Edition. Washington D.C.

Bear F. E. (Ed.) (1976). Chemistry of the Soils. 2nd Edition. Third Indian Reprint, Oxford & IBM Publishing Co., New Delhi.

Bergmeyer H. U. and Bernt E. (1974). **Method. Of Enzym. Analysis, 2:** 574-576.

Bernfeld P. (1955). **Meth. Enzyme 1:** 149-158.

Bradford M. M. (1976). **Anal. Biochem., 72:** 248-254.

Brady N. C. (1984). The Nature and Property of Soils. MacMillan, New York.

Brown (1945). **J. Biol. Chem., 158:** 601.

Bucher T. and Pfleiderer G. (1955). **Methods in Enzymology, 1:** 435-440.

Burton K. (1956). **Biochem. J., 62:** 315.

Burton K. (1968). Determination of DNA concentration with diphenylamine. In Grassman L. and Moldave K (Eds) **Methods in Enzymology, 12B:** 163-166.

Chaykin S. (1975). Biochemistry Laboratory Technique, Wiley Eastern, New Delhi.

Chen P.S. (Jr), Toribara T. Y. and Warner H. (1956). Micro determination of phosphorus. **Anal. Chem., 28:** 1756-1758.

Clark J. M. Jr. and Switzer R. L. (1977). Experimental Biochemistry, 2nd Edition, W. H. Freeman San Francisco.

Clegg K. M. (1956). The application of the anthrone reagent to the estimation of starch in cereals. **J. Sci. Food Agric.,** 7:40-44.

Colowick S. P. and Kaplan N. O. (1955). Methods in Emzymology, Academic Press, New York.

Curry Peter J. (1977). Esterase polymorphism in a field population of the African migratory locust *Locusta migratoria migratorisides.* **J. Insect Physiol., 23:** 405-414.

Davis B. J. (1964). Disc electrophoresis II. Method & application to human serum proteins. **Ann. N. Y. Acad. Sci,** 121:404-427

David Glick (ed.) (1958). Biochemical Preparations, Wiley, New York.

Dawson R. M. C., Elliot D. C., Elliot W. H. and Jones K. M. (1959). Data for Biochemical Research, Oxford University Press, London.

Dubois M., Gilles K., Hamilton J. K., Rebers P. A. and Smith F. (1951). A colorimetric method for the determination of sugars. **Nature, 168:** 167.

Dubois M., Gilles K. A., Hamilton J. K.. Rebers P. A. and Smith F. A. (1956). **Method. Enzymol., 8: 93.**

Dubois M., Gilles K. A., Hamilton J. K.. Rebers P. A. and Smith F. A. (1956). Colorimetric method for determination of sugars and related substances. **Analyt. Chem., 28:** 350-356.

Ellman G. L, Courtney K. D., Anderi V. and Ferthastone R. M. (1961). **Biochem. Pharmac., 7:** 88-95.

Fiske C. H. and Subbarao Y. P. (1925). The colorimetric determination of phosphorus. **J. Biol. Chem., 66:** 375-400.

Folch J., Ascoli I., Lees M., Meath J. A. and LeBaron F. N. (1951). **J. Biol. Chem., 191:** 833.

Folch J., Lees M. and Stanley G. H. S. (1957). A simple method for the isolation and purification of total lipids from animal tissues. **J. Biol. Chem., 226:** 497-509.

Frais F. (1972). Practical Biochemistry, Butterworths, London.

Frame, Russell and Wilhelmi (1943). J. **Biol. Chem., 149:** 255.

Garnall, A. G., Bardawill C. J. and David M. M. (1949). Determination of serum proteins by means of the Biuret reaction. **J. Biol. Chem., 177:** 751-766.

Gomez, K. A. and Gomez A. A. (1984). Statistical Procedure for Agricultural Research. John Wiley & Sons, New York.

Gopalakrishna R. and Nagarjuna (1979). **Indian J. Biochem. Biophys., 16:** 686.

Guhathakurta B. and Dutta C. C. (1974). **Appli. Microbiol., 27:** 634.

Hayaishi O. (1955). **Method. Enzymol., 1:** 672.

Hedge J. E. and Hofreiter B. T. (1962). In: Carbohydrates Chemistry, (17th Edition), Academic Press, New York.

Hofman E. (1965). In: Methods of Enzymatic Analysis (ed. Bergmeyer H.), Academic Press, New York.

Hoppe K. T., Hadley N. F. and Trelease R. N. (1975). Changes in lipid and fatty acid composition of eggs during development of the Beet Armyworm, *Spodoptera exigna.* **J. Insect Physiol., 21:** 1427-1430.

Horborne J. B. (1976). In: Phytochemical Methods. Chapman and Hall, London.

Izumi S., Tomiro S. and Chino H. (1980). Purification and molecular properties of vitellin from silkworm **Bombyx mori. Insect Biochem., 10:** 199-208.

Jayaraman K., Lakshmanan M., Gnanam A. and Jayaraman J. (1972). Experiments in Microbiology and Molecular Biology, Higginbothams, Madras.

Jayaraman J. (1999). Laboratory Manual in Biochemistry. New Age International Publishers, New Delhi.

Kessler B. and Engelberg N. (1962). **Biochem. Biophys. Acta, 55:** 70.

Kornberg A. (1955). **Method. Enzymol., 1:** 441.

Kothari C. R. (1989). Research Methodology - Methods and Techniques, Wiley Eastern, New Delhi.

Koundal K. R., Siddique E. A. and Mehta S. L. (1977). Changes in soluble proteins and isozymes in rice *(Oryza sativa)* during germination. **Ind. J. Expt. Biol., 15:** 388-390.

Lee Y. C. and Scocca J. R. (1972). **J. Biol. Chem., 247:** 5753.

Lehninger A. L. (1975). Biochemistry, 2nd Edition. Kalyani Publishers, New Delhi.

Levin R. I. and Rubin D. S. (1991). Statistics for Managements111 Edition. Prentice-Hall of India, New Delhi.

Lodha M. L., Mali P. C., Agrawal A. K. and Mehta S. L. (1974). **Phytochemistry, 13:** 539.

Lowry O.H., Rosebrough A. L. Farr and Randall R. J. (1951). Protein measurement with the Folin Phenol reagent. **J. Biol. Chem., 193:** 265.

Mahadevan A. and Sridhar R. (1986). In: Methods of Physiological Plant Pathology, (3rd Edition), Sivakami Publications, Chennai.

Markha M. (1955). In Methods of Plant Analysis, Vol. 4 (K. Parch & M. V. Travy eds.), Springler-Verlog, Berlin: 246.

Martin D. W., Mayes P. A. and Rodwell V. B. (1983). Harper's Review of Biochemistry, 19th Edition, Lange Medical Publication, Maruzen.

Mattoo A. K. and Shah Z. M. (1974). Z **alle Mikrobiol., 14:** 581-591.

Merchant D. J., Kahn R. H. and Murphy W. H. (1969). Handbook of Cell and Organ Culture (2nd ed.), Burgess, Minneapolis.

Miller G. L (1972). **Anal. Chem.,** 31:426.

Moore S. and Stein W. H. (1948). Photometric method for use in the chromatography of amino acids. **J. Biol. Chem., 176:** 367-388.

Moore S. and Stein W. H. (1948). In: Methods Enzymol., Academic Press, New York.

Naito K. A., Suzuki L. H. and Tsuji H. (1979). **Physiol. Plant, 46:** 50-53.

Nelson N. (1944) A photometric adaptation of the Somogyi method for the determination of glucose. **J. Biol. Chem., 153:** 375-380.

Opie L. H. and Newsholme E. A. (1967). **Biochem. J., 103:** 391-399.

Oser B. L. (Ed.) (1979). Hawk's Physiological Chemistry. Tata McGraw-Hill, Mumbai.

Park J. T. and Johnson M. J. A sub micro determination of glucose. J. **Biol. Chem.,** 181:149-151.

Piper C. S. (1950). Soil and Plant Analysis, Interscience Publishers Inc., New York.

Plummer D. T. (1990). In: An Introduction to Practical Biochemistry (3rd Ed.), Tata McGraw-Hill Publishing Company Limited, New Delhi.

Quickenden and Roemhild (1969). Carbohydrates in grasshopper eggs. **J. Insect Physiol., 15:** 1215-1223.

Reed G. F. and Cummings R. W. (1945). Soil Reaction: glass electrode and colorimetric method for determination of pH values of Soils. **Soil Science, 59(1):** 97-104.

Reinsfeld R. A.; Lewis U. J. and Williams D. E. (1962). **Nature (London), 195:** 281-283.

Rendina G. (1971). Experimental Methods in Modern Biochemistry. W. B. Sannders, Philadelphia.

Roman W. and Bugs J. (1961). Clin. Enzyme, 2: 121.

Sadasivam S. and Manickam A. (1992). In: Biochemical Methods for Agricultural Sciences. Wiley eastern Limited, New Delhi.

SchneiderW. C. (1957). In: **Methods Enzymol., Vol. 3.** Academic Press, New York.

Shaw C. F. (1928). A definition of terms used in soil literature. **1st International Cong. Soil Sci. Proc. and papers, 5:** 38-64.

Shenoy G. V. S. (1983). Quantitative Techniques for Managerial Decision Making, Wiley Eastern.

Sibley J. A. and Lehninger A. L. (1949). **J. Biol. Chem., 177:** 859-872.

Smille R. M. and Krotkov G. (1960). Can. J. Bot, **38:** 31.

Somogyi M. (1945). **J. Biol. Chem., 160:** 62.

Strohler A. N. and Strohler A. H. (1983). Modern Physical Geography. Second Edition. John Wiley & Sons, New York.

Stryer L. (1975). Biochemistry, W. H. Freeman, San Francisco.

Talwar G. P. (ed.) (1980). Text Book of Biochemistry and Human Biology, Prentice Hall of India.

Tashiro Y. and Trevithick J. R. (1977). **Can. J. Biochem., 55:** 249.

Thimmaiah S. K. (1999). Standard Methods of Biochemical Analysis. Kalyani Publishers, New Delhi.

Trevelyan W. E., Procter D. P. and Harrison J. S. (1950). Detection of sugars on paper chromatograms. **Nature, London, 166:** 444-445.

Veerabhadrappa P. S., Marcas S. R. and Shadaksharaswamy (1976). Identification of the soluble esterases in the beetle **Haltica caerulea** Olivier. **Ind. J. Expt. Biol., 14:** 553-556.

Walker D. G. and Holland G. (1965). **Buiochem. J., 97:** 845-854.

Ward C. W. and Schofield P. J. (1967). **Comp. Biochem. Physiol., 22:** 3-52.

Weber K. and Osborn M. (1969). **J. Biol. Chem., 244:** 4406-4412.

Yemm E. W. and Cocking E. C. (1955). The determination of amino acids with ninhydrin. **Analyst, 80:** 209-213.

Young M. K. and Raisz L. G. (1956). An anthrone procedure for determination of insulin in biological fluids. **Proc. Soc. Exp. Biol. Mol., 80:** 771-774.